Advances in Intelligent Systems and Computing

Volume 237

Series Editor

Janusz Kacprzyk, Warsaw, Poland

For further volumes:
http://www.springer.com/series/11156

Ajith Abraham · Pavel Krömer
Václav Snášel
Editors

Innovations in Bio-inspired Computing and Applications

Proceedings of the 4th International Conference on Innovations in Bio-inspired Computing and Applications, IBICA 2013, August 22–24, 2013 - Ostrava, Czech Republic

Editors
Ajith Abraham
Machine Intelligence Research Labs (MIR Labs)
Scientific Network for Innovation and Research Excellence
Auburn Washington
USA

Václav Snášel
VŠB-TUO
Department of Computer Science
Faculty of Ele. Eng. and Computer Science
Ostrava-Poruba
Czech Republic

Pavel Krömer
VŠB-TUO
Department of Computer Science
Faculty of Ele. Eng. and Computer Science
Ostrava-Poruba
Czech Republic

ISSN 2194-5357 ISSN 2194-5365 (electronic)
ISBN 978-3-319-01780-8 ISBN 978-3-319-01781-5 (eBook)
DOI 10.1007/978-3-319-01781-5
Springer Cham Heidelberg New York Dordrecht London

Library of Congress Control Number: 2013945389

Printed on acid-free paper

Springer is part of Springer Science+Business Media (www.springer.com)

Preface

This volume of Advances in Intelligent Systems and Computing contains accepted papers presented at IBICA 2013, the 4th International Conference on Innovations in Bio-Inspired Computing and Applications. The first three events, IBICA 2009, IBICA 2011, and IBICA 2012 were hosted in India and China, with great success. The aim of IBICA 2013 was to provide a platform for world research leaders and practitioners, to discuss the *full spectrum* of current theoretical developments, emerging technologies, and innovative applications of Bio-inspired Computing. Bio-inspired Computing is currently one of the most exciting research areas, and it is continuously demonstrating exceptional strength in solving complex real life problems. The main driving force of the conference is to further explore the intriguing potential of Bio-inspired Computing. IBICA 2013 was held in Ostrava, Czech Republic. Ostrava is the capital of the Moravian-Silesian Region and the third largest city in the Czech Republic as to the area and population. Ostrava has a convenient strategic position - it is situated 10 kilometres south of the Polish state border and 50 kilometres west of the Slovak border. Its distance from the country's capital Prague is 360 km, 170 km from Brno, 90 km from Katowice in Poland and just 310 km from Vienna, Austria.

The organization of the IBICA 2013 conference was entirely voluntary. The review process required an enormous effort from the members of the International Technical Program Committee, and we would therefore like to thank all its members for their contribution to the success of this conference. We would like to express our sincere thanks to the host of IBICA 2013, VŠB – Technical University of Ostrava, and to the publisher, Springer, for their hard work and support in organizing the conference. Finally, we would like to thank all the authors for their high quality contributions. The friendly and welcoming attitude of conference supporters and contributors made this event a success!

August 2013

Ajith Abraham
Pavel Krömer
Václav Snášel

Organization

General Chair

Václav Snášel	VŠB - Technical University of Ostrava

Program Chair

Ajith Abraham	MIR Labs, USA

Publicity Chair

Pavel Krömer	VŠB - Technical University of Ostrava

International Program Committee

Nitin Agarwal	University of Arkansas at Little Rock, USA
Javier Bajo	Pontifical University of Salamanca, Spain
Guido Barbian	Unversity of Luneburg, Germany
Anna Bartkowiak	University of Wroclaw, Poland
Adil Baykasoglu	University of Gaziantep, Turkey
Marenglen Biba	University of New York Tirana, Albania
Abdelhamid Bouchachia	Alps-Adriatic University of Klagenfurt, Austria
Andre C.P.L.F. de Carvalho	Universidade de Sao Paulo, Brazil
Chuan-Yu Chang	National Yunlin University of Science & Technology, Taiwan
Ching-Hsiang Chang	Chang Jung Christian University, Taiwan
Shuo-Tsung Chen	Tunghai University, Taiwan
Chao-Ho Chen	National Kaohsiung University of Applied Sciences, Taiwan
Ying-ping Chen	National Chiao Tung University, Taiwan
Ching-Han Chen	National Central University, Taiwan
Rung-Ching Chen	Chaoyang University of Technology, Taiwan
Tzu-Fu Chiu	Aletheia University, Taiwan

Sung-Bae Cho	Yonsei University, Korea
Tung-Hsiang Chou	National Kaohsiung First University of Science and Technology, Taiwan
Hsin-Hung Chou	Chang Jung Christian University, Taiwan
Ping-Tsai Chung	Long Island University, USA
Zhihua Cui	Taiyuan University of Science and Technology, China
Alfredo Cuzzocrea	University of Calabria, Italy
Mauro Dragoni	FBK Trento, Italy
Nashwa El-Bendary	Cairo University, Egypt
Alexander A. Frolov	Russian Academy of Sciences, Russia
Petr Gajdos	VSB-Technical University of Ostrava, Czech Republic
Neveen Ghali	Cairo University, Egypt
Aboul Ella Hassanien	Cairo University, Egypt
Mong-Fong Horng	National Kaohsiung University of Applied Sciences, Taiwan
Sun-Yuan Hsieh	National Cheng Kung University, Taiwan
Hui-Huang Hsu	Tamkang University, Taiwan
Hsun-Hui Huang	Tajen University, Taiwan
Huang-Nan Huang	Tunghai University, Taiwan
Tien-Tsai Huang	Lunghwa University of Science and Technology, Taiwan
Dusan Husek	Academy of Sciences of the Czech Republic, Czech Republic
Tae Hyun Hwang	University of Minnesota, USA
Jason J. Jung	Yeungnam University, Korea
Dongwon Kim	Korea University, Korea
Po-Chang Ko	National Kaohsiung University of Applied Sciences, Taiwan
Mario Koeppen	Kyushu Institute of Technology, Japan
Pavel Kromer	VSB-Technical University of Ostrava, Czech Republic
Yasuo Kudo	Muroran Institute of Technology, Japan
Miloš Kudělka	VSB-Technical University of Ostrava, Czech Republic
Yeong-Lin Lai	National Changhua University of Education, Taiwan
Chih-Chin Lai	National University of Kaohsiung, Taiwan
Jouni Lampinen	University of Vaasa, Finland
Wei-Po Lee	National Sun Yat-Sen University, Taiwan
Chang-Shing Lee	National University of Tainan, Taiwan
Chung-Hong Lee	National Kaohsiung University of Applied Sciences, Taiwan

Lin-Yu Tseng	Providence University, Taiwan
Eiji Uchino	Yamaguchi University, Japan
Julien Velcin	Universite de Lyon 2,
Lipo Wang	Nanyang Technological University, Singapore
Dajin Wang	Montclair State University, USA
Chia-Nan Wang	National Kaohsiung University of Applied Sciences, Taiwan
Junzo Watada	Waseda University, Japan
Katarzyna Wegrzyn-Wolska	ESIGETEL, France
Tzuu-Shaang Wey	Kun Shan University, Taiwan
K.W. Wong	City University of Hong Kong, Hong Kong
Chih-Hung Wu	National University of Kaohsiung, Taiwan
Fatos Xhafa	Universitat Politecnica de Catalunya, Spain
Huayang Xie	Victoria University of Wellington, New Zealand
Li Xu	Fujian Normal University, China
Ruqiang Yan	University of Massachusetts, USA
Chang-Biau Yang	National Sun Yat-Sen University, Taiwan
Horng-Chang Yang	National Taitung University, Taiwan
Hong Yu	University of Wisconsin-Milwaukee, USA
Ivan Zelinka	VSB-Technical University of Ostrava, Czech Republic
Ning Zhang	University of Manchester, UK
Xiangmin Zhou	CSIRO, Australia
Bing-Bing Zhou	University of Sydney, Australia

Sponsoring Institutions

VŠB - Technical University of Ostrava

Acknowledgement

This international conference was supported by the Bio-Inspired Methods: research, development and knowledge transfer project, reg. no. CZ.1.07/2.3.00/20.0073 funded by Operational Programme Education for Competitiveness, co-financed by ESF and state budget of the Czech Republic.

INVESTMENTS IN EDUCATION DEVELOPMENT

Contents

Power Output Models of Ordinary Differential Equations by Polynomial and Recurrent Neural Networks

Ladislav Zjavka and Václav Snášel

VŠB-Technical University of Ostrava, IT4 Innovations Ostrava, Czech Republic
{ladislav.zjavka,vaclav.snasel}@vsb.cz

Abstract. The production of renewable energy sources is unstable, influenced a weather frame. Photovoltaic power plant output is primarily dependent on the solar illuminance of a locality, which is possible to predict according to meteorological forecasts (Aladin). Wind charger power output is induced mainly by a current wind speed, which depends on several weather standings. Presented time-series neural network models can define incomputable functions of power output or quantities, which direct influence it. Differential polynomial neural network is a new neural network type, which makes use of data relations, not only absolute interval values of variables as artificial neural networks do. Its output is formed by a sum of fractional derivative terms, which substitute a general differential equation, defining a system model. In the case of time-series data application an ordinary differential equation is created with time derivatives. Recurrent neural network proved to form simple solid time-series models, which can replace the ordinary differential equation description.

Keywords: power plant output, solar illuminance, wind charger, differential polynomial neural network, recurrent neural network.

1 Introduction

Power production estimations of renewable sources are necessary as the supplies are very variable [7]. The electrical energy accumulation is an ambitious problem to solve, there is better to consume it direct by customers. A following day output production is a sufficient estimation used by the electrical network operator [8]. The photovoltaic power plant (PVP) or wind charger supply of electricity is difficult to predict using deterministic methods as weather conditions can change from day to day or within short time periods. Hence the power output model should be updated to take into account a dynamic character of applied meteorological variables. Neural networks are able to deal with some problems, which other method solutions fail. They can define simple and reliable models, which exact solution is problematic. Recurrent neural network (RNN) is often used to define models of time-series data applications, which is possible to describe by ordinary differential equations. It applies as inputs also its neuron outputs from a previous time estimate. Analogous to other common neural network solutions, there is not possible to get specifications of RNN models in the form of a math description. The model appears to the users as a "black box".

A. Abraham et al. (eds.), *Innovations in Bio-inspired Computing and Applications*,
Advances in Intelligent Systems and Computing 237,
DOI: 10.1007/978-3-319-01781-5_1, © Springer International Publishing Switzerland 2014

$$y = a_0 + \sum_{i=1}^{m} a_i x_i + \sum_{i=1}^{m}\sum_{j=1}^{m} a_{ij} x_i x_j + \sum_{i=1}^{m}\sum_{j=1}^{m}\sum_{k=1}^{m} a_{ijk} x_i x_j x_k + \ldots \tag{1}$$

m – number of variables $X(x_1, x_2, \ldots, x_m)$ $A(a_1, a_2, \ldots, a_m), \ldots$ - *vectors of parameters*

Differential polynomial neural network (D-PNN) is a new neural network type, which results from the GMDH (Group Method of Data Handling) polynomial neural network (PNN), created by a Ukrainian scientist Aleksey Ivakhnenko in 1968, when the back-propagation technique was not known yet. General connection between input and output variables is possible to express by the Volterra functional series, a discrete analogue of which is Kolmogorov-Gabor polynomial (1). This polynomial can approximate any stationary random sequence of observations and can be computed by either adaptive methods or system of Gaussian normal equations [5]. GMDH decomposes the complexity of a process into many simpler relationships each described by low order polynomials (2) for every pair of the input values. Typical GMDH network maps a vector input $\boldsymbol{x}$ to a scalar output y, which is an estimate of the true function $f(\boldsymbol{x}) = y^t$.

$$y = a_0 + a_1x_i + a_2x_j + a_3x_ix_j + a_4x_i^2 + a_5x_j^2 \tag{2}$$

D-PNN combines the PNN functionality with some math techniques of differential equation (DE) solutions. Its models are a boundary of neural network and exact computational techniques. D-PNN forms and resolves an unknown general DE description of a searched function approximation. A DE is substituted producing sum of fractional polynomial derivative terms, forming a system model of dependent variables. In contrast with the ANN functionality, each neuron can direct take part in the total network output calculation, which is generated by the sum of active neuron output values. Its function approximation is based on any dependent data relations.

2 Modeling and Forecasting of Energy Production

The potential benefits of having energy production predictability are obvious useful in automatic power dispatch, load scheduling and energy control. The chance to forecast the energy production up to 24 hours can become of the utmost importance in decision-making processes, with particular reference to grid connected photovoltaic plants. Several approaches to forecasting load, wind speed or solar irradiation can be found. They can include neural network regression methods (Auto-Regressive Moving Average Model) and time series analysis models (Nonlinear Autoregressive with Exogenous Input). However most of the existing methodologies show some drawbacks such as high average accuracy error and dependence on the particular design of the PVP. Variability of weather, in particular solar irradiation, is maybe the main difficulty faced by PVP operators so that good forecasting tools are required for the appropriate integration of renewable energy into the power system. Neural networks are able to model the nonlinear nature of dynamic processes, reproduce an empirical, possibly nonlinear, relationship between some inputs and one or more outputs. They are applied for such purpose regarding to its approximation capability of any continuous nonlinear function with arbitrary accuracy that offer an effective alternative to more traditional statistical techniques [3]. Measurements of environmental parameters are generally provided in the form of time series which are suitable to use artificial neural networks with tapped delay lines. Regarding the training window width, the typical way to provide data for solar energy climatology has monthly, annual or 10-days granularity. Several solar parameters might be considered e.g. clearness, visibility index, cloud coverage and sunshine duration however the solar radiation is the most important parameter in the prediction and modeling of renewable PVP energy systems [2].

The wind speed model can apply 2 different inputs: lagged values of the average or maximum wind speed. The wind energy and speed change are not continual throughout the entire year, for this reason might be used the wind velocity meteorological maps providing regional assessments and interpretations. The difficulty in predicting this meteorological parameter arises from the fact that it is a result of the complex interactions among large-scale forcing mechanisms of pressure and temperature differences, local characteristics of the surface, etc. Wind energy is strong especially during winter, in the period with the highest demand. End-users recognize the contribution of wind prediction for a secure and economic operation of the power system. The power models should take into account technical parameters, i.e. hub height, turbine type, etc. Wind energy is possible to forecast using neuro-fuzzy, cognitive mapping or other soft computing techniques [1].

3 General Differential Equation Composition

The basic idea of the D-PNN is to compose and substitute a general sum partial differential equation (3), which is not known in advance and can describe a system of dependent variables, with a sum of fractional relative multi-parametric polynomial derivative terms (4).

$$a+\sum_{i=1}^{n} b_i \frac{\partial u}{\partial x_i}+\sum_{i=1}^{n}\sum_{j=1}^{n} c_{ij}\frac{\partial^2 u}{\partial x_i \partial x_j}+\ldots=0 \qquad u=\sum_{k=1}^{\infty} u_k \tag{3}$$

$u = f(x_1, x_2, \ldots, x_n)$ – searched function of all input variables
a, $B(b_1, b_2, \ldots, b_n)$, $C(c_{11}, c_{12}, \ldots)$ – polynomial parameters

Partial DE terms are formed according to the adapted integral analogues method, which is a part of similarity model analysis. It replaces mathematical operators and symbols of a DE by ratio of corresponding values. Derivatives are replaced by their integral analogues, i.e. derivative operators are removed and simultaneously with all operators are replaced by similarly or proportion signs in equations to form dimensionless groups of variables [4].

$$u_i=\frac{\left(a_0+a_1x_1+a_2x_2+a_3x_1x_2+a_4x_1^2+a_5x_2^2+\ldots\right)^{m/n}}{b_0+b_1x_1+\ldots}=\frac{\partial^m f(x_1,\ldots,x_n)}{\partial x_1 \partial x_2 \ldots \partial x_m} \tag{4}$$

n – combination degree of a complete polynomial of n-variables
m – combination degree of denominator variables

The fractional polynomials (4) define partial relations of n-input variables. The numerator of a DE term (4) is a polynomial of all n-input variables and partly defines an unknown function u of eq. (3). The denominator is a derivative part, which includes an incomplete polynomial of the competent combination variable(s). The root function of numerator takes the polynomial into competent combination degree to get the dimensionless values [4]. In a case of time-series data application an ordinary differential equation is formed with only time derivatives. The partial DE (3) might become form of (5).

$$a + bf + \sum_{i=1}^{m} c_i \frac{df(t, x_i)}{dt} + \sum_{i=1}^{m} \sum_{j=1}^{m} d_{ij} \frac{d^2 f(t, x_i, x_j)}{dt^2} + \ldots = 0 \qquad (5)$$

$f(t, \mathbf{x})$ – function of time t and independent input variables $\mathbf{x}(x_1, x_2, \ldots, x_m)$

Blocks of the D-PNN (Fig.1.) consist of derivative neurons, one for each fractional polynomial derivative combination, so each neuron is considered a summation DE term (4). Each block contains a single output polynomial (2), without derivative part. Neurons do not affect the block output but participate direct in the total network output sum calculation of a DE composition. Each block has *1* and neuron *2* vectors of adjustable parameters ***a***, resp. ***a, b***.

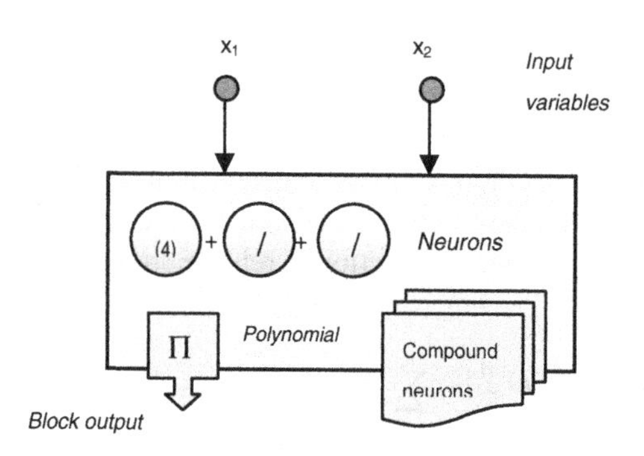

Fig. 1. D-PNN block of basic and compound neurons

In the case of 2 input variables the 2nd odder partial DE can be expressed in the form (6), which involve all derivative terms of variables applied by the GMDH polynomial (2). D-PNN processes these 2-combination square polynomials of blocks and neurons, which form competent DE terms of eq. (5). Each block so include *5* basic neurons of derivatives x_1, x_2, x_1x_2, x_1^2, x_2^2 of the 2nd order partial DE (6), which is most often used to model physical or natural systems.

$$F\left(x_1, x_2, u, \frac{\partial u}{\partial x_1}, \frac{\partial u}{\partial x_2}, \frac{\partial^2 u}{\partial x_1^2}, \frac{\partial^2 u}{\partial x_1 \partial x_2}, \frac{\partial^2 u}{\partial x_2^2}\right) = 0 \qquad (6)$$

where $F(x_1, x_2, u, p, q, r, s, t)$ is a function of 8 variables

4 Differential Polynomial Neural Network

Multi-layered networks forms composite polynomial functions (Fig.2.). Compound terms (CT), i.e. derivatives in respect to variables of previous layers, are calculated according to the composite function partial derivation rules (7)(8). They are formed by products of partial derivatives of external and internal functions.

$$F(x_1, x_2, \dots, x_n) = f(y_1, y_2, \dots, y_m) = f(\phi_1(X), \phi_2(X), \dots, \phi_m(X)) \tag{7}$$

$$\frac{\partial F}{\partial x_k} = \sum_{i=1}^{m} \frac{\partial f(y_1, y_2, \dots, y_m)}{\partial y_i} \cdot \frac{\partial \phi_i(X)}{\partial x_k} \qquad k=1, \dots, n \tag{8}$$

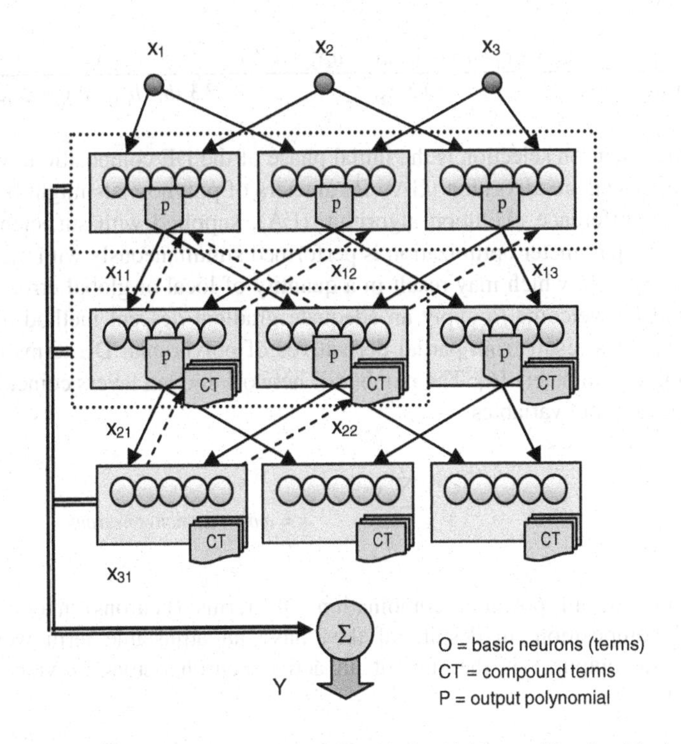

Fig. 2. 3-variable multi-layered D-PNN with 2-variable combination blocks

Thus blocks of the 2nd and following hidden layers are additionally extended with compound terms (neurons), which form composite derivatives utilizing outputs and inputs of back connected previous layer blocks. The 1st block of the last (3rd) hidden layer forms neurons e.g. (9)(10)(11) [9].

$$y_1 = \frac{\partial f(x_{21}, x_{22})}{\partial x_{21}} = w_1 \frac{\left(a_0 + a_1 x_{21} + a_2 x_{22} + a_3 x_{21} x_{22} + a_4 x_{21}^2 + a_5 x_{22}^2\right)^{1/2}}{3/2 \cdot (b_0 + b_1 x_{21})} \tag{9}$$

$$y_2 = \frac{\partial f(x_{21}, x_{22})}{\partial x_{11}} = w_2 \frac{(a_0 + a_1x_{21} + a_2x_{22} + a_3x_{21}x_{22} + a_4x_{21}^2 + a_5x_{22}^2)^{1/2}}{3/2 \cdot x_{22}} \cdot \frac{(x_{21})^{1/2}}{3/2 \cdot (b_0 + b_1x_{11})} \tag{10}$$

$$y_3 = \frac{\partial f(x_{21}, x_{22})}{\partial x_1} = w_3 \frac{(a_0 + a_1x_{21} + a_2x_{22} + a_3x_{21}x_{22} + a_4x_{21}^2 + a_5x_{22}^2)^{1/2}}{3/2 \cdot x_{22}} \cdot \frac{(x_{21})^{1/2}}{3/2 \cdot x_{12}} \cdot \frac{(x_{11})^{1/2}}{3/2 \cdot (b_0 + b_1x_1)} \tag{11}$$

The square (12) and combination (13) derivative terms are also calculated according to the composite function derivation rules.

$$y_4 = \frac{\partial^2 f(x_{21}, x_{22})}{\partial x_{11}^2} = w_4 \frac{(a_0 + a_1x_{21} + a_2x_{22} + a_3x_{21}x_{22} + a_4x_{21}^2 + a_5x_{22}^2)^{1/2}}{1.5 \cdot x_{22}} \cdot \frac{x_{21}}{3.8 \cdot (b_0 + b_1x_{11} + b_2x_{11}^2)} \tag{12}$$

$$y_5 = \frac{\partial^2 f(x_{21}, x_{22})}{\partial x_{11} \partial x_{12}} = w_5 \frac{(a_0 + a_1x_{21} + a_2x_{22} + a_3x_{21}x_{22} + a_4x_{21}^2 + a_5x_{22}^2)^{1/2}}{1.5 \cdot x_{22}} \cdot \frac{x_{21}}{3.3 \cdot (b_0 + b_1x_{11} + b_2x_{12} + b_3x_{11}x_{12})} \tag{13}$$

The best-fit neuron selection is the initial phase of the DE composition, which may apply a proper genetic algorithm (GA). Parameters of polynomials might be adjusted by means of difference evolution algorithm (EA), supplied with sufficient random mutations. The parameter optimization is performed simultaneously with the GA term combination search, which may result in a quantity of local or global error solutions. There would be welcome to apply an adequate gradient descent method too, which parameter updates result from partial derivatives of polynomial DE terms in respect with the single parameters [6]. The number of network hidden layers coincides with a total amount of input variables.

$$Y = \frac{\sum_{i=1}^{k} y_i}{k} \qquad k = \textit{amount of active neurons} \tag{14}$$

Only some of all potential combination DE terms (neurons) may participate in the DE composition, in despite of they have an adjustable term weight (w_i). D-PNN's total output Y is the sum of all active neuron outputs, divided by their amount k (14).

$$E = \sqrt{\frac{\sum_{i=1}^{M} \left(y^d - y_i\right)^2}{M}} \rightarrow \min \tag{15}$$

The root mean square error (RMSE) method (15) was applied for the polynomial parameter optimization and neuron combination selection. D-PNN is trained only with a small set of input-output data samples likewise the GMDH algorithm does [6].

5 Power Plant Output Model Experiments

D-PNN and RNN (Fig.7.) apply time-dependent series of the solar illuminance 3 variables to estimate a power plant output at the end-time (3rd) variable. Both networks were trained with previous 1 or 2 day 10-minute data series (i.e. samples), which provide the solar illuminance and corresponding power output values in the same location (Ostrava). Fig.3.-Fig.6. show the comparison of normalized power plant output day estimations, following the illuminance values. Both networks produce very similar results, despite their functionalities differs essentially. The power model outcome graph curves of Fig.5c. are nearly identical goings.

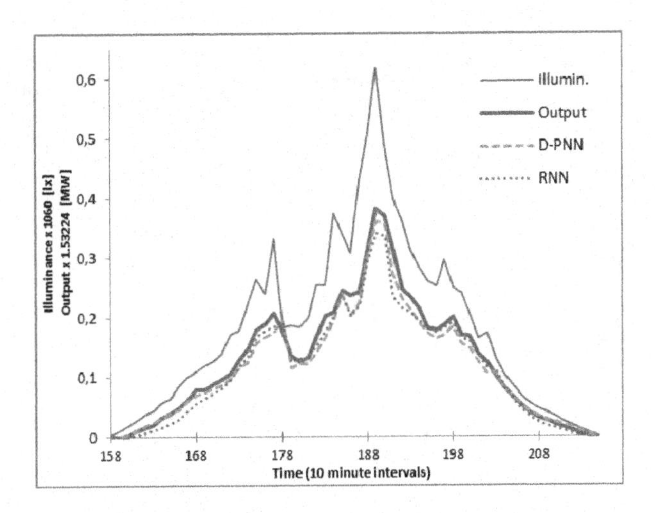

Fig. 3. RMSED-PNN = *0.0151*, $RMSE_{RNN}$ = *0.0184*

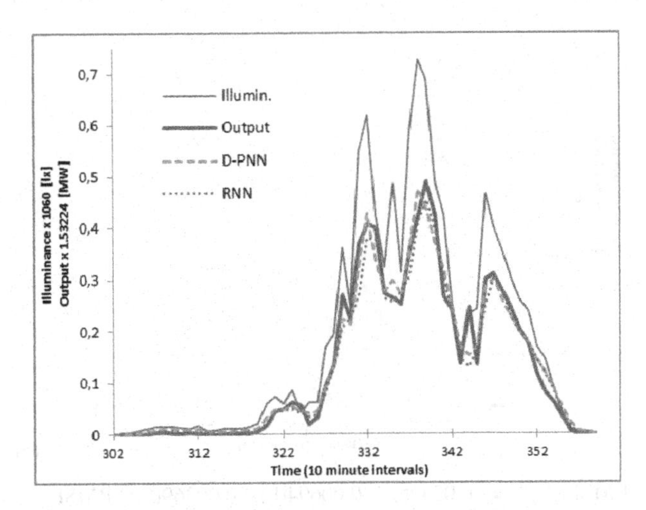

Fig. 4. RMSED-PNN = *0.0258*, $RMSE_{RNN}$ = *0.0275*

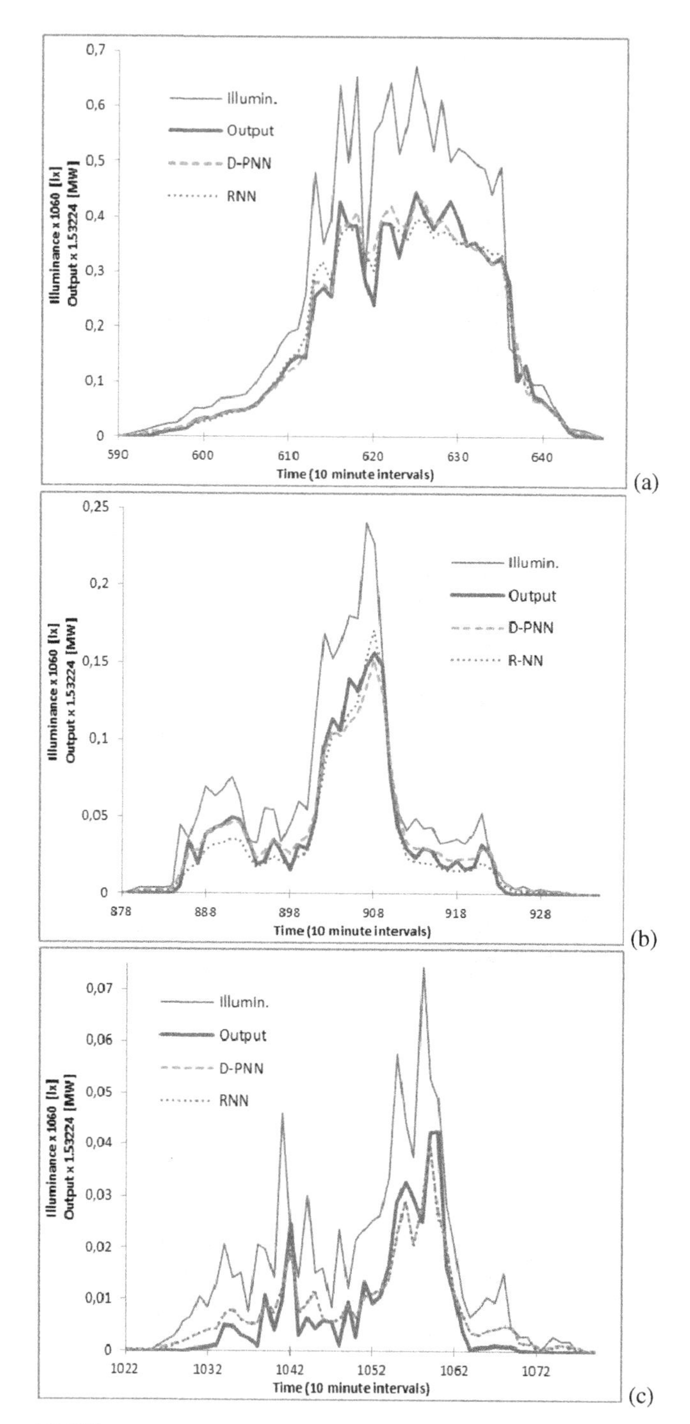

Fig. 5. a-c. $RMSE_{D\text{-}PNN}$ = 0.0242(a), 0.00694(b), 0.00369(c); $RMSE_{RNN}$ = 0.0234(a), 0.00747(b), 0.00374(c)

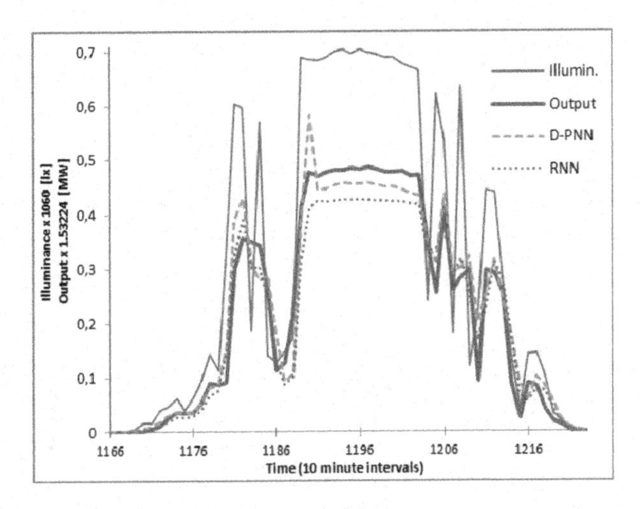

Fig. 6. RMSED-PNN = *0.0415*, $RMSE_{RNN}$ = *0.0440*

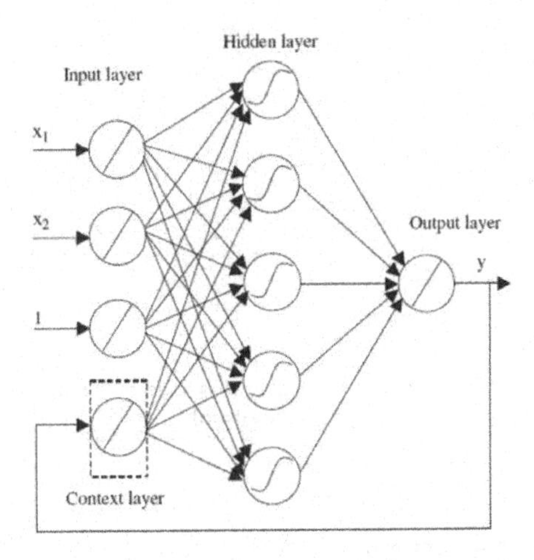

Fig. 7. Recurrent neural network

6 Wind Speed Model Experiments

The wind speed induces mainly a wind charger power output. There is usually available only its very rough forecast in a location. It could be modeled with reference to other weather variables forecasts, as meteorological predictions of this very complex dynamic system are sophisticated and not any time faithful, using simple neural network models. The fittest variables, which wind speed depends on, seem to be temperature, relative humidity and sea level pressure. The D-PNN and RNN models (Fig.8a-c.) apply 3 time-series of 3 state variables of 1 site locality, i.e. 9 input vector variables totally. Both networks were trained with previous 1 or 2 day hourly data series (24 or 48 hours, i.e. data samples), which are free online available [10].

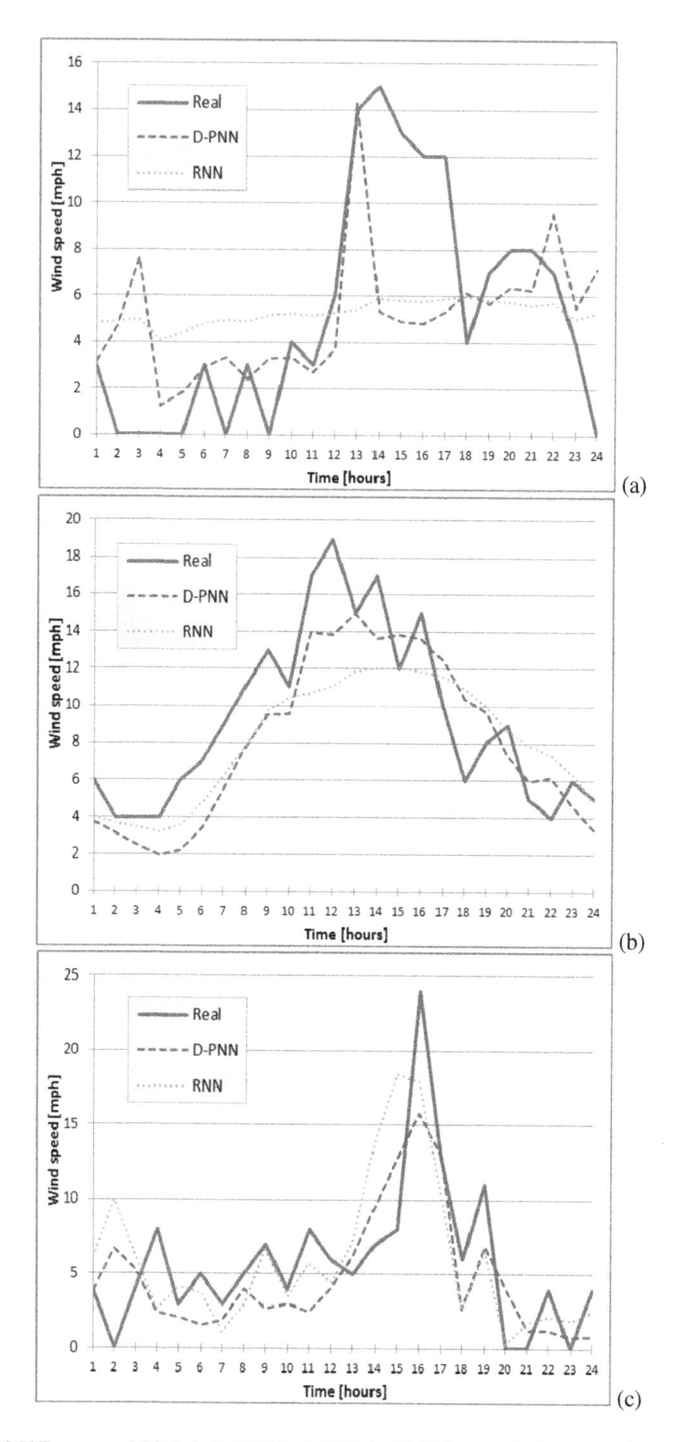

Fig. 8. a-c. $RMSE_{D\text{-}PNN}$ = 4.287(a), 2.665(b), 3.576(c); $RMSE_{RNN}$ = 4.493(a), 3.188(b), 4.116(c)

7 Conclusion

The study compares 2 neural network models, which results are very similar, despite the fact their operating principles differs by far. This comparison indicates, both method outcomes are of a very good level, extracting from the provided input data a maximum of useful information. Both networks update the models daily, to respect a dynamic character of applied meteorological variables. D-PNN is a new neural network type, which function approximation is based on generalized data relations. Its relative data processing is contrary to the common soft-computing method approach, which applications are subjected to a fixed interval of absolute values. Its operating principle differs by far from other common neural network techniques.

Acknowledgement. This work has been elaborated in the framework of the IT4Innovations Centre of Excellence project, reg. no. CZ.1.05/1.1.00/02.0070 supported by Operational Programme 'Research and Development for Innovations' funded by Structural Funds of the European Union and by the Ministry of Industry and Trade of the Czech Republic, under the grant no. FR-TI1/420 , and by SGS, VŠB – Technical University of Ostrava, Czech Republic, under the grant No. SP2012/58.

References

1. Atsalakis, G., Nezis, D., Zopounidis, C.: Neuro-Fuzzy Versus Traditional Models for Forecasting Wind Energy Production. In: Advances in Data Analysis, Statistics for Industry and Technology, pp. 275–287 (2010)
2. Cococcioni, M., D'Andrea, E., Lazzerini, B.: 24-hour-ahead forecasting of energy production in solar PV systems. In: 11th International Conference on Intelligent Systems Design and Applications (2011)
3. Hippert, H.S., Pedreira, C.E., Souza, R.C.: Neural Networks for Short-Term Load Forecasting: A Review and Evaluation. IEEE Transactions on Power Systems 16(1) (2001)
4. Kuneš, J., Vavroch, O., Franta, V.: Essentials of modeling. SNTL Praha (1989) (in Czech)
5. Nikolaev, N.Y., Iba, H.: Adaptive Learning of Polynomial Networks. Springer (2006)
6. Nikolaev, N.Y., Iba, H.: Polynomial harmonic GMDH learning networks for time series modelling. Neural Networks 16, 1527–1540 (2003)
7. Prokop, L., Mišák, S., Novosad, T., Kromer, P., Platoš, J., Snášel, V.: Artificially Evolved Soft Computing Models for Photovoltaic Power Plant Output Estimation. In: IEEE International Conference on Systems, COEX Seoul, Korea (2012)
8. Prokop, L., Mišák, S., Novosad, T., Kromer, P., Platoš, J., Snášel, V.: Photovoltaic Power Plant Output Estimation by Neural Networks and Fuzzy Inference. In: Intelligent Data Engineering and Automated Learning - 13th International Conference, Natal, Brazil (2012)
9. Zjavka, L.: Recognition of Generalized Patterns by a Differential Polynomial Neural Network. Engineering, Technology & Applied Science Research 2(1) (2012)
10. National Climatic Data Center of National Oceanic and Atmospheric Administration (NOAA), `http://cdo.ncdc.noaa.gov/qclcd_ascii/`

An Experimental Analysis of Reservoir Parameters of the Echo State Queueing Network Model*

Sebastián Basterrech[1], Václav Snášel[1], and Gerardo Rubino[2]

[1] VŠB–Technical University of Ostrava, Czech Republic
{Sebastian.Basterrech.Tiscordio,Vaclav.Snasel}@vsb.cz
[2] INRIA-Rennes, Bretagne, France
Gerardo.Rubino@inria.fr

Abstract. During the last years, there has been a growing interest in the Reservoir Computing (RC) paradigm. Recently, a new RC model was presented under the name of *Echo State Queueing Networks* (ESQN). This model merges ideas from Queueing Theory and one of the two pioneering RC techniques, *Echo State Networks*. In a RC model there is a dynamical system called *reservoir* which serves to expand the input data into a larger space. This expansion can enhance the linear separability of the data. In the case of ESQN, the reservoir is a Recurrent Neural Network composed of spiking neurons which fire positive and negative signals. Unlike other RC models, an analysis of the dynamics behavior of the ESQN system is still to be done. In this work, we present an experimental analysis of these dynamics. In particular, we study the impact of the spectral radius of the reservoir in system stability. In our experiments, we use a range of benchmark time series data.

1 Introduction

The Recurrent Neural Networks are a large class of computational models used in Computational Neuroscience and in several applications of Machine Learning. The main characteristic of this type of Neural Networks (NNs) is, in graph terms, the existence of at least one circuit in the network. Unlike feedforward neural networks, the recurrent topology enable to store historical information in internal states. Hence, recurrent NNs are a powerful tool for forecasting and time series processing applications. However, in spite of these important abilities no efficient algorithms exist to train recurrent NNs [1]. Reservoir Computing (RC) models emerge as an alternative to recurrent NNs overcoming important limitations on the training procedures while introducing no significant disadvantages. The RC approach has grown rapidly due to its success in solving learning tasks and other computational applications [1].

* This article has been elaborated in the framework of the project *New creative teams in priorities of scientific research*, reg. no. CZ.1.07/2.3.00/30.0055, supported by Operational Programme Education for Competitiveness and co-financed by the European Social Fund and the state budget of the Czech Republic.

A. Abraham et al. (eds.), *Innovations in Bio-inspired Computing and Applications*,
Advances in Intelligent Systems and Computing 237,
DOI: 10.1007/978-3-319-01781-5_2,

Recently, a new RC model which includes ideas coming from Queueing Theory was presented under the name of *Echo State Queueing Networks* [2]. The model has been proven to be efficient in temporal learning tasks (for instance: Internet Traffic prediction problems). A RC model uses a dynamical system called *reservoir* which has fixed parameters in the learning process. Unlike traditional RC model, an analysis of the stability or chaoticity in the ESQN system is still to be done. In this work, we present an experimental analysis of the some ESQN parameters. In particular, we analyze the impact of the spectral radius and the sparsity of the reservoir on the system stability behavior as well as on the model accuracy. Another important reservoir parameter is the reservoir size which was analyzed in [2]. In our experiments, we use three benchmarks widely analyzed in the RC literature.

The paper is organized as follow. We begin by presenting RC models. Section 2.1, introduces the ESQN model. Next, in Section 3 we present the benchmarks and we show the effect of reservoir parameters on the model performance. We conclude and present future works in Section 4.

2 Reservoir Computing Models

Since the early 2000s, RC has gained prominence in the NNs community [3,4]. The two pioneering Reservoir Computing techniques were: *Echo State Network* (ESN) [3] and *Liquid State Machine* (LSM) [5]. A RC model is a recurrent NN with a specific topology wherein the input and output layer haven't got circuits. The connections are allowed in the hidden layer which in this context is called *reservoir*. In order to reach a linear separability of the input data, the reservoir expand the input information in a larger space [6]. From this point of view, the reservoir acts as the expansion function used in *Kernel Methods* (see for example the *Support Vector Machine* [7]). In the ESN literature, the recurrent network (hidden layer) is referred as the *reservoir* while in the LSM literature is named the *liquid*. A distinctive principle of RC models is that the weights involved in circuits are deemed fixed during the learning process. Since the recurrent network provides all temporal information about the past, the model is built to produce the readout (model output) using a *memory-less* function. This strong assumption is based on the empirical observation that under certain algebraic hypothesis, updating only the memory-less structure can be enough to achieve good performances [1].

Formally, a RC model consists of an input layer with N_a units, a hidden recurrent structure of size N_x and an output layer of N_b neurons. Several network topologies have been proposed, we present an example in Figure 1. The connections between input and hidden neurons are given by a $N_\mathrm{x} \times N_\mathrm{a}$ weight matrix $\mathbf{w}^{\mathrm{in}}$, the connections among the hidden neurons are represented by a $N_\mathrm{x} \times N_\mathrm{x}$ weight matrix $\mathbf{w}^{\mathrm{r}}$ and a $N_\mathrm{b} \times N_\mathrm{x}$ weight matrix $\mathbf{w}^{\mathrm{out}}$ represents the connections between hidden and output neurons. Denote by $\mathbf{a}(t) = (a_1(t), \ldots, a_{N_\mathrm{a}}(t))$ the input vector and by $\mathbf{x}(t) = (x_1(t), \ldots, x_{N_\mathrm{x}}(t))$ the reservoir state at any time t. In order to obtain a meaningful linear structure in the reservoir projection, the dimension of $\mathbf{x}(t)$ should be much larger than the input data dimension $((N_a \ll N_x))$. The reservoir state $\mathbf{x}(t)$ is generated by a recurrent neural network $\phi(\mathbf{a}(t), \mathbf{w}^{\text{in}}, \mathbf{w}^{\text{r}}) : \mathbb{R}^{N_a} \to \mathbb{R}^{N_x}$ which transforms $\mathbf{a}(t)$ into a vector $\mathbf{x}(t)$. Let x_0 be a initial reservoir state, the reservoir state is given by:

$$\mathbf{x}(t) = \phi(\mathbf{x}(t-1), \mathbf{a}(t)). \tag{1}$$

As a consequence, the activation state of the reservoir at time t depends on the input data $\mathbf{a}(t)$ and on the preceding inputs $\mathbf{a}(t-1), \mathbf{a}(t-2), \ldots$, potentially the current input can depend on an infinite history of inputs. The model output usually are generated by a linear regression, that is:

$$\mathbf{y}(t) = \mathbf{w}^{\text{out}}\mathbf{x}(t). \tag{2}$$

For brevity, we omit the bias value added to expressions (1) and (2).

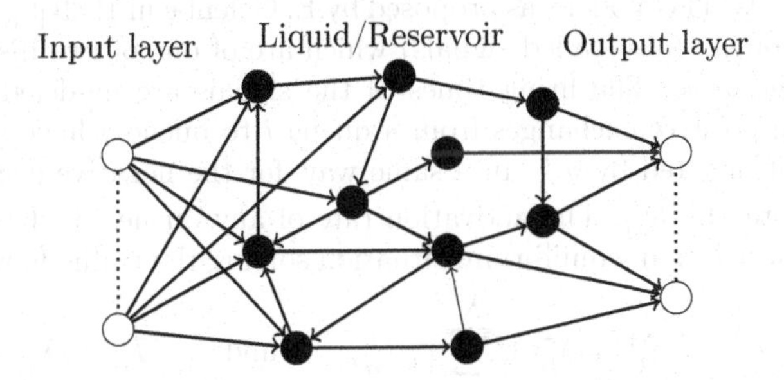

Fig. 1. Example of a standard topology of the Reservoir Computing model.

The LSM model has been proposed by Wolfgang Maass [5]. The model comes from the interest in making a conceptual representation of the cortical microstructures in the brain. There are several biological properties that motivate the LSM model [1,5]. The LSM approach suggests new ways of modeling cognitive processing in neural systems. In this model, the recurrent structure is built using network with *Leaky Integrate and Fire* (LIF) neurons [8].

Herbert Jaeger introduced the ESN model [3]. In this model, the reservoir is composed by neurons with $\tanh(\cdot)$ activation function. Thus, the state of each neuron in the reservoir is given by

$$x_m(t) = \tanh\left(w_{m0}^{\text{in}} + \sum_{i=1}^{N_a} w_{mi}^{\text{in}} a_i(t) + \sum_{i=1}^{N_x} w_{mi}^{\text{r}} x_i(t-1)\right), \qquad m = [1, N_x]. \tag{3}$$

The network outputs (*readouts*) $\mathbf{y}(t) = (y_1(t), \ldots, y_{N_b}(t))$ are often generated by a linear regression:

$$y_m(t) = w_{m0}^{\text{out}} + \sum_{i=1}^{N_a} w_{mi}^{\text{out}} a_i(t) + \sum_{i=1+N_a}^{N_a+N_x} w_{mi}^{\text{out}} x_i(t), \qquad m \in [1, N_b]. \tag{4}$$

An important property of the ESN reservoir is the *Echo State Property* (ESP) in [3]. In essence, the ESP guarantees that under certain conditions the state of the reservoir becomes asymptotically independent of the initial conditions, meaning that the effects of a finite sequence of inputs eventually vanish [1]. In practical applications, the ESP is almost always ensured when the ESN reservoir weights are appropriately scaled. A similar property was presented for LSMs [5].

2.1 Description of the Echo State Queuing Network model

Recently a new RC model called the *Echo State Queueing Network* (ESQN) was introduced [2]. In this model the reservoir dynamics are inspired from the behavior in steady-state of a queueing network type called *Generalized Queueing Network* (G-Network). A G-Network model is a mathematical object that can be classified into the area of Spiking Neuron Networks (SNNs) or into the family of Queueing Models. The model was proposed by E. Gelenbe in 1989 [9]. The queues exchange customers (also called signals) which are of one of two disjoint types: *positive* and *negative*. The firing times of the signals are modeled as Poisson processes. The positive exchanges from a queue i to queue j have associated a positive weight denoted by w^+_{ij}. In a same way, for the negative exchanges, the weight is denoted by w^-_{ij}. The activation rate of any queue i is denoted by x_i, and it is computed in an equilibrium situation solving the traffic flow equations:

$$x_i = \frac{T^+_i}{r_i + T^-_i}, \qquad T^+_i = \lambda^+_i + \sum_{j=1}^{N} x_j w^+_{ij}, \qquad \text{and} \qquad T^-_i = \lambda^-_i + \sum_{j=1}^{N} x_j w^-_{ij},$$

where λ^+_i and λ^-_i are the Poisson rates of external signals (positive and negative, respectively) and r_i which is the server rate of the queue i. A mathematical restriction associated to the model stability is that : $0 < x_i < 1$, for any i. In order to solve the traffic flow is necessary to use a fixed point procedure. A summary about how to compute $\mathbf{x}, \boldsymbol{T}^+$ and $\boldsymbol{T}^-$ is presented in [10]. Under certain hypothesis, it is guaranteed that the system has a unique equilibrium point [9, 11]. The G-Network model was successful used in supervised learning tasks as well as in biological modeling problems. In these fields the model has been presented under the name of *Random Neural Networks*. For more details about the G-Network model see [9, 11].

The ESQN model is a discrete or continuous mapping between an input space in $\mathbb{R}^{N_\mathrm{a}}$ into an output space in $\mathbb{R}^{N_\mathrm{b}}$. The architecture of an ESQN consists of an input layer, a reservoir and a readout layer. The outputs of the model are obtained by first expanding the input data using a nonlinear transformation (the reservoir) and then taking a simple function such as regression model. Let us index the input neurons from 1 up to N_a, and the reservoir neurons from $N_\mathrm{a}+1$ up to $N_\mathrm{a} + N_\mathrm{x}$. At any time t when an input $\mathbf{a}(t)$ is offered to the network, we assign the rates of the external positive customers with that input, that is: $\lambda^+_i(t) = a_i(t)$. Traditionally, when the G-Network is used for learning tasks there aren't negative signals from the environment, that is: $\lambda^-_i(t) = 0$, for all $i = 1, \ldots, N_\mathrm{a}$ and for all time t [11, 12]. The load or activity rate of a neuron i, in the stable case ($a_i(t) < r_i$) is given by: $x_i(t) = a_i(t)/r_i$, $i = 1, \ldots, N_a$. Thus, in the ESQN model at any time t, input neurons behave as a $M/M/1$ queue. Note that, following the RC approach the input and reservoir weights as well as the service rates are fixed in time, meaning that no temporal references are used.

The state of the ESQN reservoir is a simply vector $\mathbf{x}$ of dimension N_{x}. Each time that an input data $\mathbf{a}(t)$ is presented, the new reservoir state is computed using

$$x_i(t) = \frac{\sum_{j=1}^{N_{\mathrm{a}}} \frac{a_j(t)}{r_j} w_{ij}^{+} + \sum_{j=N_{\mathrm{a}}+1}^{N_{\mathrm{a}}+N_{\mathrm{x}}} x_j(t-1) w_{ij}^{+}}{r_i + \sum_{j=1}^{N_{\mathrm{a}}} \frac{a_j(t)}{r_j} w_{ij}^{-} + \sum_{j=N_{\mathrm{a}}+1}^{N_{\mathrm{a}}+N_{\mathrm{x}}} x_j(t-1) w_{ij}^{-}}, \tag{5}$$

for all $i \in [N_{\mathrm{a}}+1, N_{\mathrm{a}}+N_{\mathrm{x}}]$. When this is seen as a dynamical system, on the left we have the state values at t, and on the r.h.s. the state values at $t-1$. In the same way that others RC models, the readout part is computed using the linear regression given by the expression 4.

3 Experimental Analysis of the ESQN Reservoir Dynamics

3.1 Benchmark Descriptions

We use three time series data, one is a simulated data widely used in the NN literature and the other ones are two real data set. As usual, in order to evaluate the accuracy of the model we use a training data set and a validation data set. The quality measure of the model accuracy is the Normalized Mean Square Error (NMSE) [13]. The preprocessing data step consisted of the data rescaling in the interval $[0, 1]$. The learning method used was offline ridge regression algorithm which a regularization parameter is empirically adjusted for each data set. The first benchmark is the fixed 10th order NARMA time series, generated by

$$b(t+1) = 0.3b(t) + 0.05b(t)\sum_{i=0}^{9} b(t-i) + 1.5v(t-9)v(t) + 0.1,$$

where the distribution of $v(t)$ is $Unif[0, 0.5]$. The length of the training data is 1990 and 390 is the length of the validation data. The second data set consists of the Internet Traffic data, it is from the United Kingdom Education and Research Networking Association (UKERNA). The last one, is from an Internet Service Provider (ISP). Both data set were exhaustively analyzed in [2,14]. We evaluate the results collecting the data each 5 minutes for the ISP data and using a day scale for the UKERNA data The training data corresponds to the initial 66% of the data, that is 14772 samples for ISP and 51 samples for UKERNA data (using a day time scale).

For the Fixed 10th NARMA data, we use a network with 10 input neurons corresponding to the time lag from $t-10$ up to $t-1$. We follow the architecture

Table 1. NMSE obtained from 20 independent trials with random weight initialization and the corresponding Confidence Interval (CI) (95%). The reservoir size was 80 units for the first data set and 40 units for the other ones.

Series	NMSE	CI (95%)
NARMA	0.1004	± 0.0025
ISP	0.0100	$\pm 1.2436 \times 10^{-4}$
UKERNA	0.2030	± 0.0335

used in [14], then for the UKERNA day scale, the number of input neurons is 3 and the time lag is $\{t-1, t-6, t-7\}$. In the ISP 5 minutes scale case, we test with 7 input neurons corresponding to lag $\{t-1, t-2, t-3, t-4, t-5, t-6, t-7\}$.

We initialize the positive and negative weights of the ESQN model using a uniform distribution $Unif[0, 0.2]$. The initial reservoir state was randomly chosen in $[0, 1]^{N_x}$. In order to analyze the impact of the spectral radius and the sparsity on the accuracy, we made a grid in $[0, 1]^3$ where the variables are $\rho(\mathbf{w}^+)$, $\rho(\mathbf{w}^-)$ and the sparsity of $\mathbf{w}^+$ (denoted as $s(\mathbf{w}^+)$). We considered the same sparsity of both reservoir matrices ($s(\mathbf{w}^+) = s(\mathbf{w}^-)$). For each pair $(\rho(\mathbf{w}^+), s(\mathbf{w}^+))$ many associated NMSE values are obtained (each one corresponding to different $\rho(\mathbf{w}^-)$ values). In an analogous way, for each pair $(\rho(\mathbf{w}^-), s(\mathbf{w}^+))$, we obtain many NMSE values. The figures presented in the next section show the average of these NMSE values.

3.2 Result Analysis

The accuracy of the model for data set is showed in Table 1. Additionally, these accuracies are illustrated in the Figures 2(a) and 2(b). These model performances can be compared with the accuracy of other methods studied in [13, 14].

Figures 3(a), 3(b) illustrate the impact of the spectral radius $\rho(\mathbf{w}^+)$ and sparsity $s(\mathbf{w}^+)$ on the model accuracy. We present the NMSE average (computed for different $\mathbf{w}^-$ values) for each pair $(\rho(\mathbf{w}^+), s(\mathbf{w}^+))$. Figure 3(c) shows the impact of the $\rho(\mathbf{w}^-)$ and $s(\mathbf{w}^-)$ on the model accuracy. We can see that the spectral radius is an important variable and the sparsity does not seem to influence the accuracy.

Figure 3(d) illustrates the accuracy according to the $\rho(\mathbf{w}^+)$ and $s(\mathbf{w}^+)$ using UKERNA data set and 150 units in the reservoir. In a similar way, Figure 3(e) presents the NMSE according to the $\rho(\mathbf{w}^+)$ and $s(\mathbf{w}^+)$ when there are 80 units in the reservoir. In this case, when $\rho(\mathbf{w}^+)$ is close to 0 the accuracy of the model decreases. It is difficult to determinate the influence of the sparsity in the model accuracy. These two graphics also show the relevance of the reservoir size on the accuracy. The behavior for the negative weights was similar to that of the experiments with positive weights.

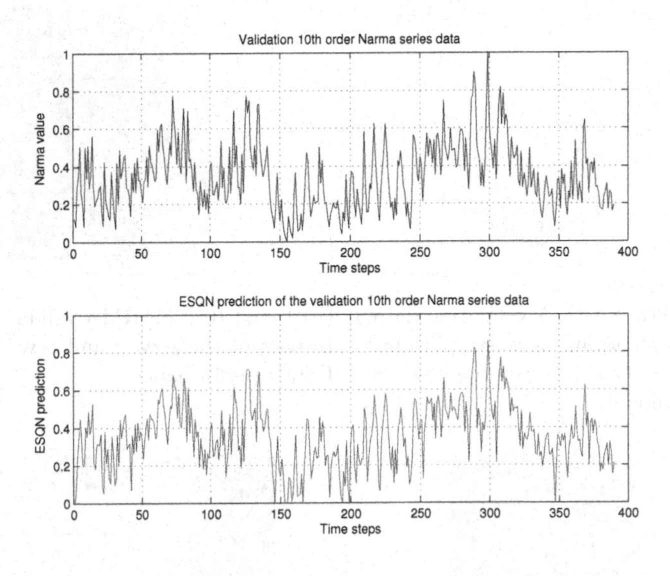

(a) Fixed 10th order NARMA times series.

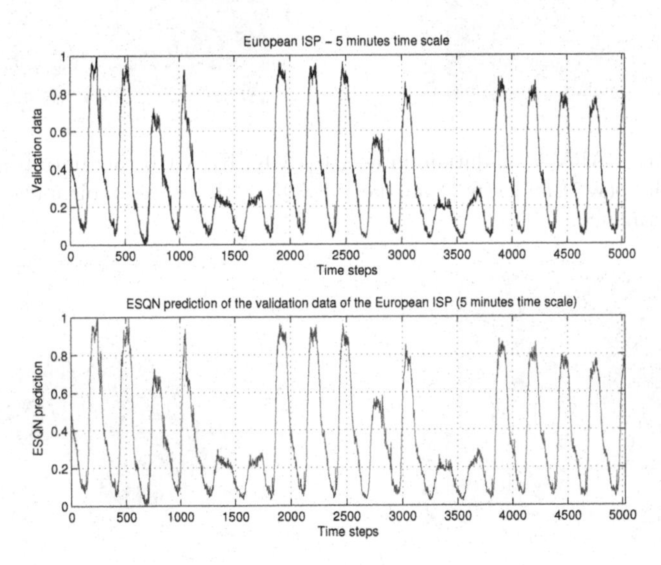

(b) Example of ESQN prediction for 5000 instances in the ISP data set.

Fig. 2. Comparison between the validation data and the ESQN model prediction

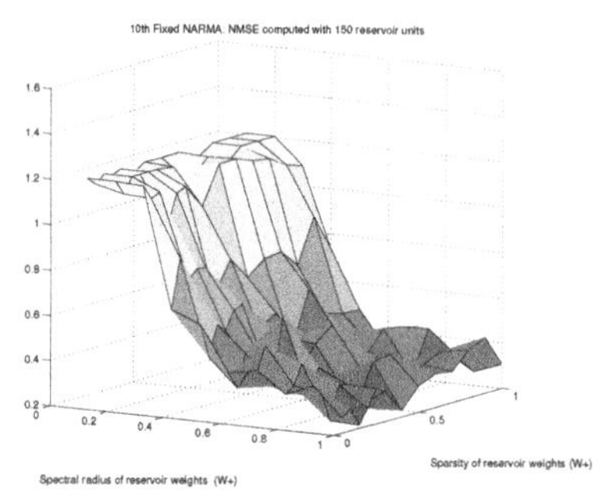

(a) Fixed 10th NARMA validation data. The NMSE according the spectral radius and the sparsity of the reservoir $\mathbf{w}^+$ of the ESQN model.

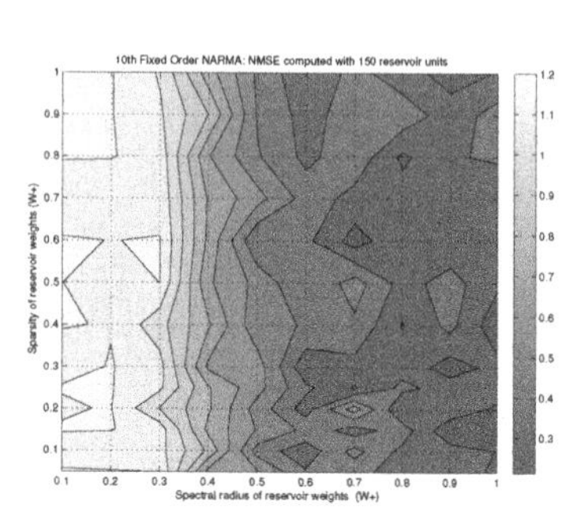

(b) Fixed 10th NARMA validation data. Impact of the $\rho(\mathbf{w}^+)$ and $s(\mathbf{w}^+)$ on the ESQN performance

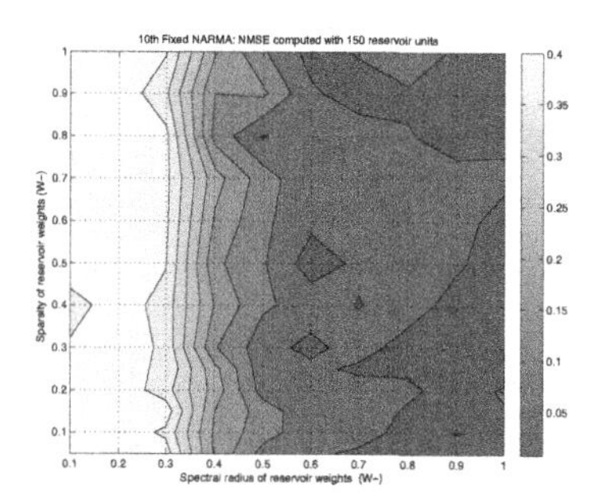

(c) Fixed 10th NARMA validation data. Impact of the $\rho(\mathbf{w}^-)$ and $s(\mathbf{w}^-)$ on the ESQN performance

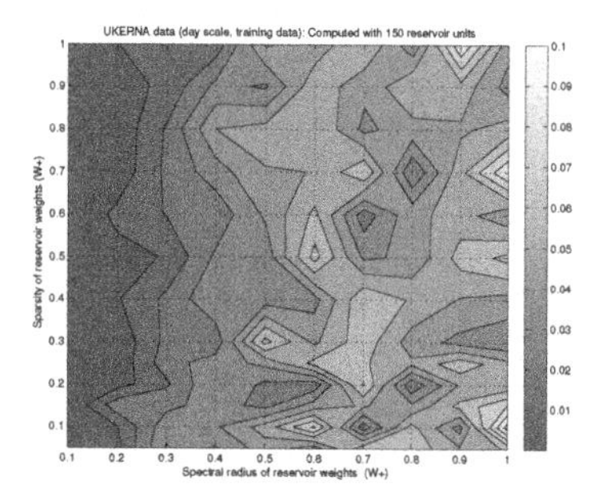

(d) UKERNA training data (day scale). ESQN with 150 reservoir units

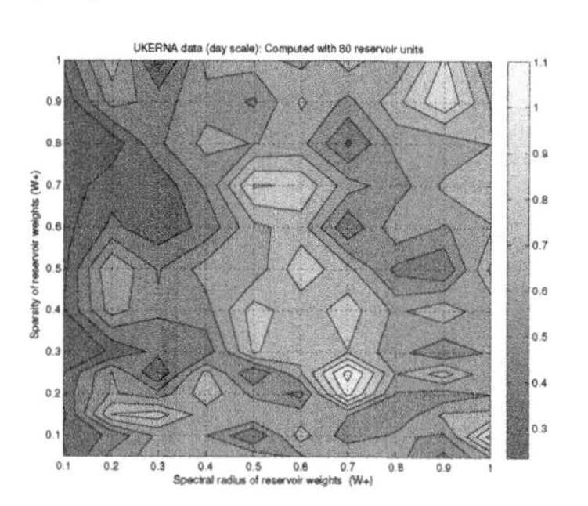

(e) UKERNA validation data (day scale). ESQN with 80 reservoir units

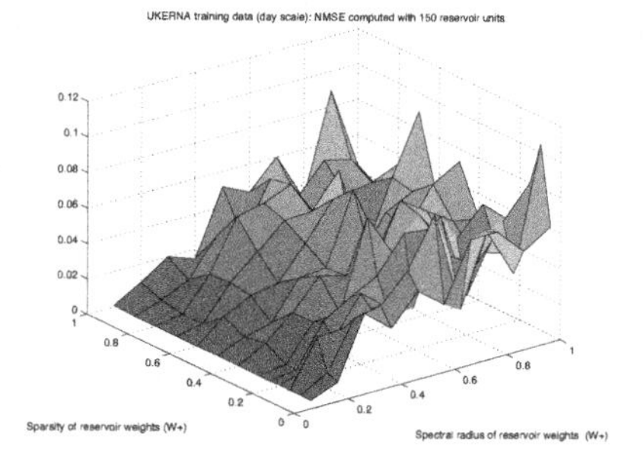

(f) UKERNA training data (day scale). ESQN with 150 reservoir units

Fig. 3. The figures show the influence of the sparsity and the spectral radius of the ESQN reservoir

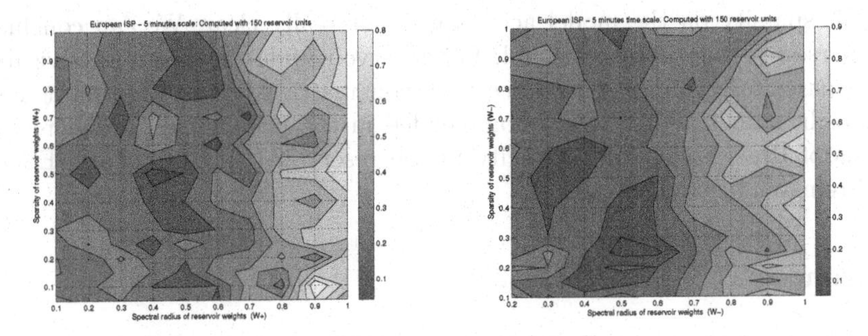

(a) ISP validation data (5 minutes scale). ESQN with 150 reservoir units

(b) ISP validation data (5 minutes scale). ESQN with 150 reservoir units

Fig. 4. The figures show the influence of the sparsity and the spectral radius of the ESQN reservoir using ISP data

Figure 4(a) illustrates the impact of the $\rho(\mathbf{w}^{+})$ and $s(\mathbf{w}^{+})$ on the model performance using the ISP data. In the same way, Figure 4(b) shows the impact of the $\rho(\mathbf{w}^{+})$ and $s(\mathbf{w}^{+})$ when we use the ISP data. We can see that the spectral radius causes an impact on the model performance for the ISP data, when it is close to 0 the performance decreases. The influence of the reservoir sparsity in the model for this data set is not relevant. We can state that the spectral radius behavior of positive weights and negative weights in the reservoir present similar characteristics.

4 Conclusions and Future Works

In many RC applications it has been shown that the reservoir size influences the model accuracy [13, 15]. As in all machine learning devices, there is a tradeoff to reach in the number of model parameters. If it is too small, we can have poor results due to not exploiting enough the benefits of separating the input data into a larger space. If it is too large, the *over-fitting* phenomenon can be presented. The reservoir size impact for the ESQN model case was discussed in [2]. Another important RC parameter is the spectral radius. This parameter influences the memory capability of the ESN model [3]. The role of the spectral radius is more complex when the reservoir is built with spiking neurons in the LSM model [6]. In this paper, we discuss the impact of the spectral radius of the ESQN reservoir on the model accuracy. In spite of the fact that the ESQN reservoir is built based on a Spiking Neuron Network (G-Networks), we can state that the spectral radius has an important influence on the memory capability of the model. According to our experimental results a spectral radius $\rho(\mathbf{w}^{\mathrm{r}})$ close to 1 is appropriate for learning tasks that require long memory, and a value close to 0 is adequate for tasks requiring short memory. This result is similar to that found in the ESN literature [1]. On the other hand, the relevance of the reservoir sparsity on the model accuracy is not really clear. We can conclude that sparse reservoir matrices enable fast reservoir updates, then it is advisable to use sparse reservoirs. For instance, reservoir matrices with 15% or 20% non-zero weight values as it is also recommended in the ESN model. A theoretical analysis of the stability or chaoticity in the reservoir dynamics of the ESQN model remains to be done.

References

1. Lukoševičius, M., Jaeger, H.: Reservoir computing approaches to recurrent neural network training. Computer Science Review, 127–149 (2009)
2. Basterrech, S., Rubino, G.: Echo State Queueing Network: a new Reservoir Computing learning tool. In: IEEE Consumer Comunications & Networking Conference, CCNC 2013 (January 2013), doi:10.1109/CCNC.2013.6488435
3. Jaeger, H.: The "echo state" approach to analysing and training recurrent neural networks. Technical Report 148, German National Research Center for Information Technology (2001)
4. Jaeger, H., Maass, W., Príncipe, J.C.: Special issue on echo state networks and liquid state machines - editorial. Neural Networks (3), 287–289 (2007)
5. Maass, W., Natschläger, T., Markram, H.: Real-time computing without stable states: a new framework for a neural computation based on perturbations. Neural Computation, 2531–2560 (November 2002)
6. Verstraeten, D., Schrauwen, B., D'Haene, M., Stroobandt, D.: An experimental unification of reservoir computing methods. Neural Networks (3), 287–289 (2007)
7. Cortes, C., Vapnik, V.: Support-Vector Networks. Mach. Learn. 20(3), 273–297 (1995)
8. Maass, W.: Noisy spiking neurons with temporal coding have more computational power than sigmoidal neurons. Technical Report TR–1999–037, Institute for Theorical Computer Science. Technische Universitaet Graz, Graz, Austria (1999)
9. Gelenbe, E.: Random Neural Networks with Negative and Positive Signals and Product Form Solution. Neural Computation 1(4), 502–510 (1989)
10. Timotheou, S.: The random neural network: A survey. The Computer Journal 53(3), 251–267 (2010)
11. Gelenbe, E.: Learning in the Recurrent Random Neural Network. Neural Computation 5(1), 154–511 (1993)
12. Basterrech, S., Mohamed, S., Rubino, G., Soliman, M.: Levenberg-Marquardt Training Algorithms for Random Neural Networks. Computer Journal 54(1), 125–135 (2011)
13. Rodan, A., Tiňo, P.: Minimum Complexity Echo State Network. IEEE Transactions on Neural Networks, 131–144 (2011)
14. Cortez, P., Rio, M., Rocha, M., Sousa, P.: Multiscale Internet traffic forecasting using Neural Networks and time series methods. Expert Systems (2012)
15. Basterrech, S., Fyfe, C., Rubino, G.: Self-Organizing Maps and Scale-Invariant Maps in Echo State Networks. In: 2011 11th International Conference on Intelligent Systems Design and Applications (ISDA), pp. 94–99 (November 2011)

Measuring Phenotypic Structural Complexity of Artificial Cellular Organisms

Approximation of Kolmogorov Complexity with Lempel-Ziv Compression

Stefano Nichele and Gunnar Tufte

Norwegian University of Science and Technology,
Department of Computer and Information Science,
Sem Selandsvei 7-9, 7491, Trondheim, Norway
{nichele,gunnart}@idi.ntnu.no

Abstract. Artificial multi-cellular organisms develop from a single zygote to different structures and shapes, some simple, some complex. Such phenotypic structural complexity is the result of morphogenesis, where cells grow and differentiate according to the information encoded in the genome. In this paper we investigate the structural complexity of artificial cellular organisms at phenotypic level, in order to understand if genome information could be used to predict the emergent structural complexity. Our measure of structural complexity is based on the theory of Kolmogorov complexity and approximations. We relate the Lambda parameter, with its ability to detect different behavioral regimes, to the calculated structural complexity. It is shown that the easily computable Lempel-Ziv complexity approximation has a good ability to discriminate emergent structural complexity, thus providing a measurement that can be related to a genome parameter for estimation of the developed organism's phenotypic complexity. The experimental model used herein is based on 1D, 2D and 3D Cellular Automata.

Keywords: Developmental Systems, Emergence, Structural Complexity, CAs.

1 Introduction

Artificial developmental systems take inspiration from biological development, where a unicellular organism, i.e. a zygote, develops to a multi-cellular organism by following the instructions encoded in its genome. The genome contains the building instructions and not a description of what the organism will look like. Several artificial developmental systems take inspiration from cellular models [2-4], where the construct and constructor element is a cell. Thus, each cell in the system is a building block of the system and encapsulates the genome information that regulates the cellular actions, e.g. growth, differentiation, apoptosis. The emergent phenotype can result in a very simple or extremely complex structure. Our goal is to measure the complexity of the phenotypic structure and relate it to the genome information. The notion of

A. Abraham et al. (eds.), *Innovations in Bio-inspired Computing and Applications*,
Advances in Intelligent Systems and Computing 237,
DOI: 10.1007/978-3-319-01781-5_3, © Springer International Publishing Switzerland 2014

structural complexity used is based on the theory of Komogorov complexity [5, 8]. As stated by the Incomputability Theorem (proof in [5]), Kolmogorov complexity is incomputable. Compression algorithms are often used as an approximation of the Kolmogorov complexity [11, 20]. In the experimental work herein, Lempel-Ziv algorithm is used to estimate the phonotypic structural complexity. In the genotype space, genomes are characterized and described by the Lambda genome parameter [6]. Such parameter has shown interesting abilities to discriminate genotypes in different behavioral classes, e.g. fixed, chaotic, random [7]. It is thus investigated if λ is useful to relate genotype composition to the structural complexity of the emergent phenotypes.

The article is laid out as follows: Section 2 presents background and motivation. In Section 3 cellular automata are formally defined. Section 4 describes measures of structural complexity and the notion of Kolmogorov complexity and approximations. Section 5 introduces the developmental model and Lambda genome parameter. Section 6 describes the experimental setup and results. Section 7 includes analysis and discussion and Section 8 concludes the work.

2 Background and Motivation

In the field of Artificial Embryogeny, the goal is often to exploit emergent complexity out of the parallel local interactions of a myriad of simple components. To evaluate such systems' ability, a clear notion of complexity is necessary. Consider the notion of "edge of chaos" [6], a critical region of a parameter space where the system is between order and randomness. In ordered regimes there are only a few distinct possible configurations whether with total randomness the system exhibits the same statistical distribution of behaviors for any initial condition. Therefore, it is in the edge of chaos that systems exhibit high complexity to support advanced features favorable to perform computation. Such emergent complexity, if meaningfully measured, could be predicted using genome parameters. Langton [6] introduced the Lambda parameter to differentiate behavioral regimes where different levels of complexity could emerge.

A developmental mapping may be represented by a function that maps elements in the genotype space with elements in the phenotype space. Such mapping may have regions where small distances between genotypes are preserved into small differences between resulting phenotypes, whether in some other regions distances are hardly preserved at all. In practice, small mutations can have a huge impact on the emergent phenotype. Therefore, a genome parameter predicting phenotypic behavior is useful as a guidance tool to keep resulting phenotypes within a complexity regime, reducing phenotypic difference as long as the genome parameter is kept within defined bounds.

3 Cellular Automata

Cellular automata (CA), originally studied by Ulam [18] and von Neumann [19] in the 1940s, are idealized versions of parallel and decentralized computing systems, based on a myriad of small and unreliable components called cells. Even if a single cell itself can do very little, the emergent behavior of the whole system is capable to obtain complex dynamics. In cellular computing each cell can only communicate with a few other cells, most or all of which are physically close (neighbors). One implication of this principle is that there is no central controller; no one cell has a global view of the entire system. The metaphor with biology can be exploited on cellular systems because the physical structure is similar to the biological multi-cellular organisms.

Formally, a cellular automaton is a countable discrete array of cells i with a discrete-time update rule Φ that executes in parallel on local neighborhoods of a specified radius r. In every time step the cells allow values in a finite alphabet A of symbols: $\sigma^i_t \in \{0, 1, ..., k\text{-}1\} \equiv A$. The local update rule is $\sigma^i_{t+1} = \Phi(\sigma^{i-r}_t, ..., \sigma^{i+r}_t)$. At time t, the state s_t of the cellular automaton is the configuration of the finite or infinite spatial array: $s_t \in A^N$, where A^N is the set of all possible cell value permutations on a lattice of size N. The CA global update rule $\Phi: A^N \rightarrow A^N$ executes Φ in parallel to all sites in the lattice: $s_t = \Phi\, s_{t-1}$. For finite N, the boundary cells are usually dealt with by having the whole lattice wrap around into a torus, thus boundary cells are connected to "adjacent" cells on the opposite boundary. In this paper, 1D, 2D and 3D cellular automata with cyclic boundary conditions are considered.

4 Measuring Structural Complexity

Several complexity measures are proposed in literature, both to quantify genotype and phenotype complexity, e.g. [17]. For genotypes, size may not be an important factor. Even in nature, some unicellular eukaryotic organisms have much larger genomes than humans. Another possibility is to evaluate genotype complexity based on the number of activated genes. Such an activity measure may strongly relate on initial conditions, resulting in a non-precise complexity measure. However, emergent complexity appears at the phenotype level. Important factors can be related to cell organization or functions that the organism is able to perform. Within such an approach Kolmogorov complexity complies well to be able to capture such features.

4.1 Kolmogorov Complexity

The notion of complexity is used differently in distinct fields of computer science. Kolmogorov complexity could be used for understanding emergent complexity in artificial developmental systems.

Let us consider the following strings representing two different states of a 1 dimensional cellular automaton of size 20 at time step t:

a = "01010101010101010101" *b = "01234567894978253167"*

We can intuitively see that string *b* is more complex than string *a*. String *a* is just a repetition of *"01"* whether string *b* does not seem to show any repeating pattern, i.e. string *a* is less complex because we can represent it with a shorter description than for string *b*. Kolmogorov complexity represents the length of the shortest description of a string. In his work, Kolmogorov made use of a Universal Turing Machine to define complexity in an unambiguous way.

Definition *(Kolmogorov Complexity)*: Fix a Turing Machine U. We define the Kolmogorov function, $C(x)$ as the length of the smallest program generating x. This is shown in Equation (1).

$$C(x) = min_p \{|p| : U(p) = x\} \quad (1)$$

It is proven by the *Invariance Theorem* [5] that the particular choice of the universal machine only affects $C(x)$ by a constant additive factor and in particular, $\forall x$, $C(x) \leq |x| + c$. Kolmogorov complexity is incomputable in theory and thus, some approximations are needed.

4.2 Incomputability Theorem

If the problem of computing the Kolmogorov complexity of a string x is to be handled, the way to proceed is to run all the programs which compute x as output and then find the shortest among them, thus testing all the possible programs. Unfortunately, there is no way of knowing if a program halts or not, hence the undecidability of the halting problem [9] implies the incomputability of Kolmogorov complexity. Fortunately, in practice we are not interested in the exact value of the Kolmogorov complexity. Data compression algorithms could be used, to some extent, to approximate it. In fact, strings that are hardly compressible have a presumably high Kolmogorov complexity. Complexity is then proportional to the compression ratio. As stated earlier, the Kolmogorov complexity of a string x is always less than or equal to the length of the string x itself plus a small constant: $C(x) \leq |x| + O(1)$. Yet, as proven by the *Incompressiblity Lemma* [5], there are some strings that are not compressible, i.e. random strings. Formally, a string x is c-incompressible if $C(x) \geq |x| - c$.

4.3 Lempel-Ziv Compression Algorithm

Compression algorithms have been widely used as approximations of Kolmogorov complexity. For example, Lehre, Hartmann and Haddow [10-11], successfully computed approximations of Kolmogorov complexity as measures of genotype and phenotype complexity, using Lempel-Ziv compression algorithm. Zenil and Villareal-Zapata [20] studied one-dimensional cellular automata rules' behavior using approximations of Kolmogorov Complexity. Compression algorithms tend to compress repeated patterns and structures, thus being able to detect structural features in phenotype states. In the experiments herein, we use Deflate [12] algorithm, which is a variation of LZ77 [13]. Deflate is a loseless data compression algorithm that combines LZ77 and Huffman coding [14]. This choice is based on the fact that Deflate is a computationally inexpensive operation and, as long as the state compression process is precisely defined, it is independent of the dimensionality of the state.

If 1D cellular automaton is considered, the correspondent string representing the state of the system at a certain time step could be compressed directly. For a 2D cellular automaton of size 3 by 3, as an example, single rows are concatenated together to compose the state string r:

0	1	0
1	1	2
1	0	0

$\rightarrow r =$ *"010112100"*

The same procedure is applied for a 3D cellular automaton, where all the rows are listed for all the depth levels. Such measure is dimensionality-independent, since it can be used for 1D, 2D or 3D cellular automata. The state string r can now be compressed using the Deflate algorithm, which produces the compressed state string t.

$$t = Deflate\ (r)$$

The next step would be to calculate the length q of the compressed string t.

$$q = Length\ (t)$$

q can then be used to compare the approximate complexity of the states. However, the value has to be normalized in order to compare complexities for different dimensionalities and grid sizes. It is necessary to find lower and upper bounds for the compressed string length, in order to scale the value of q. To do that, it is possible to consider the least and the most complex states. Again, for a 3 by 3 CA, the bounds are:

$$r_{min} = \text{"000000000"} \qquad r_{max} = \text{"012345678"}$$

r_{min} yields the lowest compressed size q_{min} for states of the given dimension and size. Likewise, r_{max} which has no identical symbols, yields the highest compressed size q_{max}:

$$q_{min} = Length(Deflate(r_{min})) \qquad q_{max} = Length(Deflate(r_{max}))$$

The normalized structural complexity measure c of the state s is then:

$$c = (q - q_{min}) / (q_{max} - q_{min})$$

One last remark on the structural complexity calculation is about position and orientation of structures in a state. In fact, it may be better if such measure is transformation invariant. States that represent the same structure, only in a different orientation or position in the grid, should have equal structural complexity. Let us consider the following example:

0	0	0
0	0	0
2	1	2

$\xrightarrow{T}$

2	0	0
1	0	0
2	0	0

In such case, it is evident that the state on the left has equivalent structure to the state on the right, since the transformation T rotates the state 90 degrees. Even though, the measured structural complexity of the row-concatenated-state is different. As such, the following measures are evaluated, as specified in Table 1:

Table 1. Complexity measures

1	Simple Deflate compression	The CA state is represented as a concatenated string and directly compressed.
2	Average of all rotations	The CA state is rotated in all the possible orientations and the correspondent state strings are compressed. The average is computed.
3	Average of all translations	The CA state is shifted in all the possible positions and the correspondent state strings are compressed. The average is computed.
4	Rotations + translations	Both point 2 and 3. The CA state is rotated in all the possible orientations. Each of them is shifted in all the possible positions and the correspondent state strings are compressed. The overall average is computed.

5 Cellular Developmental Model

The developmental model used in this work is an embryomorphic system [1] based on cellular automata, where the goal is the self-assembly of cells from a single zygote which holds the complete genotype information. A CA can be considered as a developing organism, where the genome specification and the gene regulation information control the cells' growth, differentiation and apoptosis. The global emergent behavior of the system is then represented by the emerging phenotype, which is subject to size, shape and structure modifications along the developmental process.

The experimental work is conducted on cellular automata with different dimensionalities, 1 – 3, and neighborhood configurations, 3 – 7.

All CAs have cyclic boundary conditions. Each cell has 3 possible states (cell type 0: empty/dead cell, cell type 1 and cell type 2). The grid is initialized with a cell of type 1 (zygote) in the middle of the grid and develops according to a genotype based on a cellular developmental table that fully specifies all the possible regulatory input combinations, i.e. all 3^n neighborhood configurations are explicitly represented (n represents the neighborhood size). To ensure that cells will not materialize where there are no other cells around, a restriction has been set in the developmental table: if all the neighbors of an empty cell are empty, the cell will be empty also in the following development step. A more detailed description of the developmental model is given in [15-16].

During the development process, a unicellular organism grows to a multi-cellular organism. Two different life phases are identified: the transient phase and the attractor. The transient phase begins with the initial state of the CA (zygote) and ends when the organism reaches its adult form and an attractor begins. Note that this definition is not biologically correct. The attractor represents the time lapse between two repetitions of the same state, i.e. the same state is encountered twice. A complete trajectory is then defined as the sum of transient phase and attractor.

The Lambda Genome Parameter obtained from the genome information can be used to estimate the dynamic behavior of the system and thus can be related to the emergent complexity of the phenotype. Langton [6] studied the parameter λ as a measure of the activity level of the system. λ has shown to be particularly well suited to discriminate genotypes that will develop phenotypes in different behavioral classes, e.g. fixed, chaotic, random [7]. Lambda is calculated out of the regulative outcome of the developmental table, i.e. the output at time *t+1* based on a specific neighborhood configuration at time *t*. According to Langton's definition, a quiescent state must be chosen. We choose the empty cell (type 0) as the quiescent state. λ is then calculated according to Equation 2, where *n* represents the number of transitions to the quiescent state, *K* is the number of cell types (three in our case) and *N* is the neighborhood size, as defined in Table 2.

$$\lambda = \frac{K^N - n}{K^N} \tag{2}$$

Langton observed that the basic functions required for computation (transmission, storage and modification of information) are more likely to be achieved in the vicinity of phase transitions between ordered and disordered dynamics (edge of chaos). He hypothesized that it is easier to find genotypes capable of complex computation in a region where the value of λ is critical.

In the experiments herein, genomes are generated in the whole Lambda spectrum, from 0 to 1, using a similar method to Langton's random table method [6], i.e. for every entry in the developmental table, with probability *(1-λ)* the cell type at the next developmental step is quiescent (type 0); with probability *(λ)*, the cell type at the next developmental step is generated by a uniform random distribution among the other cell types (type 1 or 2).

Previous work [16] has shown that Lambda is able to discriminate genotypes that will end up with very long or extremely short trajectories and attractors. In this paper we investigate relationship between λ, as a genotype measurement, and emergent structural complexity of the corresponding phenotypes.

6 Experimental Setup

Two different sets of experiments are conducted. First, the four different complexity measures in Table 1 are tested. 100 developmental tables are generated for each λ value with granularity 0.01. The correspondent genotypes are developed starting from a single cell on different cellular architectures (1, 2 or 3D with 3, 5 or 7 neighbors), as specified in Table 2, experiment 1. The structural complexity is then measured for the whole trajectory and for the attractor.

Plots from Figure 1.1 to 1.7 show the results for each configuration in Table 2 – experiment 1, where the four lines represent the complexity measures in Table 1. The x-axes plots the whole Lambda spectrum whether the y-axes is the measured structural complexity.

Table 2. Experiment configurations

Dimensionality	*Size*	*Cells*	*Neighborhood radius*
Experiment 1:			
1D	9	9	3
1D	9	9	5
1D	16	16	5
1D	8	8	7
2D	3x3	9	5
2D	4x4	16	5
3D	2x2x2	8	7
Experiment 2:			
1D	25	25	3
1D	27	27	3
1D	25	25	5
1D	27	27	7
2D	5x5	25	5
3D	3x3x3	27	7

The results show that there is a clear relation between the genome parameter value and all the complexity measures, independently from the dimensionality, neighborhood configuration and grid size. Moreover, it is clear that such complexity measures, in relation to λ, are able to characterize both trajectory structural complexity and attractor structural complexity. In most of the cases, the four lines are almost always overlapping, except for some trajectories with λ value between 0 and 0.4, which represents the ordered behavioral regime. In conclusion, it is not necessary to perform expensive rotation and translation before the actual compression. Thus, in the remainder of the paper, the term structural complexity refers to the approximation of Kolmogorov complexity using Deflate as an implementation of Lempel-Ziv.

In the second set of experiments, more accurate tests are performed. 1000 developmental tables are generated for each λ value with granularity 0.01. The correspondent genotypes are again developed starting from a zygote on different cellular architectures as shown in Table 2, experiment 2. The structural complexity for trajectory and attractor is measured in relation with Lambda and compared with the correspondent trajectory and attractor length, measured as the number of development steps. Results are shown in Figure 2.1 to 2.6.

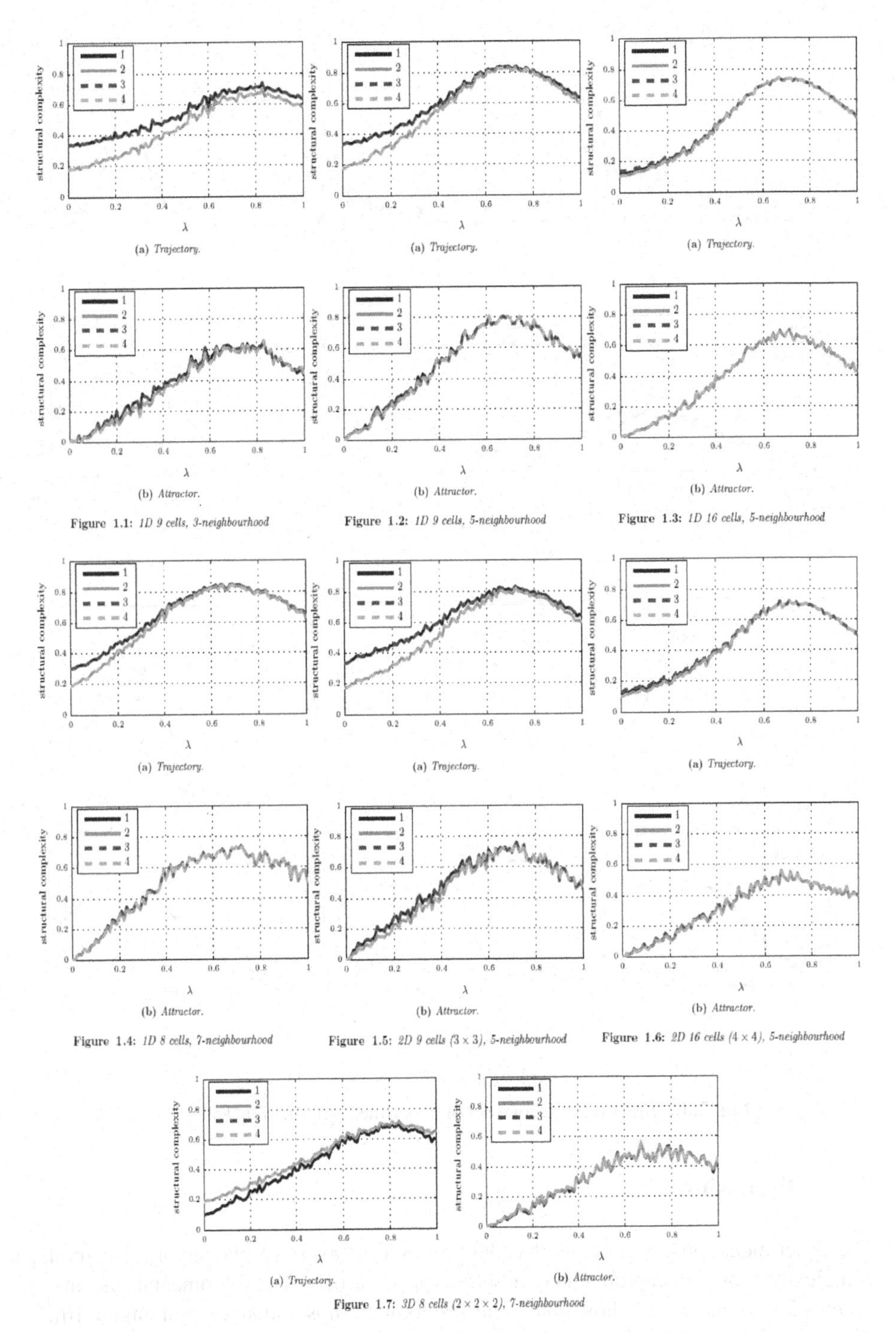

Fig. 1. Results for the first set of experiments defined in Table 2. Lambda on x and structural complexity on y. Lines 1, 2, 3 and 4 represent the measures in Table 1.

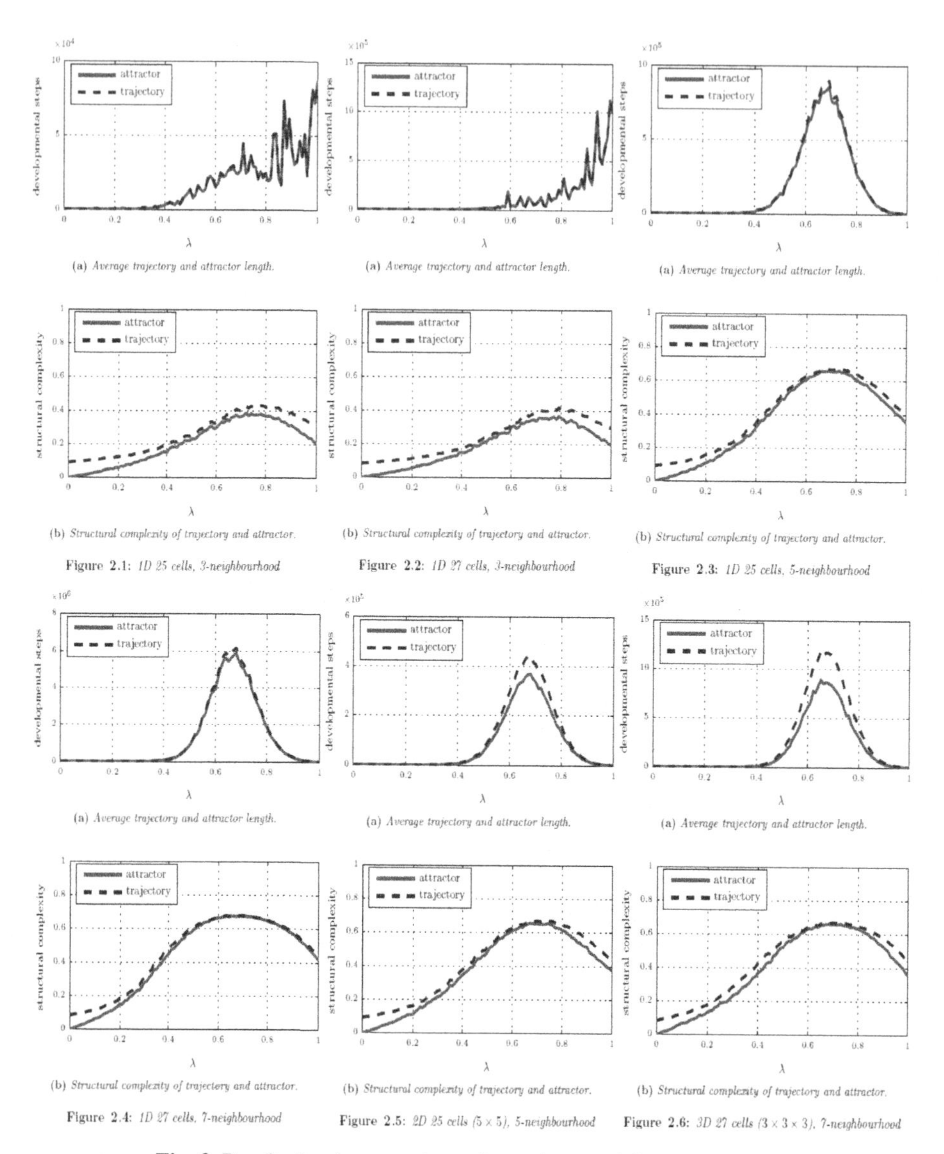

Fig. 2. Results for the second set of experiments defined in Table 2

7 Discussion

The experiments presented show that the proposed measure of phenotypic structural complexity is able to capture emergent properties of artificial developmental systems. Figures 2.1(a) and 2.2(a) show consistent results with those obtained by Langton [6], where Lambda is not able to accurately describe the search space for 1D CA with rather small neighborhood radius and 3 cell types. Remarkably, the structural complexity describes well the parameter space, with low complexity when Lambda is close to 0 and higher complexity where λ reaches the critical value around 0.66. This can be observed in Figure 2.1(b) and 2.2(b). If only the plots (b) are analyzed, from 2.1(b) to 2.6(b), it is possible to spot that the structural complexity curve has the same shape for any configuration.

For 1D CA with small neighborhood the curve is flattened whether for 1D with bigger neighborhoods, 2D and 3D is wider. Overall, the maximum structural complexity that emerges is always slightly over 0.6, meaning that adding dimensions to the developing structure and keeping the total number of cells constant does not increase the structural complexity of the developed organisms. In that sense, 1D, 2D and 3D organisms with same size have the same relative potential to show complex structures. This is an interesting result if one considers adding a new dimension to an EvoDevo system to achieve higher structural complexity. Again, it is possible to observe this result comparing Figure 2.3(b) and 2.5(b) where the developing structures are 1D and 2D respectively, both with 25 cells and 5 neighbors. Same result for Figure 2.4(b) and 2.6(b), using 1D and 3D CA, both with 27 cells and 7 neighbors.

Comparing Figure 2.4(b) and 2.5(b), it is possible to observe that moving from a 1D CA to a 2D CA the parameter region with higher structural complexity is larger with a single dimension. In fact, the shape is more stretched and almost flat on the peak, whether with two dimensions becomes spikier. This seems to be an effect of the enlarged neighborhood. Looking carefully at plot in Figure 2.6(b), this effect caused by an increased neighborhood is still noticeable, even if mitigated by the addition of dimensions. The same behavior is not present in Figure 2.3(b), where development happens on a 1D automaton and structural complexity is analogous to 2D automaton with same neighborhood configuration and same number of cells, as represented in Figure 2.5(b).

Overall, it appears that extending the neighborhood setting results in a wider area where higher structural complexity is reachable. On the other hand, moving from a one-dimensional structure to a two or three-dimensional structure produces more sudden increases in structural complexity for parameter values between 0.3 and 0.5.

8 Conclusion

This paper investigated the emergent structural complexity of artificial cellular organisms at the phenotype level, using approximations of Kolmogorov complexity. Since Kolmogorov complexity in not computable in theory, Deflate compression algorithm based on Lepmpel-Ziv has been used. Such complexity measure is well suited for understanding emergent properties of artificial developmental systems. In particular, it has been shown that structural complexity is strongly related to Lambda genome parameter and its ability to detect different behavioral regimes. This makes it possible to understand if genome information could be used to predict the emergent structural complexity of developing phenotypes. Moreover, the measurement we have used is dimensionality independent and has been experimented on 1D, 2D and 3D CA.

Another observed result is that structural complexity has shown to be powerful enough to characterize the parameter space even when the dimensionality, number of states per cell and neighborhood size are rather small. In such cases, it would not be possible to obtain predictions about trajectory and attractor length at the genotype stage, thus being uncertain about the emergent behavioral regime of the system. As a future work, it may be possible to exploit the potential of Lambda genome parameter to guide evolution towards desirable levels of phenotypic structural complexity.

Acknowledgment. Thanks to John Anthony for carrying out part of the experimental work during his master thesis project.

References

1. Doursat, R., Sánchez, C., Dordea, R., Fourquet, D., Kowaliw, T.: Embryomorphic engineering: emergent innovation through evolutionary development. In: Doursat, R., Sayama, H., Michel, O. (eds.) Morphogenetic Engineering. UCS Series, pp. 275–311. Springer (2012)
2. Kumar, S., Bentley, P.J.: Biologically inspired evolutionary development. In: Tyrrell, A.M., Haddow, P.C., Torresen, J. (eds.) ICES 2003. LNCS, vol. 2606, pp. 57–68. Springer, Heidelberg (2003)
3. Miller, J.F., Banzhaf, W.: Evolving the program for a cell: from French flag to boolean circuits. In: Kumar, S., Bentley, P.J. (eds.) On Growth, Form and Computers, pp. 278–301. Elsevier Limited, Oxford (2003)
4. Tufte, G.: Evolution, development and environment toward adaptation through phenotypic plasticity and exploitation of external information. In: Bullock, S., Noble, J., Watson, R., Bedau, M.A. (eds.) Artificial Life XI, pp. 624–631. MIT Press, Cambridge (2008)
5. Li, M., Vitanyi, P.M.B.: An introduction to Kolmogorov complexity and its applications, 2nd edn. Springer, New York (1997)
6. Langton, C.G.: Computation at the edge of chaos: phase transitions and emergent computation. In: Forrest, S. (ed.) Emergent Computation, pp. 12–37. MIT Press (1991)
7. Wolfram, S.: Universality and complexity in CA. Physica D 10(1-2), 1–35 (1984)
8. Kolmogorov, A.N.: Three approaches to the quantitative definition of information. Problems Inform. Transmission 1(1), 1–7 (1965)
9. Turing, A.: On computable numbers, with an application to the Entscheidungsproblem. Proceedings of the London Mathematical Society, Series 2 42, 230–265 (1936)
10. Hartmann, M., Lehre, P.K., Haddow, P.C.: Evolved digital circuits and genome complexity. In: NASA/DoD Conference on Evolvable Hardware 2005, pp. 79–86. IEEE Press (2005)
11. Lehre, P.K., Haddow, P.C.: Developmental mappings and phenotypic complexity. In: Proceedings of 2003 IEEE CEC, vol. 1, pp. 62–68. IEEE Press (2003)
12. Deutsch, L.P.: DEFLATE Compressed data format specification version 1.3 (1996)
13. Ziv, J., Lempel, A.: A universal algorithm for sequential data compression. IEEE Transactions on Information Theory 23(3), 337–343 (1977)
14. Huffman, D.A.: A method for the construction of minimum-redundancy codes. Proceedings of the I.R.E., 1098–1102 (1952)

15. Tufte, G., Nichele, S.: On the correlations between developmental diversity and genomic composition. In: 13th Annual Genetic and Evolutionary Computation Conference, GECCO 2011, pp. 1507–1514. ACM (2011)
16. Nichele, S., Tufte, G.: Genome parameters as information to forecast emergent developmental behaviors. In: Durand-Lose, J., Jonoska, N. (eds.) UCNC 2012. LNCS, vol. 7445, pp. 186–197. Springer, Heidelberg (2012)
17. Kowaliw, T.: Measures of complexity for artificial embryogeny. In: GECCO 2008, pp. 843–850. ACM (2008)
18. Ulam, S.: Los Alamos National Lab 1909-1984, vol. 15(special issue), pp. 1–318. Los Alamos Science, Los Alamos (1987)
19. von Neumann, J.: Theory and organization of complicated automata. In: Burks, A.W. (ed.), pp. 29–87 (2, part one). Based on transcript of lectures delivered at the University of Illinois (December 1949)
20. Zenil, H., Villareal-Zapata, E.: Asymptotic behaviour and ratios of complexity in cellular automata. arXiv/1304.2816 (2013)

Fuzzy Rules and SVM Approach to the Estimation of Use Case Parameters

Svatopluk Štolfa, Jakub Štolfa, Pavel Krömer,
Ondřej Kobĕrský, Martin Kopka, and Václav Snášel

Department of Computer Science
VSB - Technical University of Ostrava
17. listopadu 15, Ostrava-Poruba, Czech Republic
{svatopluk.stolfa,jakub.stolfa,pavel.kromer,
ondrej.kobersky,vaclav.snasel}@vsb.cz, martin.kopka@c4u.cz

Abstract. Many decisions that are needed for the planning of the software development project are based on previous experience and competency of project manager. One of the most important questions is how much effort will be necessary to complete the task. In our case, the task is described by the use case and manger has to estimate the effort to implement it. However, such estimations are not always correct, not estimated extra work has to be done sometimes. Our intent is to support manager's decision by the estimation tool that uses know parameters of the use cases to predict other parameters that has to be estimated. This paper focuses on the usage of our method on the real data and evaluates its results in real development. The method uses parameterized use case model trained from the previously done use cases to predict extra work parameter. Estimation of test use cases is done several times according to the managers needs during the project execution.

Keywords: software development, estimation, SVM, fuzzy rules.

1 Introduction

Estimation of the work effort, risk factors and other parameters is one of the most important tasks in the software planning. These parameters influence of the cost of the project is very important for the project managers that are pushed to estimate the cost at the very beginning of the project. From the systematic point of view, the decision support process can be established when the knowledge about previous similar projects is available. It is important to mention that presented decision support approach takes place after the analysis phase of the software development process. In many companies such estimations are done repeatedly during the development process and the estimation can be re-evaluated several times.

The idea is to support the estimation with the use of methods that are not common in this area and can bring a new view to the area. Our approach can help manager to

A. Abraham et al. (eds.), *Innovations in Bio-inspired Computing and Applications*,
Advances in Intelligent Systems and Computing 237,
DOI: 10.1007/978-3-319-01781-5_4, © Springer International Publishing Switzerland 2014

estimate the project complexity and the risk of additional work for new projects based on the analysis of use cases. The method uses knowledge base of previous (historical) use cases and provides a decision support in the form of a probability given by estimation of extra work in the project and is based on the usage and fuzzy rules [26].

This paper describes an experiment that extends our previous work on software project decision approach by the results from the real usage and SVM is used as a second prediction method [18]. The estimation experiment was performed on the use case data set from one big long term project in real company. The company has its own standardized way how to describe use cases and stores a lot of use case parameters. Thus, the company has an ideal experimental environment for testing of our approach. Our experiment was performed as a simulation of real development. Data set was divided into sets according to the periods of real development and test data sets (use cases) were selected by the project manager. The goal of our experiment was to test, whether our approach could be really supportive for the parameters estimation in real usage.

1.1 State of the Art

Many formal methods were published in the area of estimation for software development project. In early 90s, Heemstra presented the basic ideas why, when and how to estimate projects in paper "Software cost estimation. In Information and Software Technology" [11]. This paper speaks about importance of estimation of the projects. Many companies do not make estimations at all. Even if they make an estimation, it does not correspond to reality. We are focused on the methodologies of project effort estimation based on use cases. Therefore, next paragraphs describe methods based on use case approach that are mainly used in the use case driven development.

The COCOMO methodology computes effort as a function of program size and set of cost drivers on separate project phases. The name of the model, which was originally developed by Dr. Barry Boehm and published in 1981 [12], is Constructive Cost Model. Dr. Barry Boehm found out a formula computing the classification of separate cost drivers.

Another methodology [14] computes the project estimation using complexity of use cases and its transactions applying the set of adjustment factors.

One of the publications about prediction or estimation of software projects is Defect Inflow Prediction in Large Software Projects [13]. It presents a method to construct prediction models of trends in defect inflow in large software projects. They are focused on two types of the prediction models. It is short-term and long-term prediction model. First one predicts the number of defects, which are discovered in the code up to three weeks. Long-term prediction models predict the defect inflow for the whole project. In our approach, we focus on predicting parameters of the software project or more precisely use case and its extra work parameter.

Usually, the project estimation methodologies and approaches based on the software project parameter analysis rely on known static relationship between parameters. In contrast, the presented approach is based on data mining and machine learning. The method learns the relations and rules from provided knowledge base and in reality creates a model of particular company with its business processes, software development processes, team management etc.

Our approach is based on our previous research that is described in the paper [26]. The method is fully automated and no external knowledge in the form of parameter ranges and so on is required.

The paper is organized as follows: Section 2 introduces the methods used for the estimation; Section 3 describes the logic of our approach: the parameterization of the recent use cases, the parameterization and adding of the new use case, the processing of the data set by fuzzy rules or by SVM, and the estimation of the extra work parameter; Section 4 depicts realized experiment and results; concluding Section 5 provides a summary and discusses the planned future research.

2 Brief Introduction of Methods

2.1 Evolutionary Fuzzy Rules

Evolutionary fuzzy rules [24, 17] (FRs) are fuzzy classifiers heavily inspired by the area of information retrieval (IR). In the IR, extended Boolean IR model utilizes fuzzy set theory and fuzzy logic to facilitate flexible and accurate search [25]. It uses extended Boolean queries that contain search terms, operators, and weights and evaluates them against an internal representation (index) of a collection of documents. FRs use similar data structures, basic concepts, and operations and apply them to general data processing such as classification, prediction, and so forth.

The database used by the FR is a real valued matrix. Each row of the matrix corresponds to a single data record which is interpreted as a fuzzy set of features. Such a general real valued matrix D with m rows (data records) and n columns (data attributes, features) can be mapped to an IR index that describes a collection of documents.

The FR has the form of a weighted symbolic expression roughly corresponding to an extended Boolean query in the fuzzy IR analogy. The predictor consists of weighted feature names and weighted aggregation operators. The evaluation of such an expression assigns a real value from the range [0,1] to each data record. Such a valuation can be interpreted as an ordering, labeling, or a fuzzy set induced on the data records.

The FR is a symbolic expression that can be parsed into a tree structure. The tree structure consists of nodes and leafs (i.e. terminal nodes). In the fuzzy rule, three types of terminal nodes are recognized:

1. feature node which represents the name of a feature (a search term in the IR analogy). It defines a requirement on a particular feature in the currently processed data record.
2. past feature node which defines a requirement on certain feature in a previous data record. The index of the previous data record (current - 1, current - 2 etc.) is a parameter of the node.

3. past output node which puts a requirement on a previous output of the predictor. The index of the previous output (current - 1, current - 2) is a parameter of the node.

An example of FR written down using a simple infix notation is given bellow:

feature1:0.5 and:0.4 (feature2[1]:0.3 or:0.1 ([1]:0.1 and:0.2 [2]:0.3))

where feature1:0.5 is a feature node, feature2[1]:0.3 is a past feature node, and [1]:0.5 is a past output node. Different node types can be used when dealing with different data sets. For example, the past feature node and past output node are useful for the analysis of time series and data sets where the ordering of the records matters, but their usage is pointless for the analysis of regular data sets. The feature node is the basic building block of predictors developed for any type of data.

Operator nodes supported currently by the fuzzy rule are and, or, not, prod, and sum nodes but more general or domain specific operators can be defined. Both nodes and leafs are weighted to soften the criteria they represent.

The operators and, or, not, prod, and sum can be evaluated using fuzzy set operations. Fuzzy set operations are extensions of crisp set operations on fuzzy sets [7]. They are defined using the characteristic functions of operated fuzzy sets [22]. In [7] L. Zadeh defined basic formulas to evaluate the complement, union, and intersection of fuzzy sets but besides these standard fuzzy set operations, whole classes of prescriptions for the complements, intersections, and unions on fuzzy sets were defined [19].

This study employed standard t-norm (1) and t-conorm (2) for the implementation of and and or operators and fuzzy complement for the evaluation of the not operator (3). Product t-norm (4) was used to evaluate the prod operator and dual product t-conorm (5) was used to evaluate the sum operator.

$$t(x,y) = min(x,y) \tag{1}$$

$$s(x,y) = max(x,y) \tag{2}$$

$$c(x) = 1 - x \tag{3}$$

$$t_{prod}(x,y) = xy \tag{4}$$

$$s_{prod}(x,y) = a + b - ab \tag{5}$$

The FR is a simple version of a general fuzzy classifier. In contrast to more complex fuzzy rule-based systems that usually constitute traditional fuzzy classifiers, it consists of a single expression that describes soft requirements on data records in terms of data features. The fuzzy rule modelling selected attribute of a data set is in this study created by the means of genetic programming [16]. The tree structures of parsed fuzzy rules are evolved by iterative application of crossover, mutation, and selection operators in order to find an accurate model of the training data. The IR measure F-Score was used as fitness function [24].

For more information about fuzzy rules, their structure, weighting, evaluation, and evolution see [24, 17].

2.2 Support Vector Machines

Support vector machines (SVM) represent a family of supervised machine learning tools based on statistical learning theory originally proposed by Vapnik [23, 18]. SVM were designed to find an optimal directed hyperplane separating two non-overlapping classes of data with the help of support vectors (i.e. the points in the data closest to the separating hyperplane) [18]. However, later extensions enabled the SVM to learn and classify multiple classes of data, overlapping classes, and noisy data by the introduction of slack variables ξ_i, ξ_i^* that enable soft-margin classifiers [20, 18].

The SVM uses a linear separating hyperplane to construct a classifier with maximum margin by the means of constrained non-linear optimization [23]. Data that is not lineary separable can be processed by the SVM with the help of kernel substitution (kernel trick), i.e. a transformation of input data to a high-dimensional feature space where it might be lineary separable [21, 18]. The SVM combine both, success in practical applications and well-established theoretical background.

The basic SVM for binary classification aims to learn a decision function [18]

$$f(x) = sign(w \cdot x + b) \tag{6}$$

where $\cdot$ is dot product, x is the set of input data vectors (points) $x_1, x_2, \dots, x_m$ and $f(x)$ is the vector of corresponding labels $y_1, y_2, \dots, y_m$, subject to $y_i = \pm 1$. In a geometrical representation, the hyperplanes $w \cdot x + b = 1$ and $w \cdot x + b = -1$ are called canonical hyperplanes and the area between them margin band. Maximizing the margin (i.e. finding optimal hyperplane) involves maximization of the function

$$W(\alpha) = \sum_{i=1}^{m} \alpha_i - \frac{1}{2} \sum_{i,j=1}^{m} \alpha_i \alpha_j y_i y_j K(x_i \cdot x_j) \tag{7}$$

subject to

$$\alpha_i \geq 0, \ \sum_{i=1}^{m} \alpha_i y_i = 0 \tag{8}$$

where K is a kernel used for mapping of input data to high-dimensional feature space (kernel substitution) and α_i, α_j are Lagrange multipliers. Bias b is given by [18]

$$b = -\frac{1}{2}\left[\max_{\{i|y_i=-1\}}\left(\sum_{j=1}^{m} \alpha_j y_j K(x_i, y_i)\right) + \min_{\{i|y_i=1\}}\left(\sum_{j=1}^{m} \alpha_j y_j K(x_i, y_i)\right)\right] \tag{9}$$

The learning of a SVM can be formulated as a quadratic programming problem and implemented using quadratic programming methods [18].

3 Logic of the Approach

This section contains information about the logic of the approach. The approach is based on the approach described in our previous paper [26].

The approach is focused on estimation of future parameters of the use case based on the known parameters in the beginning of the project. For the estimation this approach uses fuzzy rules or SVM methods. Section 4 Experiments and Results shows successfulness of the particular methods.

The figure 1 depictures simplified view of process of current approach. In the left side of the figure you can see values parameterizations of the particular use cases. This set of the use cases is divided into two parts.

- The first part is fulfilled by use cases of recent projects – these use cases have fulfilled all the parameters, known on beginning and also parameters known after realization. For example extra work parameter.
- The second part is fulfilled only by one use case of a new project. This use case have fulfilled all the parameters known on the beginning and the question is how will be its remaining parameters which you can find out only after its realization..

If the new project has more use cases, the approach is repeated for every use case sequentially. The process continues with decision if to use processing by SVM (action 2) or by fuzzy rules (action 1). This is a step how we predict parameters of a use case that we can know only after project - extra work.

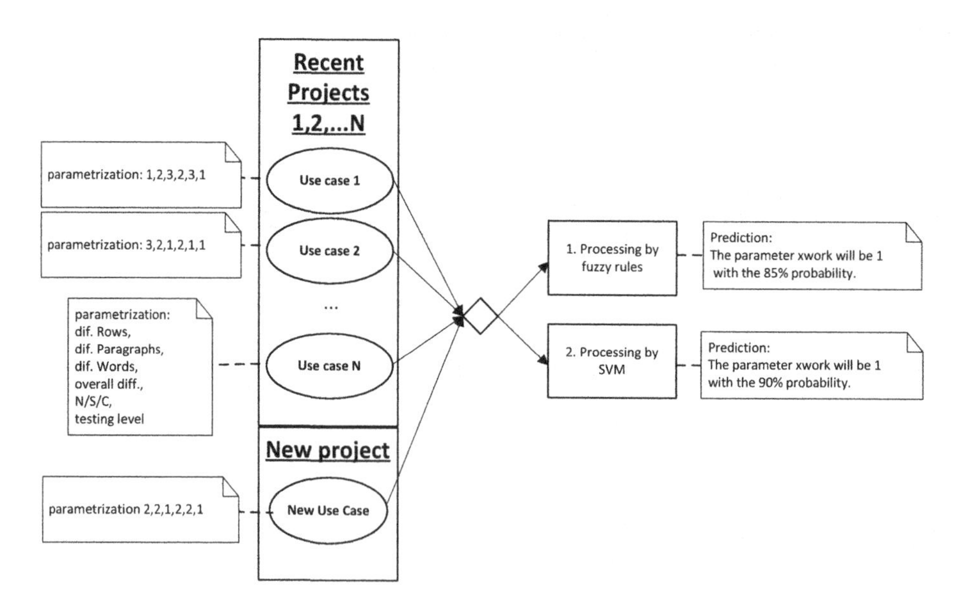

Fig. 1. The logic of the approach

In short, there are 2ways how to use our approach:

1. Predicting parameters by SVM.
2. Predicting parameters by fuzzy rules

Results of tests of our approach are described in the section 4 Experiments and Results.

4 Experiments and Results

The experiment was performed on one data set from one big project. The data set describes development of software project by the means of use case scenarios and its attributes in real software company. The project has more than 1000 use cases of different difficulty. Use cases were implemented between January 2008 and January 2013. Experimental data set was divided into six groups according to the use case development time line. It means that first group of use cases was developed at the beginning of the project starting from January 2008 and the last group was implemented at the end of the data time range that ends on January 2013.

Parameters of FR and SVM used in this study are shown in Table 1. The parameters selection was based on best practices, previous experience, and grid search for SVM parameters.

Table 1. Algorithms and settings

Algorithm	Parameters
FR	Population size 100, crossover probability P_C 0.8, mutation probability P_M 0.02, 100000 generations, fitness function F-Score with $\beta = 1$ [24]
SVM	C-SVM with radial basis function kernel $e^{-\gamma\|u-v\|^2}$, $\varepsilon = 1 \cdot 10^{-3}$, cost $C = 1 \cdot 10^2$, and $\gamma = 2$

The data sets were divided into training and test subsets. Prediction models were trained from the data sets in six steps. In the first step, prediction model was trained from the first part of the data set and consequently used to predict the presence of extra work for the test sample set that always contains 10 use cases that has to be predicted. Next, the prediction model was trained from first and second data sets and used to predict presence of extra work parameter for the second test sample that contained 10 use cases that has to be predicted. Then, the training of the prediction models continues with the addition of data sets 3, 4, 5, and 6. Prediction test were performed in each step when the new data set was added to the training of the prediction model. The results (precision) of training and test data sets using FRs and SVM methods are shown in the following tables (Tables 2-5). The FRs were evolved using genetic programming which is a stochastic method. Because of that, 30 independent models were evolved for each subset in each scenario.

The tables show classification precision, sensitivity and Specificity of training and test data sets for SVM and FR methods.

Table 2. Training precision, sensitivity, and specificity for SVM method

TRAIN	All	Pos	Neg	Precision	Sensitivity	Specificity
s1	561	191	370	68,0927	0,7	0,67984934
s2	671	223	448	69,0015	0,69230769	0,68987342
s3	781	260	521	71,959	0,59808612	0,76398601
s4	891	294	597	72,9517	0,61471861	0,76969697
s5	1001	328	673	72,8272	0,609375	0,76912752
s6	1111	352	759	74,2574	0,62790698	0,77725674

Table 3. Testing precision, sensitivity, and specificity for SVM method

TEST	All	Pos	Neg	Precision	Sensitivity	Specificity
s1	10	5	5	40	0	0,44444444
s2	10	4	6	90	1	0,85714286
s3	10	2	8	90	1	0,88888889
s4	10	4	6	100	1	1
s5	10	2	8	70	0,33333333	0,85714286
s6	10	4	6	90	0,8	1

Table 4. Training precision, sensitivity, and specificity for FR method

TRAIN	All	Pos	Neg	Precision	Sensitivity	Specificity
s1	561	191	370	69,9822	0,573775	0,748373
s2	671	223	448	69,1456	0,542174	0,760677
s3	781	260	521	65,1302	0,53798	0,779098
s4	891	294	597	66,8088	0,562	0,777772
s5	1001	328	673	64,5721	0,529243	0,77121
s6	1111	352	759	71,1041	0,582913	0,79196

Table 5. Testing precision, sensitivity, and specificity for FR method

TEST	All	Pos	Neg	Precision	Sensitivity	Specificity
s1	10	5	5	36	0,190476	0,242857
s2	10	4	6	67,3333	0,499444	0,619365
s3	10	2	8	56,6667	0,385873	0,658148
s4	10	4	6	72,3333	0,669021	0,714444
s5	10	2	8	52	0,189259	0,718333
s6	10	4	6	62,6667	0,499577	0,602024

Tables show that the SVM was able to learn the data well. The average learning precision of FR-based models was worse than those of SVM. Precision of the first training set is 68%, but the test showed only 40%. Training precision for second and all other data sets is between 69%-74%. Precision of test data for those data sets is between 90%-100% with one exception in test s5 with precision 70%. FR results show that the precision for training sets is between 64%-71%. Test precision for those data sets is between 36%-72%. The FR-based models are less numerically stable but the best models achieved better prediction than SVM-based models. Moreover, they are directly interpretable due to their symbolic nature and can be used to provide feedback to the software development process. The SVM-based models on the other hand are stable and their learning is faster.

5 Conclusion and Future Work

The main question of this experiment was to answer the question, whether our approach is able to predict the extra work parameter for use cases that were selected by the manager of the project. Manager has selected use cases from each period of time. The selection was made for all data sets that represent a periods of development time. Managers' selection was made on the basis of his preference, for which use cases from each period of time he would like to know the prediction most.

The result show that for the first set of test use cases the prediction precision was only 40% and 36% for SVM and FR methods respectively. This is not a satisfactory result for manager. When second data set was added to the training set, the result precision was 90% and 67% for SVM and FR respectively. These results started to be more satisfactory for the manager and predictions for next data sets started to be good. However, the test for the fifth data set showed high decrease of probability. The precision for the training data set shows no high deviation from other tests, but the precision of the test set is significantly lower than other results. According to us, this deviation is caused by the selection of the use cases that are somehow not really very well predictable by people and the experiment shows that there are not really precisely predictable by our method as well.

The result of the tests shows also a better precision of the SVM method than FR method. Comparison of these methods was not a goal of this experiment and will be more tested in the future.

We have performed a very constrained experiment that showed the applicability of the automated methods for the prediction of software development parameters. The problem is that our presented method is still not good enough to be a right supportive tool. Our intention is to adjust the approach and perform more comprehensive test to improve and confirm precision of predictions. The goal is also to predict as many other parameters as possible, like e.g. effort, time, resources etc. There is still a lot of work to do, first of all, we would like to properly set, describe the fitting and extend the set of parameters that are inputs for the predictions extend the method and start to use it in real environment.

Acknowledgement. This work was supported by the European Regional Development Fund in the IT4 Innovations Centre of Excellence project (CZ.1.05/1.1.00/02.0070), by the BioInspired Methods: research, development and knowledge transfer project, reg. no. CZ.1.07/2.3.00/20.0073 funded by Operational Programme

Education for Competitiveness, co-financed by ESF and state budget of the Czech Republic and by the internal grant agency of VSB-TU of Ostrava -SP2013/207 "Application of artificial intelligence in process-knowledge mining, modelling and management".

References

1. Crestani, F., Pasi, G.: Soft information retrieval: Applications of fuzzy set theory and neural networks. In: Kasabov, N., Kozma, R. (eds.) Neuro-Fuzzy Techniques for Intelligent Information Systems, pp. 287–315. Springer, Heidelberg (1999)
2. Klir, G.J., Yuan, B.: Fuzzy Sets and Fuzzy Logic. Theory and Applications. Prentice Hall, Upper Saddle River (1995)
3. Kraft, D.H., Petry, F.E., Buckles, B.P., Sadasivan, T.: Genetic Algorithms for Query Optimization in Information Retrieval: Relevance Feedback. In: Sanchez, E., Shibata, T., Zadeh, L.A. (eds.) Genetic Algorithms and Fuzzy Logic Systems. World Scientific, Singapore (1997)
4. Krömer, P., Platoš, J., Snášel, V., Abraham, A., Prokop, L., Mišák, S.: Genetically evolved fuzzy predictor for photovoltaic power output estimation. In: 2011 Third International Conference on Intelligent Networking and Collaborative Systems (INCoS), pp. 41–46. IEEE (2011)
5. Krömer, P., Snášel, V., Platoš, J.: Learning patterns from data by an evolutionary-fuzzy approach. In: Corchado, E., Snášel, V., Sedano, J., Hassanien, A.E., Calvo, J.L., Ślęzak, D. (eds.) SOCO 2011. AISC, vol. 87, pp. 127–135. Springer, Heidelberg (2011)
6. Snášel, V., Krömer, P., Platoš, J., Abraham, A.: The evolution of fuzzy classifier for data mining with applications. In: Deb, K., et al. (eds.) SEAL 2010. LNCS, vol. 6457, pp. 349–358. Springer, Heidelberg (2010)
7. Zadeh, L.A.: Fuzzy sets. Information and Control 8, 338–353 (1965)
8. Kohonen, T.: The Self-Organizing Map. Proceedings of the IEEE 78(9) (September 1990)
9. Kohonen, T., Oja, E., Simula, O., Visa, A., Kangas, J.: Engineering Applications of the Self-Organizing Map. Proceedings of the IEEE 84(10) (October 1996)
10. Vesanto, J., Alhoniemi, E.: Clustering of the Self-Organizing Map. IEEE Transactions on Neural Networks 11(3) (May 2000)
11. Heemstra, F.J.: Software cost estimation. Information and Software Technology 34(10) (October 1992)
12. Boehm, B.: Software Engineering Economics. Prentice Hall (1981)
13. Staron, M., Meding, W.: Defect Inflow Prediction in Large Software Projects. e-Informatica Software Engineering Journal 3(1) (2009)
14. Ochodek, M., Nawrocki, J., Kwarciak, K.: Simplifying effort estimation based on Use Case Points. Information and Software Technology 53(3), 200–213 (2011)
15. Štolfa, J., Štolfa, S., Kobĕrský, O., Kopka, M., Kožuszník, J., Snášel, V.: Methodology for Estimating Working Time Effort of the Software Project. In: 2012 Databases, Texts, Specifications, and Objects (DATESO), pp. 25–37 (2012)
16. Affenzeller, M., Winkler, S., Wagner, S., Beham, A.: Genetic Algorithms and Genetic Programming: Modern Concepts and Practical Applications. Chapman & Hall/CRC (2009)
17. Beshah, T., Ejigu, D., Kromer, P., Snasel, V., Platos, J., Abraham, A.: Learning the classification of traffic accident types. In: 2012 4th International Conference on Intelligent Networking and Collaborative Systems (INCoS), pp. 463–468 (September 2012)

18. Campbell, C., Ying, Y.: Learning with support vector machines. Synthesis Lectures on Artificial Intelligence and Machine Learning 5(1), 1–95 (2011)
19. Feuring, T.: Fuzzy-systeme. Institut für Informatik. Westfälische Wilhelms Universität, Münster (1996)
20. Hamel, L.H.: Knowledge Discovery with Support Vector Machines. Wiley-Interscience, New York (2009)
21. Herbrich, R.: Learning Kernel Classifiers: Theory and Algorithms (Adaptive Computation and Machine Learning). The MIT Press (December 2001)
22. Jantzen, J.: Tutorial On Fuzzy Logic. Technical Report 98-E-868 (logic), Technical University of Denmark, Dept. of Automation (1998)
23. Kecman, V.: Support vector machines an introduction. In: Wang, L. (ed.) Support Vector Machines: Theory and Applications. STUDFUZZ, vol. 177, pp. 1–47. Springer, Heidelberg (2005)
24. Krömer, P., Platoš, J., Snášel, V., Abraham, A.: Fuzzy classification by evolutionary algorithms. In: IEEE International Conference on Systems, Man, and Cybernetics, pp. 313–318. IEEE System, Man, and Cybernetics Society (2011)
25. Pasi, G.: Fuzzy sets in information retrieval: State of the art and research trends. In: Bustince, H., Herrera, F., Montero, J. (eds.) Fuzzy Sets and Their Extensions: Representation, Aggregation and Models. STUDFUZZ, vol. 220, pp. 517–535. Springer, Heidelberg (2008)
26. Štolfa, J., Koběrský, O., Kopka, M., Krömer, P., Štolfa, S., Kožuszník, J., Snášel, V.: Value estimation of the use case parameters using SOM and fuzzy rules. In: The International ACM Conference of Emergent Digital EcoSystems, MEDES (2012)

Multi Objective Optimization Strategy Suitable for Virtual Cells as a Service

Ibrahim Kabiru Musa and Walker Stuart

School of Computer Science and Electronic Engineering
University of Essex, Colchester, UK, CO4 3SQ
ikmusa@essex.ac.uk

Abstract. Performance guarantee and management complexity are critical issues in delivering next generation infrastructure as a service (IAAS) cloud computing model. This is normally attributed to the current size of datacenters that are built to enable the cloud services. A promising approach to handle these issues is to offer IAAS from a subset of the datacenter as a, biologically inspired, virtual service cell. However, this approach requires effective strategies to ensure efficient use of datacenter resources while maintaining high performance and functionality for the service cells. We present a multi-objective and multi-constraint optimization (MOMCO) strategy based on genetic algorithm to the problem of resource placement and utilization suitable for virtual service cell model. We apply a combination of NSGA-II with various crossover strategies and population sizes to test our optimization strategy. Results obtained from our simulation experiment shows significant improvement on acceptance rate over non optimized solutions.

Keywords: Cloud computing, IAAS, biologically-inspired, NGSA-II.

1 Introduction

Cloud computing, as an emerging paradigm, provides a cost effective and flexible model for performing computationally and data intensive tasks. Examples of such tasks include large-scale scientific data analysis, numerical analysis [3], multimedia applications [1], high performance scientific computing [2, 3], and internet-enabled content storage and delivery [2]. In this model, users are allowed to use resources on Pay-Per-Use basis and thereby eliminating any upfront investment [3] [2]. Infrastructure As A Service (IAAS) is a common model of delivering cloud computing. IAAS offers low-level components as Virtual Machines (VMs), which can be booted with a user-defined hard-disk image, and network resources as services. To meet the requirements of scalability, resources deployed in datacenter to enable IAAS can be extremely large. A typical cloud datacenter may comprise thousands of network and IT resources [4]. The magnitude of these resources and the technologies required to enable the cloud services, such as virtualization, raises numerous management and performance challenges including efficient resource utilization, flexibility, and scalability [3]. Numerous proposals, including live migration [5], to meet these challenges are available in the literature.

A. Abraham et al. (eds.), *Innovations in Bio-inspired Computing and Applications*,
Advances in Intelligent Systems and Computing 237,
DOI: 10.1007/978-3-319-01781-5_5,

Inspired by the concept of biological cell [6], a new approach to meet these challenges proposed in [7] envisions template based converged (network and IT) virtual service cell (vCell) similar to a biological cell. We refer to this cell as virtual service cell [8] offered in Virtual Cells As A Service (vCAAS) model of IAAS. The organelles, substrates, and enzymes in biological cells (e.g. mitochondria, endoplasmic reticulum (ER), the Golgi complex, trans Golgi network) interact, based on the chemical reactions between the interacting component and the task to perform, to realize specific cell functions [9]. This type of interaction is similar, in context, to the kind of interaction in a vCell. The intracellular membrane-bound structures and their interaction here are the VMs and network links respectively. In vCAAS, the vCell is totally isolated from other tenants in the datacenter and the owner controls the entire components in the service cell from a template via a service console [8].

In this paper, we extend our previous work [8] with optimization algorithms for solving vCAAS delivery problems. We specifically focus on requests acceptance and resource utilization problems in vCAAS. Our work implements the solution using non-dominated sorting genetic algorithm-II (NSGA-II). Mathematical models for the problems are formulated and the optimization strategy tested in a simulation experiment. We argue that virtual service cell model, as an emerging paradigm, can benefit immensely from existing natural mechanisms for managing complexity. The rest of the paper is formulated as follows. Section 2 gives a brief description of our proposed approach, section 3 presents an overview of related works, section 4 proposes the multi objective multi constraint solution to the problem of request acceptance and resource placement in vCAAS, section 5 presents the result of our experiment, and section 6 concludes the paper and gives an insight into our future work.

2 Virtual Infrastructure Cell

vCAAS envisages a request for large number of virtual resources as a vCell. Request for vCell is submitted in a template. Each vCell is characterized with a service level agreement (SLA) contained in the template which defines important components of the vCell such as number of VMs, type and size of the VMs, storage, and network requirements for intra vCell communication. Similar to biological cell, each vCell is self-contained and has its own control and management services. The fundamental assumptions made in vCAAS are that VMs interact to execute a task thereby incurring communication overheads. Furthermore, vCAAS Virtual Infrastructure Provider (vCAAS-VIP) may allocate VMs from multiple physical hosts to achieve business and functional objectives and a virtual machine utilizes any available resource in the resident vCell to execute a task.

Various actors interact to deliver vCAAS. Physical resources owned by vCAAS physical Infrastructure Providers (vCAAS-PIP) are accessed and virtualized by vCAAS-VIP - also acting as a broker. vCell creation starts with a submission of requests to vCAAS-VIP. vCAAS-VIP then query all vCAAS-PIP and select one. To meet the requirements of the vCell request, vCAAS-VIP evaluate existing resource pool and consults vCAAS-PIP to appraise available resources suitable for the submitted request. vCAAS-VIP then virtualize and offer the vCell to satisfy initial request.

The proposed service cell model offers many advantages to next generation cloud computing. This approach reduces the size of management space to only vCell, allows components in the same vCell share available resources, and provide secured infrastructure [7]. Also, using the template mechanism, similar to biological cell, vCAAS offers better performance in service creation or replication of existing virtual service cell [8] than traditional cloud provisioning.

3 Related Works

The application of genetic mechanisms in systems to solve diverse problem types is common in the literature. Example of successful applications include Condor-G [10] and recently in Cloud Computing [11]. Enormous literature [12-16] exist on the application of genetically inspired heuristics and Meta heuristics to solve known NP-Complete problems. Liu and Tan [15] implemented evolutionary particle swarm optimization to solve bin packing problem that adopts random speed to position the particles. Population based local refinement method mimicking evolutionary selection process is demonstrated in [16] to solve two dimensional bin packing problem. In both [15] [16] mentioned hybrid variations of genetic applications. The use of non-dominated sorting genetic algorithm-II (NSGA-II) is known to perform efficiently in solving non-convex, disconnected, and non-uniform Pareto fronts [17].

In solving a problem using Multi objective evolutionary algorithm (MOEA), [14] identified individuals, population, fitness, and generic operators as major components which can be used to perform selection, re-combinatory, and mutation. In MOEA, coded representations (chromosomes) that represent decision variables in an optimization problem are used as input to crossover and mutation operators [14, 18, 19]. The right size of initial population to generate new solution is critical to the success of evolutionary algorithm. [19] proposed a technique for calculating the minimum population to start the evolution and Cluster analysis [20] has been used in many genetic algorithms to reduce the size of the population before reproduction is carried out. The cluster analysis partitions a collection of p elements in q groups of homogenous elements, where $q < p$ and characteristic or centroid element for the cluster is then selected.

Our work builds on the above scholarships and proposes an optimization strategy suitable for next generation converged (computation, storage, and network) cloud service [7, 8].

4 Multi Objective Optimization for vCAAS

The requirement to achieve many objectives and constraints in vCAAS model is a feature that characterized such problem as MOMCO. vCAAS optimization requires achieving maximum request acceptance and high physical host utilization through effectient VM placements (Figure 1). The problem can be reduced [21] to bin packing problem - a well-known NP-complete problem. Using Cook's theorem [21], the problem of vCAAS resource provisioning can also be viewed as NP-complete with the added requirement that all vCell must be accepted or request for a vCell is rejected. Other additional constraints are communication overhead and the unknown size of request common in cloud environment.

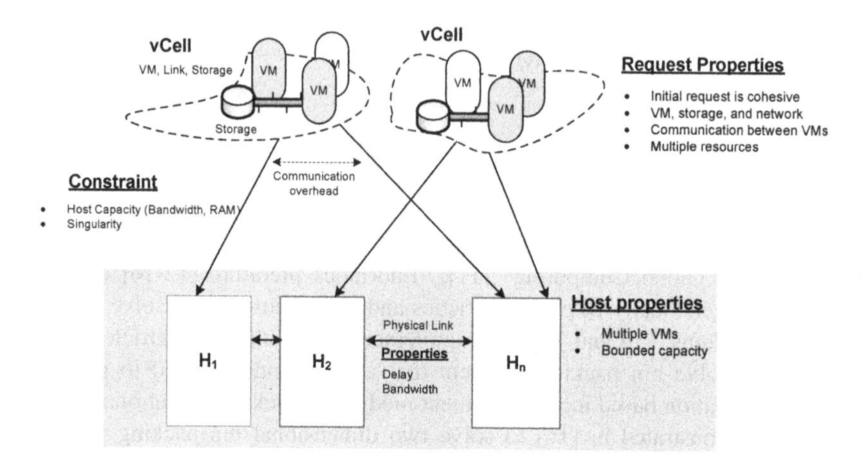

Fig. 1. Optimization problem in vCAAS. H:host, VM:virtual machine

Based on the description of vCAAS presented in section 2, we formulate a multi objective optimization strategy for vCAAS delivery. We begin with a single objective formulation of the problem. As mentioned in section 2, the most important objective to optimize, from vCAAS-VIP perspective, is acceptance of requests. Hence we start with the single objective as:

$$\max \sum_{r \in R} \sum_{v \in r} v \tag{1}$$

$$s.t. \ \sum_{v \in r} v = m \qquad \text{m= |r| (Cohesiveness)} \tag{2}$$

$$\sum_{j \in r} v_{r,j} = 1 \quad r \in I \text{ (Singularity)} \tag{3}$$

$$\sum_{r \in R} W_r X_{r,j} \leq WH_r \ \text{ (Hosts capacity)} \tag{4}$$

$$X_{r,j} = \begin{cases} 1 & if \ \ r \ is \ provisioined \ from \ Host \ J \\ 0 & otherwise \end{cases} \tag{5}$$

v is the VM request, WH is host capacity, and W_r is size of request, r. Equation 1 defines single objective VM provisioning problem. The objective is to maximize the number of accepted Service Cell request R given the finite resources available (WH). The constraints (Eqns. 2, 3, and 4) require that each request from vCell is serviced only once. Next we extend the above formulation to a MOMCO problem based on the requirements of vCAAS describe in section 2. Each vCell contains several VMs and placing each in different physical host generates additional communication overheads as a function of traffic and link characteristics. We denote this overhead as d_r for virtual machine interaction and formulate it as:

$$d_r = \sum_{i=1}^{M} \|h_{0,0} - h_{i,r}\| \tag{6}$$

$h_{m,n}$ is the location identifier (link and distance) of host id m and request n, M is the number of available hosts, and $h_{0,0}$ is the first suitable host for the unit (VM) of request. If we assume that host useable bandwidth, h_c, is bounded by the available link capacity, L_c, such that $h_c \leq L_c$, we can estimate the value of d_r as:

$$d_r = \sum_{k=1}^{N} \sum_{j \in H} \frac{X_{0,j}^{k} * Bw_k}{C_j} \tag{7}$$

The objective is to minimize d_r given initial bandwidth request Bw and record of VM placement. Variable $X_{0,j}^{k}$ defines the use of link between initial chosen host and host j. $X_{0,j}^{k} = 0$ if host j is used as VM container. C_j is the host capacity (and conversely the link capacity) and N is the number containers requested. Another optimization goal is to minimize U_j - unused resource in host j after VM provisioning. This is formulated as:

$$U_j = C_j - \sum_{r \in R} \sum_{j \in r} Bw_r \, X_{r,j} \tag{8}$$

U_j is expected to generate new solution with each permutation of request vector. By applying selection, crossover, and mutation the process can be guided toward the best solution.

For a set of request R, the acceptance rate, A, is determined as the ratio of requests r∈R that are satisfied to the total size of all requests. We aim to maximize A as:

$$max\, A = max \frac{\sum_{r \in R} (S_{r,j} X_{r,j}^{k})_{j \in H, k \in \Gamma}}{\sum_{r \in R} S_r} \tag{9}$$

Subject to $|\Gamma_k| = |S_{r,j} \, X_{r,j}^{k}| \quad \forall \, r \in R$

Γ is the given set of vCell requests each indexed with k, $S_{r,j}$ is the size of request r provisioned from host j and S_r is the request size for r. As defined previously, $X_{r,j}^{k}$ is a binary variable which takes 1 if request r submitted in vCell *k* is provisioned in host j and 0 otherwise. The constraint enforces that all the initial resource requests for vCell Γ_k must be accepted otherwise the request is rejected.

4.1 Formulating the GA

In this section we formulate an implementation of the propose problem using genetic algorithm (GA) and describe the strategies adopted for the problem encoding, crossover, mutation, and determining the minimum population sizes.

Encoding

We encode the problem as an integer vector of size, S, representing the request. Each index X_n is associated with a value V_n which is the index of the resource. n represents the order the requests are submitted. This way, if we generate requests as a vector $\vec{V}$ then the encoding V_0 V_3 V_1 means that the VM requests are submitted as ordered set $[\vec{V}(V_0), \vec{V}(V_3), \vec{V}(V_1)]$ for processing. Since the requests vary in size and assuming offline optimization - where all initial requests are submitted at start, altering the order at which the requests are sent by permutation.

Ordered request	V_0	V_1	V_2	V_3	V_4	V_m
Index	X_0	X_1	X_2	X_3	X_4	X_m

Population

The minimum initial population for the experiment is computed as a function of the VM request size and the number of iterations [19] and represented as:

$$\Pr(S) = (1 - (\frac{1}{2})^{N-1})^{l} \tag{10}$$

This has been shown to give the minimum probability that at least one allele is present at each locus thereby guaranteeing that every point in the search space, S, is reachable from the initial population by crossover only.

Operators

In this work we consider the partially matched crossover (PMX) strategy which has been shown in [22] to be suitable for similar problem and performs better than other techniques such as ordered crossover strategy. The technique involves initialization, applying the PMX crossover with probability, getting two cutting points, and getting the sub chains to interchange, and performing the Interchange. The mutation considered for the generating new solutions is the swapping strategy [18].

5 Results

This section describes the application of the optimization model presented previously to the virtual service cell design problem using NSGA-II. To assess the quality of the NSGA-II results, various population sizes, crossover strategies, and maximum evolution are tested.

The experiment is tested on JMetal [18] which provides a framework for multi objective optimization written in java. All experiments were implemented on a 64 bit Dell OptiPlex system with 8GB Ram and core 2 Duo processor running at 3.33GHz. We begin with investigating the improvement on acceptance due to our optimization strategy. We generate 10000 VMs request, from vCell templates of varying VMs size, to be provisioned on 500 hosts all connected on one level rack topology. In all the

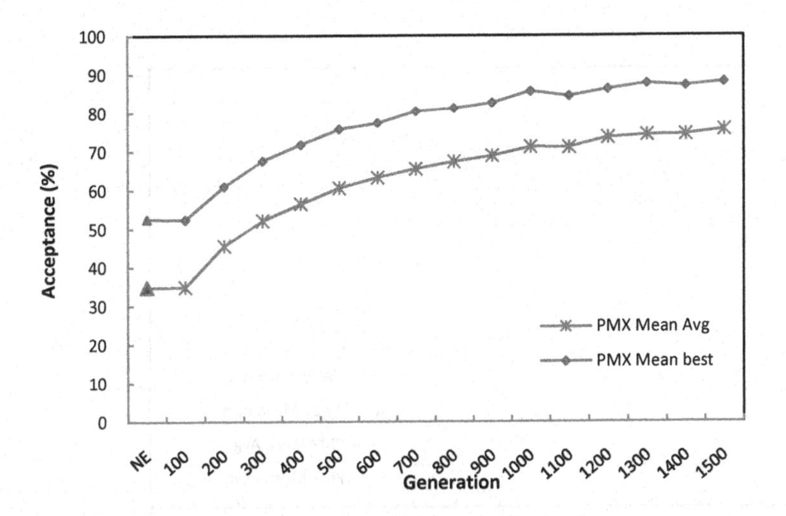

Fig. 2. Result showing acceptance rate at various generations. NE is acceptance rate without optimization.

subsequent experiment, crossover and swap mutation probability are fixed at 0.95 and 0.2 respectively. The binary tournament is used as the selection operator. In this experiment a combination of first fit and NSGA-II is implemented to calculate the number of accepted virtual cell requests.

As shown in figure 2, the First Fit greedy force algorithm (labeled NE) mean of best and average solution are 52% and 34% acceptance rate respectively. Our optimization strategy at evolutions range 100-1500 improved the initial solution by up to 90%.

We then compare the result obtained from the first experiment with a solution of different crossover strategy. Using two crossover strategies - partial match crossover (PMX) and 2 Way crossovers (2WX) - and various iteration values, the improvement on the acceptance rate is maintained by up to 50% (Figure 3) from initial non optimized best fit allocation. This improvement is true for both the crossover strategies with increasing number of generations. The improvement is incremental until a saturation point is reached at which small change on the result is obtained. However, figure 3 shows that PMX slightly performs better than 2WX for our problem, though the change is not significant as shown in the graph.

To further investigate the strength of PMX over 2WX, we conduct experiment to measure the time required (figure 4) to perform number of evolutions by each strategy. In the 25th percentile of generation, both strategies indicate same time overheads.

However, 2way crossover shows a slight improvement over PMX as evolutions approach 700. As indicated in figure 3 very little improvements are noticed on the higher percentiles. This means that despite the lower time overheads shown by 2WX, PMX still proved suitable for the problem of virtual cell composition.

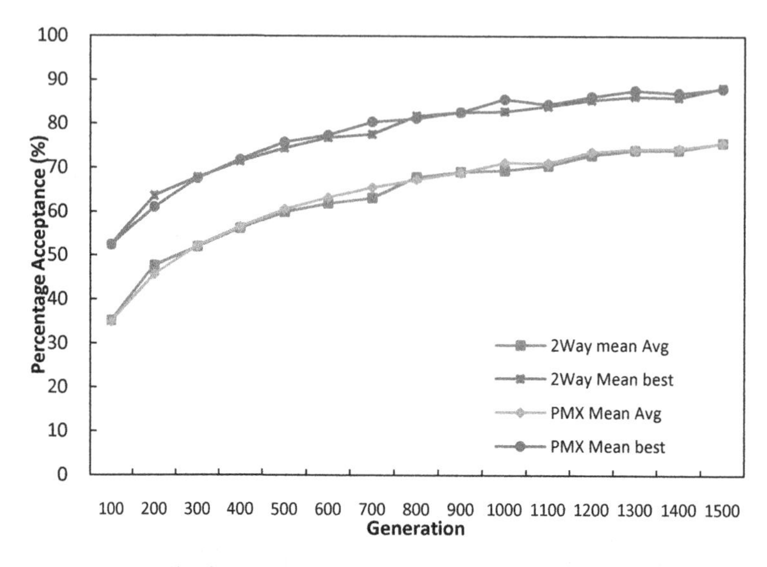

Fig. 3. Acceptance rate for varying number of generations

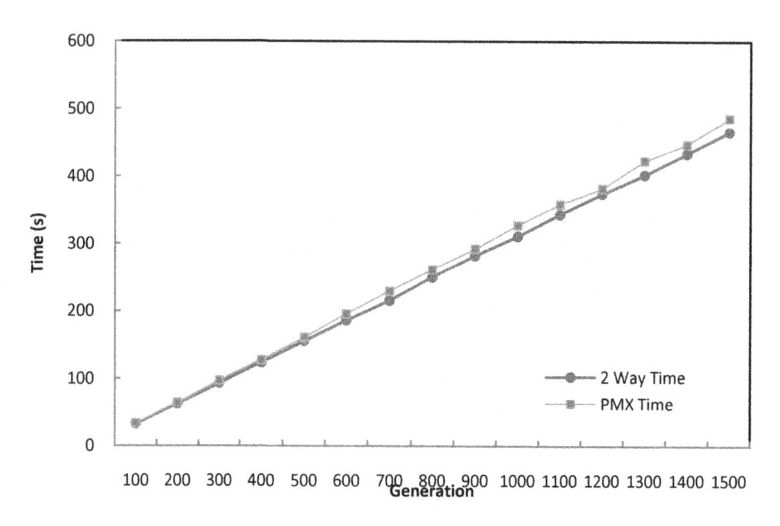

Fig. 4. Comparing the time overheads by two crossover strategies

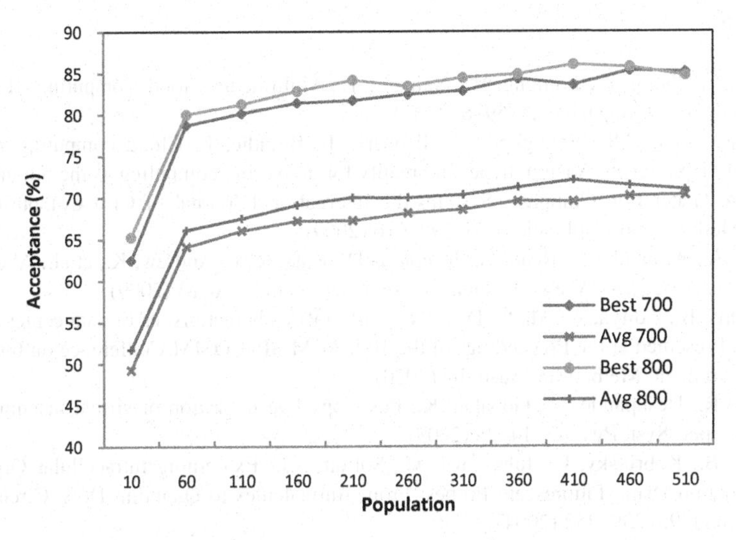

Fig. 5. Investigating the impact of varying population sizes on acceptance rate

We then investigate the impact of population on PMX. Figure 5 shows that increasing the population size has significant impact on the acceptance rate at lower values of populations size (population size<120) and little significance on the search space and the performance as the population size reaches higher values. In both cases, significant improvement is achieved by applying the optimization.

6 Conclusion

In this work we demonstrated two biologically inspired techniques – virtual service cell and genetic algorithm - applied to solve resource management problem in cloud computing. We begin with describing the analogy between components in biological cell and the virtual service cell in our proposed cloud model. Using genetic algorithm, we then applied optimization to improve resource usage and requests acceptance. Our work confirms the suitability of partial match crossover in virtual infrastructure container model. Our work explore the trade-offs associated with higher performance of virtual infrastructure container and time overheads required to achieve higher result in terms of request acceptance. In this work we focus on very simple network topology, we hope to advance the techniques to complex hierarchical network topologies in the future.

References

1. Wenwu, Z., Chong, L., Jianfeng, W., Shipeng, L.: Multimedia Cloud Computing. IEEE Signal Processing Magazine 28, 59–69 (2011)
2. Buyya, R., Yeo, C.S., Venugopal, S., Broberg, J., Brandic, I.: Cloud computing and emerging IT platforms: Vision, hype, and reality for delivering computing as the 5th utility. Future Generation Computer Systems-the International Journal of Grid Computing-Theory Methods and Applications 25, 599–616 (2009)
3. Michael, A., Armando, F., Rean, G., Joseph, A.D., Katz, R.H., Andrew, K., et al.: Above the Clouds: A Berkeley View of Cloud Computing. Commun. ACM (2009)
4. Theophilus, B., Aditya, A., Maltz, D.A.: Network traffic characteristics of data centers in the wild. Presented at the Proceedings of the 10th ACM SIGCOMM Conference on Internet Measurement, Melbourne, Australia (2010)
5. Hines, M.R., Deshpande, U., Gopalan, K.: Post-copy live migration of virtual machines. SIGOPS Oper. Syst. Rev. 43, 14–26 (2009)
6. Zorov, D.B., Kobrinsky, E., Juhaszova, M., Sollott, S.J.: Examining Intracellular Organelle Function Using Fluorescent Probes: From Animalcules to Quantum Dots. Circulation Research 95, 239–252 (2004)
7. Banerjee, P., Friedrich, R., Bash, C., Goldsack, P., Huberman, B.A., Manley, J., et al.: Everything as a Service: Powering the New Information Economy. Computer 44, 36–43 (2011)
8. Musa, I.K., Stuart, W.: A Converged Service Plane for Virtual Infrastructure Containers. IJCSI International Journal of Computer Science 10, 12 (2013)
9. Thomas, F.J.M.: The Biogenesis of Cellular Organelles. Plenum Publishers (2005)
10. James Frey, T.T., Foster, I., Livny, M., Tuecke, S.: Condor-G: A Computation Management Agent for Multi-Institutional Grids. Journal of Cluster Computing 5, 237–246 (2002)
11. Junlin, C., Wei, Z., Jing, Z., Wei, W.: Design of cloud model controller based on multi-objective optimization. In: Control and Decision Conference (CCDC), pp. 19–24 (2011)
12. Rothlauf, F.: Design of modern heuristics principles and application. In: Natural Computing. Springer, Berlin (2011)
13. Kramer, O.: Self-adaptive heuristics for evolutionary computation. SCI, vol. 147. Springer, Heidelberg (2008)
14. Donoso, Y., Fabregat, R.: Multi-objective optimization in computer networks using metaheuristics. Auerbach Publications, Boca Raton (2007)
15. Liu, D.S., Tan, K.C., Huang, S.Y., Goh, C.K., Ho, W.K.: On solving multiobjective bin packing problems using evolutionary particle swarm optimization. European Journal of Operational Research 190, 357–382 (2008)
16. Fernández, A., Gil, C., Márquez, A.L., Baños, R., Montoya, M.G., Parra, M.: A memetic algorithm for two-dimensional multi-objective bin-packing with constraints. In: Proceedings of the 13th Annual Conference Companion on Genetic and Evolutionary Computation, pp. 341–346 (2011)
17. Deb, K., Pratap, A., Agarwal, S., Meyarivan, T.: A fast and elitist multiobjective genetic algorithm: NSGA-II. IEEE Transactions on Evolutionary Computation 6, 182–197 (2002)
18. Durillo, J.J., Nebro, A.J.: jMetal: a Java Framework for Multi-Objective Optimization. In: Advances in Engineering Software, pp. 760–771 (2011)
19. Reeves, C.R.: Modern heuristic techniques for combinatorial problems. Blackwell Scientific Publications, London (1993)

20. Moulton, C.M.: Hierarchical Clustering of Evolutionary Multiobjective Programming Results to Inform Land Use Planning (2007)
21. Garey, M.D., Johnson, D.S.: Computers and Intractability: A Guide to the Theory of NP-Completeness. Freeman, CA (1979)
22. Naveen, K., Karambir, R.K.: A Comparative Analysis of PMX, CX and OX Crossover operators for solving Travelling Salesman Problem. International Journal of Latest Research in Science and Technology 1, 98–101 (2012)

An Application of Process Mining to Invoice Verification Process in SAP

Jakub Štolfa, Martin Kopka, Svatopluk Štolfa, Ondřej Koběrský, and Václav Snášel

Department of Computer Science
VSB - Technical University of Ostrava
17. listopadu 15, Ostrava-Poruba
Czech Republic
{jakub.stolfa,svatopluk.stolfa,
ondrej.kobersky,vaclav.snasel}@vsb.cz,
martin.kopka@c4u.cz

Abstract. There are many processes in companies that are enacted many times every day. The key issue of the company management is to control the company cash flow and try to optimize the cost of everyday operations. There are many ways how to support the process enactment, but at the end, when there are some data from the process usage, the analysis of the efficiency is needed. One of the ways how to analyze the process and effectively analyze the process data is to use process mining methods. In this paper, we present the usage of process mining methods to real invoicing process and show the possible impact of the results to the process or organizational improvement.

Keywords: process mining, case study, SAP, process improvement.

1 Introduction

Companies enact most of their key business processes with the support of information systems. Enacted processes are represented by workflows, partly managed by the information system, partly managed by users' decisions and activities. It is not easy to understand whether the specific process runs efficiently, because usually many various activities are processed in parallel and process definition allows plenty of process enactment variations. Our task was to analyze specific process with request to suggest steps for its simplification, curtailment and enact the process cheaper.

First of all, we have found that even when processes are enacted in standardized platform (SAP), the requested information for process analysis using process model and monitoring of its enactment is not available. We have decided to apply the process mining approach that use rich information log saved by the SAP within the process enactment. This approach enabled us to start analysis immediately without long term process modeling and implementation phases.

A. Abraham et al. (eds.), *Innovations in Bio-inspired Computing and Applications*,
Advances in Intelligent Systems and Computing 237,
DOI: 10.1007/978-3-319-01781-5_6,

Our paper is organized as follows: Section 2 introduces the state of the art; Section 3 depicts the experiment that we have performed: describes the data that we have obtained; shows the usage of the process mining methods and explains obtained results and its interpretation in real business; concluding Section 4 provides a summary and discusses the planned future research.

2 State of the Art

In last decades, systems started to be more and more process oriented [1]. The shift to process oriented systems was motivated by the idea of supporting systems to the daily business, to shift the knowledge about operations that could be described as processes from humans to systems. Process oriented systems started to be worshiped as the only way to control the processes and activities that has to be enacted. The knowledge about the processes and their enactment was transferred to the systems.

The shift from the data oriented systems to the process oriented systems brought the companies tools to control and check the enactment of the processes and resources that are involved.

However, to build a quality process oriented system means to build a system supported by workflow system. There are many reasons why this is not always possible, such as a cost, change of doing things very often etc. Thus there are many systems that are process supportive but not exactly process driven like ERP systems (SAP, CRM systems, B2B systems and many other types of information systems). Sometimes these systems are not even aware of the processes that are supported. Processes were not defined at the beginning of the systems implementation or were defined but lost in the long term usage and maintenance of the system.

The need of the proper knowledge about the business data led to the usage of Business intelligence (BI) tools. Since the systems shifted from the data oriented to the process oriented systems, new subdomain of BI, Business process intelligence (BPI), was defined. BPI and its supportive tools help users to manage process execution quality by the analysis, prediction, monitoring, control and optimization [2].

On the other hand, there is a Business process management (BPM). BPM [3] can be defined as whole business company process management and optimization. Its concern is on the process improvement and its alignment to the needs of clients. BPM lifecycle consist of design, modeling, execution, monitoring and optimization. It means that the BPM take care of the composition, enactment and analysis of the operational business processes.

Business process definitions are sometimes quite complex and allow many variations. All of these variations are then implemented to supportive systems. If you want to follow some business process in a system, you have many decisions and process is sometimes lost in variations. Modeling and simulations can help you to adjust the process, find weaknesses and bottlenecks during the design phase of the process. Sometimes you can guess or know the patterns and occurrence probabilities of variations that are used during the execution phase. However, not even modeling and simulation of the processes can tell you, how processes are really enacted in the system, what is e.g. the perceptual usage of the variations and whether some variations are enacted at all. If you want to analyze the real usage of the system, recognize its weaknesses, bottlenecks or strangeness on the real data, you have to know how the process was followed in reality. Process mining is an approach that is used for the analysis of real enactment of the processes. Process mining uses logs of real process enactments to analyze the process itself. Process mining can answer you the question, how the process was really executed, which variations were used and what are the probabilities of the enactment of each process variation. Process mining can be seen as a supportive method for the BP and BPI analysis and from the perspective of BPM, can be used as a feedback to the BPM methods [4].

2.1 Related Work – Case Studies

There is a lot of papers that describe new ways or improvements of methods, techniques and algorithms used in the process mining. Surprisingly only several papers are focused on the case studies, it means, in fact, practical usage of the process mining. Detailed overview of recent case studies up to 2011 can be found in [16]. Authors describe 11 case studies in several domains, mainly in public services [5-15].

Since then, some other papers that describe the usage of process mining where published. In [17], authors applied fuzzy mining, trace clustering and other additional methodologies among ProM to analyze block movement process in shipyards. They used real-world event logs of the Block Movement Control System (BMCS).

In financial sector, authors of [18] used process mining for the identification of financial processes to analyze the compliance to the security requirements needed by the security audit.

In [19], authors demonstrate the applicability of process mining to the discovery of processes that characterize the knowledge maintenance and the organizational perspective is used to find relations between individual performers. This study was made on the real case of a knowledge maintenance process in aviation institute.

The range of process mining case studies and variations of process mining types and methods shows the wide range of process mining applicability to answer different questions in different domains.

3 Case Study

This cases study reacts to a company request to analyze business process of the invoice verification in SAP system with aim to identify context in which the process is not effective and provide a suggestion for the process improvement. The analyzed company runs SAP system in five countries (with five different jurisdictions) and processes approx. 30 000 supplier invoices per year.

3.1 Context

Examined business process of the invoice verification is implemented in SAP ERP and SAP DMS, user activities are controlled by SAP business workflow. Users participate in the invoice verification workflow in several different roles (creator, accountant – completion, approver, and accountant – decision and posting).

Generally, it is process where the accountant should create the invoice, verify it, then send to the approvers and finally when he gets it back he does invoice posting. The case study is about verification of the idea about the process, find deviations and in case of it do retrospective view to the particular instance of invoice verification.

3.2 Data Obtaining and Data Structure

The main task during data preparation was to collect and adjust the application logs to the final structure (see Table 1) needed for the process mining analysis. Data was collected from the workflow system log containing most information about activities. It was needed then to make an adjustment, because the basic log contained the "activity" object that was split between two fields.

The second important step was the anonymization of the data (so that no real name or legible ID is present in the analysis). It is possible to find back the original IDs for essential analysis of the real meaning of results in reality, but with the mapping keys only. Following Table 1 depictures the structure of the adjusted log that we have used for the process mining – we have collected data for all fields, but only mandatory fields were used for the process mining analysis.

Table 1. Data structure of the log

IDOBJ	ID of the invoice	mandatory
IDACTIVITY	ID of the activity	mandatory
DATESTART	Date when the activity started	optional
TIMESTART	Time when the activity started	optional
DATEEND	Date when the activity ended	mandatory
TIMEEND	Time when the activity ended	mandatory
ROLE	Role of the invoice processor (creator, accountant, approver)	mandatory
USER	User account of the invoice processor (real user name)	optional
TRANSACTION	System transaction run by the user during processing the activity	mandatory
DATA	Detailed information about context of the invoice – specifically ORGID, INVOICETYPE, ...	optional

We have loaded the log with DATEEND between 1. 1. 2012 - 30. 6. 2012, totally we have loaded **37 991 records** for adjusting.

3.3 Preprocessing of Data

The main task of the data preprocessing was to map records from the application log to the parts of the process in the process mining.

The process in the process mining consists at least of:

- Case – one pass of the process
- Event – one step of the process
- Start time – start time of the task
- End time – end time of the task
- Originator – originator of the particular task

Our log is mapped in the following way:

- Case – IDOBJ
- Event – ROLE + TRANSACTION
- End time – DATEEND + TIMEEND
- Originator - USER

Connection of the ROLE and TRANSACTION allows us to recognize specific activities that are performed by one role in the process. ROLE contains the information which role is performing particular activity (for example accountant). TRANSACTION contains the information about what kind of operation is doing particular role. We have recognized following event classes by the connection of these two attributes:

- Invoice *Creation* (accountant-01) – this event in the real business process means completion of the invoice (scanned originals, completion of all important fields) or recompletion of the existing invoice.
- Invoice *Verification* (accountant-02) – this event verifies the formal correctness of the current invoice (VAT, accounts, …)
- Invoice *Approval* (approver-04) – this event approves the invoice (responsible persons approve the factual correctness of the invoice and approve or reject the invoice, or send the workflow back to the Verification/Creation)
- Invoice *Posting* (accountant-05) – alternative decision of accountant (largely the final posting of the invoice and change of the responsible approver)

There were found some unacceptable records in the log during preprocessing analysis. There were missing mandatory fields of Originator in some records of the log – the reason is that some records were not processed by manual workflow to the end, but were processed by automatic job without updating the application log. We removed these (1279) records and all related records of the particular IDOBJ or case. So at the end, the analysis was performed on **36 711 records**, or respectively events.

3.4 Summary of the Log and Adjustment of Data

The process mining tool ProM 6.2 was used for all analysis attempts [20].

The first step of the analysis was the analysis of the log summary. The log contained **7350 cases** of one process, this means **36711 events**.

The Table 2 describes occurrences, absolute or relative, of particular tasks. Task *Approval* is a most frequented task in the process.

Table 2. Occurrences of all events

Event class	Occurrences (absolute)	Occurrences (relative)
Approval	15524	42.287%
Creation	7669	20,890%
Verification	7344	20,005%
Posting	6174	16,818%

The Table 3 depictures start-event classes of the process. It means start classes that are starting events for at least one of process enactments. The Table 4 depictures end-event classes of the process. It means task classes that are end for at least one of process enactments. Both tables show us that there are event classes that should not start or end the process. It can be caused by some deviations in the process. These deviations of start and end event classes can be target for the further analysis, but in our case, we are sure that correct process starts by event Creation and ends by event Posting every time. It is controlled by the workflow in the SAP, there are no other starts and ends that could be possible. Thus, the only possible explanation in our case is that deviations of start and end event classes are caused by the selection of the time segment from the log for the analysis. Some cases started before 1.1.2012 and some cases started before 30.6.2012 but ended after that date. It caused incomplete cases that have to be filtered. Nevertheless, this can be also analyzed further by other views.

Table 3. Start events

Event class	Occurrences (absolute)	Occurrences (relative)
Creation	7218	98.204%
Posting	122	1,66%
Verification	8	0,109%
Approval	2	0,027%

Table 4. End events

Event class	Occurrences (absolute)	Occurrences (relative)
Posting	5552	75.537%
Approval	969	13,184%
Verification	496	6,748%
Creation	333	4,531%

ProM method Filter log using Simple Heuristics was used for the **filtering** of incomplete cases. There was specified that every case can start only by event class *Creation* and can end only by event class *Posting*. This adjustment of the log caused that log finally contained **5424 cases** and **30784 events** for further analysis.

3.5 Analysis of the Process Characteristics

The final log of examined process without deviations that were caused by data obtaining contained, as was mentioned before, 5424 cases and 30784 events. Occurrence of the event classes is depictured in the Table 5.

Table 5. Occurrences of all events

Event class	Occurrences (absolute)	Occurrences (relative)
Approval	13563	44.059%
Creation	5917	19,221%
Verification	5652	18,360%
Posting	5652	18,360%

Creation and *Posting* are the only start and end event classes, but for the better recognition in the analysis by different ProM methods, artificial events *Start* and *End* were added by the ProM function – Add Artificial Events.

Process Model

Process model is depictured in the figure Fig. 1. We have reconstructed process model from the log by BPMN Analysis (using Casual Net Miner) [22]. Process starts by artificial event Start and then continues by event *Creation.* Next one is *Verification.* These two events Creation and Verification can be done repeatedly. Next, there is a gateway where the *Approval* event can be skipped. *Approval* event can be done repeatedly too. Last events are *Posting* and artificial event *End.*

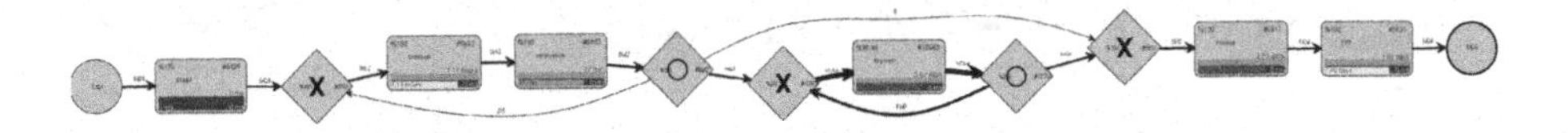

Fig. 1. Process model

The process model shows even the frequency of the transitions between events. It is depicted by numbers next to arcs or by thickness of arcs. When the arc is bolder, then the path is more frequented. We can see the high frequency of regular repeating of *Approval* event. It means that invoice is *Approved* more than once quite often.

Other interesting finding is that *Creation* and *Verification* evens are done always together sequentially.

We have discovered the most frequented paths, respectively all the used paths. Number of all used paths (patterns) is 76. The Table 6 shows summary of the most frequented and most interesting paths that were. Detailed analysis of these most frequent paths is provided in the next section.

Table 6. Interesting paths in the process model

Path	Case(s)	Occurrence (relative)	Number of events
Path 1 – 1st most frequented	2436	44.912%	7
Path 2 – 2nd most frequented	1564	28.835%	8
Path 3 – 3rd most frequented	504	9.292%	6
Path 4 – 4th most frequented	200	1,862%	9
Path 5 – 5 th most frequented – more *Posting* events	101	1,862%	9
Path 6 – most time consuming path	1	0.018%	18
Path 7 – least time consuming path	1	0.018%	10
Path 8 – without *Approval* event	1	0.018%	5

3.6 Analysis of Interesting Paths

This section describes the analysis of most interesting paths that we chose from all paths of the process model.

Path 1 – 1st Most Frequented Path

This path is used by most process enactments (cases) of the log. It is almost the half of all cases. The process model of this path is depictured on the Fig. 2. Process starts by *start* artificial event, continues by *Creation* event, *Verification* event. Next, there are **two** repetitions of ***Approval*** event and process ends by *Posting* event followed by the *end* event.

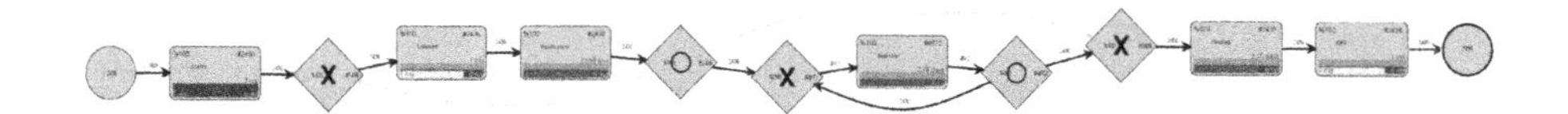

Fig. 2. Process model - Path 1

Path 2 – 2nd Most Frequented Path

This path contains 1564 cases. It is almost 29% of all cases. The process model of this path is depictured on the Fig. 3. Process starts by *start* artificial event, continues by *Creation* event, *Verification* event. Next, there are **three** repetitions of ***Approval*** event and process ends by *Posting* event followed by *end* event.

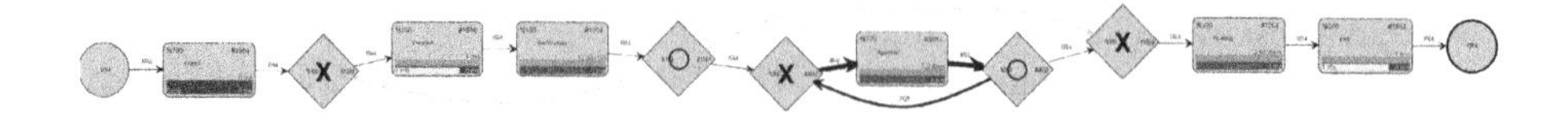

Fig. 3. Process model - Path 2

Path 3 – 3rd Most Frequented Path

This path contains 504 cases. The process model of this path is depictured on the Fig. 4. Process starts by *start* artificial event, continues by *Creation* event, *Verification* event. Next, there is **one *Approval*** event and process ends by *Posting* event followed by *end* event.

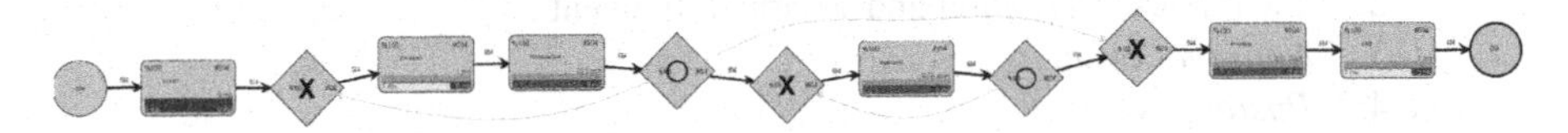

Fig. 4. Process model – Path 3

All three most frequented paths are almost similar. The only one relevant difference is a number of repeats of the *Approval* event. The most frequented path has two, second most frequented path has three and third most frequented path has one repetition of it.

Path 4 – 4th Most Frequented Path

This path contains 200 cases. The process model of this path is depictured on the Fig. 5. Process starts by *start* artificial event, continues by *Creation* event, *Verification* event. Next, there are **four *Approval*** events and process ends by *Posting* event followed by *end* event.

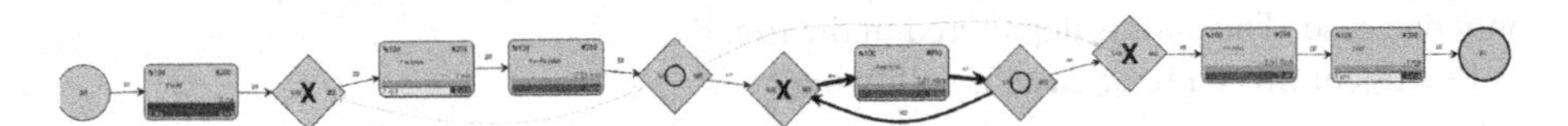

Fig. 5. Process model - Path 4

Path 5 – 5th Most Frequented – More Posting Events

We have chosen this path for detailed analysis because it is fifth most frequented path and contains repetition of *Posting* event. This is some kind of deviation that we are looking for. The path is depictured in the Fig. 6.

Process of this path, or case, is following:

1. *Start* artificial event
2. *Creation* and *Verification* event
3. Two times - *Approval* event
4. *Posting* event
5. *Approval* event
6. *Posting* event
7. *End* artificial event

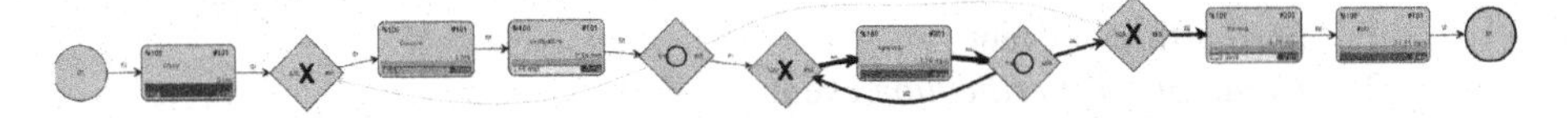

Fig. 6. Process model - Path 5

Path 6 – Most Time Consuming Path

This path is most time consuming path. This path includes only one case. This case was for some reason longest lasting. The path is depictured in the Fig. 7.

Process of this path, or case, is following:

1. *Start* artificial event
2. Two times – *Creation* and *Verification* event
3. *Approval* event
4. *Posting* event
5. seven times - *Approval* event
6. *Posting* event
7. *Approval* event
8. *Posting* event
9. *End* artificial event

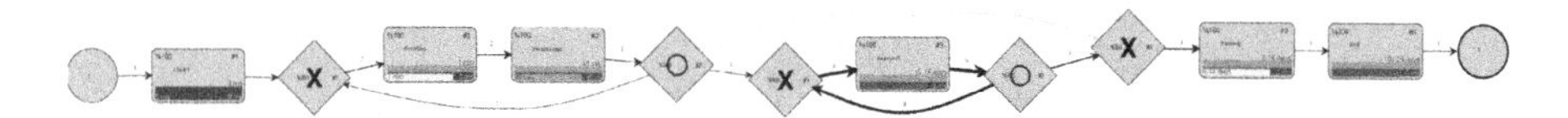

Fig. 7. Process model - Path 6

Path 7 – Least Time Consuming Path

This path is least time consuming path. Same as like most time consuming it includes only one case. The path is depictured in the Fig. 8.

Process of this path, or case, is following:

1. *Start* artificial event
2. Three times – *Creation* and *Verification* event
3. *Approval* event
4. *Posting* event
5. *End* artificial event

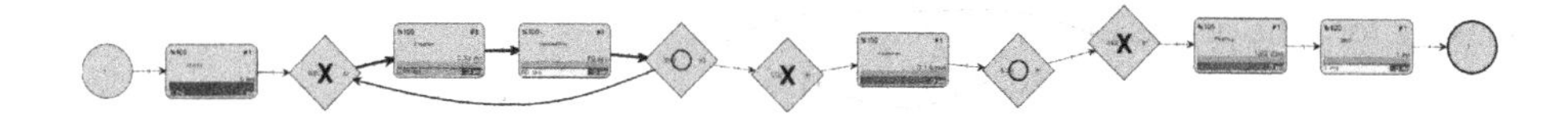

Fig. 8. Process model - Path 7

Path 8 – Without Approval Event

We have chosen this path for detailed analysis because it does not contained *Approval* event. This is some kind of deviation we were looking for. The path is depictured in the Fig. 9.

Process of this path, or case, is following:

1. *Start* artificial event
2. *Creation* and *Verification* event
3. *Posting* event
4. *End* artificial event

Fig. 9. Process model - Path 8

Time analysis of the Process

There is information about waiting time between events in the process model. It means how much time is consumed by waiting for the completion of previous event. Unfortunately we do not have information about consumed time of particular events - it was not included in the log. So the only information we know is when the events were finished.

3.7 Results and Retrospect to Real Business

We have analyzed following interesting average time parameters:

- Waiting time between *Creating* and *Verification* events is 12 minutes in average. It means that these two activities are enacted almost in the same time. The reason is that these events are performed usually by the same person (accountant).
- Average waiting time between particular *Approval* events is 5,62 days. It is relevant for managers – they will use it for the motivating actors to process the Approval task in shorter time (better monitoring and remainder tool was activated).

We have analyzed specific paths and found the following results:

- **Path 1:** The invoices from this group were found as invoices connected to the purchase order/contract and with proper receipt without differences (amount, price). Typically there are invoices for the investments, strategic raw material and overhead invoices with two persons approving the invoice. This proves that the purchasing system is well defined and settled.
- **Path 2:** Most of invoices from path 2 come from the purchases during main business process and financial services; the confirmation by three persons for specific invoice types was presented with the motivation to find some level of approving during purchasing process and to reduce the number of approvals during invoice verification. The process is mature and correct, the motivation is to move one step to purchase phase where this person is also active.
- **Path 3:** Most of invoices from path 3 come from the purchase in marketing; this testifies maturity of the purchasing along the main business process.
- **Path 4:** This group contains invoices of various types. Most of them contain short discussion between Approvals (request-answer within the meaning of explaining the differences in amount of maintenance service, providing another document …) otherwise they would fall to Path 1. This invoice approval process could be shorten only in case that all information required for invoice approval would be gathered by some "acceptation process – acceptation protocol". One invoice set in this path has 4 Approval steps based on the organizational structure that is defined in the approving instructions – shortening would mean that theses instructions would have been changed.

- **Path 5:** repetition of the "posting" step always means that the first approver of the invoice refused the responsibility for this invoice (step "not me"). Such decision has mostly reason in assignment of competencies, sometimes in mistakes in invoice classification or other typing error. We can see that although this is the 5th frequent scenario, but occurring only in 1,8% of cases – it is now on managers' decision if cost of improving will be compensated by benefits. Monitoring of this path is suggested for the future.
- **Path 6:** this is possible (but not typical) enactment for the invoice without purchase order/contract, the purchase order must have been prepared and agreed – it is strictly suggested to have the purchase number approved before the invoice comes (otherwise the invoice is not accepted), this case was combined with issues discussed in Path 5; another case from this group was the invoice from key customer created supposedly pursuant to the contract (there was a long discussion about legitimacy and correctness – but as the partner was strategic partner, the invoice was not turned back and the process was prolonged).
- **Path 7:** very specific invoices with irrelevant amount, high priority verification under control of the company owner, result is irrelevant for other analysis.
- **Path 8:** there was not found any workflow for specific invoices, there are notes at these invoices with link to the documents with confirmation (meeting notes) – it means the invoice was approved by competent person, but out of the workflow system. Such path must be analyzed separately.

4 Conclusion and Future Work

There are some conclusions that we have found about the invoice verification process using process mining techniques.

1. We have found the average values of the running activities, the result about possible time reserves was presented and accepted.
2. The most frequented paths were identified (Path 1, Path 2, Path 3) and business cases along which they occur were found. Analysis showed very frequent approving by three persons (in Path 2) and the business process owner got the material for the process improving.
3. Paths with more than three Approving steps (Path 4) contain communication of some issue during the invoice verification. These cases could be shortened if these identified most frequent issues are part of the acceptation procedure and would be solved before the invoice verification process and the acceptation protocol would be attached by default. Other set of the invoices with four approval steps have this procedure defined by the instructions (only change of the instruction can make the process shorten).
4. Path 5 provides the information about the model of competencies and it is appropriate to monitor this path.
5. The invoices without purchase order have usually longer verification time (the backward purchase order creation takes a lot of time), in some cases it was extremely long (Path 6). These exceptions should not be allowed by default.

6. Very specific cases were found in Path 7 and Path 8 – they do not have any frequent pattern and from this reason are not interesting for conclusion.

This case study opened several tasks for future work. First of all, we have not paid attention to the resources. In future work, we would like to focus also on users, their interaction during the process and construction of their social network

We have been concentrated to most frequent paths with the aim to find reserves and improving themes – in the future, we would like to focus also on deviations (process without expected start and stop event, multiply "unique tasks" as was found in Path 4) this can means some specific problem.

Our study proved that even the basic process mining methods used to analyze the enactment of the processes can be very useful for the real businesses. The processes performed by humans and supported by different systems have often wide range of variability, thus it is not always possible to predict and control the real usage of the process even with the sophisticated system. Thus we have to use information from the real enactments to see what really happened. We can reconstruct the process by the process mining tools; find deviations and other many useful pieces of information by still grooving bunch of process mining methods. This analysis can be then used to improve the process, supporting system or the organizational behavior. Our intention for the future is to study the usage of the process mining methods, find their best value for the business and try to present new methods that can answer the questions that arise from the real business needs.

Acknowledgement. This research has been supported by the internal grant agency of VSB-TU of Ostrava - SP2013/207 "Application of artificial intelligence in process-knowledge mining, modeling and management".

References

[1] Dumas, M., van der Aalst, W.M.P., ter Hofstede, A.H.M.: Process Aware Information Systems: Bridg-ing People and Software Through Process Technology. Wiley-Interscience (2005)

[2] Grigori, D., Casati, F., Castellanos, M., Dayal, U., Sayal, M., Shan, M.-C.: Business process intelligence. Computers in Industry 3, 321–343 (2004)

[3] Weske, M., van der Aalst, W.M.P., Verbeek, H.M.W.E.: Advances in business process management. Data & Knowledge Engineering 50(1), 1–8 (2004)

[4] van der Aalst, W.M.P., ter Hofstede, A.H.M., Weske, M.: Business process management: A survey. In: van der Aalst, W.M.P., ter Hofstede, A.H.M., Weske, M. (eds.) BPM 2003. LNCS, vol. 2678, pp. 1–12. Springer, Heidelberg (2003)

[5] Alves de Medeiros, A.K., Weijters, A.J.M.M., van der Aalst, W.M.P.: Genetic process mining: an experimental evaluation. Data Mining and Knowledge Discovery 14(2), 245–304 (2007)

[6] Bozkaya, M., Gabriels, J., van der Werf, J.M.E.M.: Process diagnostics: a method based on process mining. In: Kusiak, A., Lee, S. (eds.) eKNOW, pp. 22–27. IEEE Computer Society (2009)

[7] Goedertier, S., De Weerdt, J., Martens, D., Vanthienen, J., Baesens, B.: Process discovery in event logs: an application in the telecom industry. Applied Soft Computing 11(2), 1697–1710 (2011)
[8] Jans, M., van der Werf, J.M., Lybaert, N., Vanhoof, K.: A business process mining application for internal transaction fraud mitigation. Expert Systems with Applications 38(10), 13351–13359 (2011)
[9] Mans, R.S., Schonenberg, H., Song, M., van der Aalst, W.M.P., Bakker, P.J.M.: Application of process mining in healthcare—a case study in a Dutch hospital. In: Fred, A.L.N., Filipe, J., Gamboa, H. (eds.) BIOSTEC 2008. CCIS, vol. 25, pp. 425–438. Springer, Heidelberg (2008)
[10] Rebuge, Á., Ferreira, D.R.: Business process analysis in healthcare environments: a methodology based on process mining. Information Systems 37(2), 99–116 (2012)
[11] Rozinat, A., de Jong, I.S.M., Günther, C.W., van der Aalst, W.M.P.: Process mining applied to the test process of wafer scanners in ASML. IEEE Transactions on Systems, Man, and Cybernetics, Part C 39(4), 474–479 (2009)
[12] Rozinat, A., Mans, R.S., Song, M., van der Aalst, W.M.P.: Discovering simulation models. Information Systems 34(3), 305–327 (2009)
[13] Song, M., van der Aalst, W.M.P.: Towards comprehensive support for organizational mining. Decision Support Systems 46(1), 300–317 (2008)
[14] van der Aalst, W.M.P., Reijers, H.A., Weijters, A.J.M.M., van Dongen, B.F., Alves de Medeiros, A.K., Song, M., Verbeek, H.M.W.: Business process mining: an industrial application. Information Systems 32(5), 713–732 (2007)
[15] van der Aalst, W.M.P., Schonenberg, M.H., Song, M.: Time prediction based on process mining. Information Systems 36(2), 450–475 (2011)
[16] De Weerdt, J., Schupp, A., Vanderloock, A., Baesens, B.: Process Mining for the multi-faceted analysis of business processes—A case study in a financial services organization. Computers in Industry 64(1), 57–67 (2013) ISSN 0166-3615, 10.1016/j.compind.2012.09.010
[17] Lee, D., Bae, H.: Analysis framework using process mining for block movement process in shipyards. ICIC Express Letters 7(6), 1913–1917 (2013)
[18] Accorsi, R., Stocker, T.: On the exploitation of process mining for security audits: The conformance checking case. Paper presented at the Proceedings of the ACM Symposium on Applied Computing, pp. 1709–1716 (2012)
[19] Li, M.: Process mining in knowledge maintenance a case study. Advances in Information Sciences and Service Sciences 4(9), 293–301 (2012)
[20] van Dongen, B.F., de Medeiros, A.K.A., Verbeek, H.M.W., Weijters, A.J.M.M.T., van der Aalst, W.M.P.: The proM framework: A new era in process mining tool support. In: Ciardo, G., Darondeau, P. (eds.) ICATPN 2005. LNCS, vol. 3536, pp. 444–454. Springer, Heidelberg (2005)
[21] Van Der Aalst, W.M.P., Van Dongen, B.F., Günther, C.W., Mans, R.S., Alves De Medeiros, A.K., Rozinat, A., Song, M., Verbeek, H.M.W., Weijters, A.J.M.M.: Process mining with ProM. In: Belgian/Netherlands Artificial Intelligence Conference, p. 453 (2007)
[22] Van Der Aalst, W., Adriansyah, A., Van Dongen, B.: Causal nets: A modeling language tailored towards process discovery (2011)

Emergent Induction of Deterministic Context-Free L-system Grammar

Ryohei Nakano

Chubu University
1200 Matsumoto-cho, Kasugai, 487-8501 Japan
nakano@cs.chubu.ac.jp

Abstract. L-system is a bio-inspired computational model to capture growth process of plants. This paper proposes a new noise-tolerant grammatical induction LGIC2 for deterministic context-free L-systems. LGIC2 induces L-system grammars from a transmuted string mY, employing an emergent approach in order to enforce its noise tolerance. In the method, frequently appearing substrings are extracted from mY to form grammar candidates. A grammar candidate is used to generate a string Z; however, the number of grammar candidates gets huge, meaning enormous computational cost. Thus, how to prune grammar candidates is vital here. We introduce a couple of techniques such as pruning by frequency, pruning by goodness of fit, and pruning by contractive embedding. Finally, several candidates having the strongest similarities between mY and Z are selected as the final solutions. Our experiments using insertion-type transmutation showed that LGIC2 worked very nicely, much better than an enumerative method LGIC1.

Keywords: L-system, plant modeling, knowledge discovery, grammatical induction, noise tolerance.

1 Introduction

The beauty and elegance of plants have attracted humankind. In many developmental processes of living organisms, especially of plants, regularly repeated appearances of structures are readily noticeable. L-system was originally introduced by Lindenmayer as a *bio-inspired computational model* to capture such growth processes of plants [12].

The central concept of L-system is rewriting, a powerful mechanism for generating complex objects from a simple initial object. The relationship with the concept of fractals led L-systems to various practical applications such as synthesis of realistic plant images, design of geometric patterns, analysis of DNA sequences, or synthesis of musical scores [11].

The reverse process of rewriting is *grammatical induction*, a kind of knowledge discovery, which discovers quite succinct knowledge called a grammar from very long strings. Induction of L-system grammars has been an open problem little explored so far. Moreover, an extensive survey paper [3] on grammatical inference

A. Abraham et al. (eds.), *Innovations in Bio-inspired Computing and Applications*, Advances in Intelligent Systems and Computing 237,
DOI: 10.1007/978-3-319-01781-5_7, © Springer International Publishing Switzerland 2014

tells us that for many applications it is necessary to be able to cope with noise, and most grammatical inference algorithms are not robust to noise, whatever its source; thus, dealing with noisy data is one of the important open problems.

L-systems can be classified using two aspects: (1) deterministic or stochastic, and (2) context-free or context-sensitive. McCormack [6] addressed CG modeling through evolution of deterministic/stochastic context-free L-systems. Nevill-Manning [10] proposed a simple algorithm Sequitur, which uncovers structures from sequences; however, it did not work well for L-systems. Schlecht, et al. [13] proposed statistical structural inference for microscopic 3D images through learning stochastic L-systems. Damasevicius [2] addressed structural analysis of DNA sequences through evolution of stochastic context-free L-systems.

A fast induction method [9] was once proposed for deterministic context-free L-system. The method assumes a given string includes no errors, making it possible to employ the number theory. In the real world, however, any object may have some errors or transmutation. As for transmutation, there can be several types such as replacement-type, insertion-type, deletion-type, or mixed-type. To cope with a noisy string, proposed was LGIC1 (LGIC, ver.1) [7,8], enumerative induction of deterministic context-free L-system grammar from a transmuted string mY. LGIC1 worked well when the transmutation rate is small for replacement-type, but did not work well for insertion-type.

This paper proposes LGIC2 (LGIC, ver.2), noise-tolerant emergent induction of deterministic context-free L-system grammar. In LGIC2, frequently appearing substrings are extracted from mY to form grammar candidates. A grammar candidate is used to generate a string Z; however, the number of grammar candidates gets huge because the numbers of the substrings are large. Thus, introduced are a couple of pruning techniques such as pruning by frequency, pruning by goodness of fit, and pruning by contractive embedding. Finally, several candidates having the strongest similarities between mY and Z are selected as the final solutions. Our experiments using insertion-type transmutation showed that LGIC2 worked very nicely, much better than LGIC1.

2 Background

D0L-system

The simplest L-systems called D0L-systems are deterministic context-free. D0L-system is defined as $G = (V, C, \omega, P)$, where V and C denote sets of *variables* and *constants*, ω is an initial string called *axiom*, and P is a set of *production rules.* A variable is a symbol that is replaced in rewriting; a constant is a symbol that remains unchanged in rewriting and is used to control turtle graphics.

In this paper we consider the following grammar of D0L-system having two rules. Here n denotes the number of rewritings, while A and B denote variables.

$$n = ?, \quad \text{axiom} : A$$
$$\text{rule A} : A \rightarrow ????????$$
$$\text{rule B} : B \rightarrow ??????$$

We have three strings: Y, mY, and Z. Y is an original string generated using the original grammar, while mY is a given transmuted string generated by applying transmutation to Y. Z is a string generated using a grammar candidate, where a grammar candidate means a candidate set of n, rules A and B.

Transmutation

As for string transmutation, there can be replacement-type (r-type), insertion-type (i-type), deletion-type (d-type), or mixed-type (m-type). Here only i-type transmutation is considered; Since LGIC1 was rather weak on i-type [7], we want to know how LGIC2 overcomes this drawback. In i-type transmutation, a randomly selected symbol is inserted at a designated position of a string.

As for how transmutation occurs, we consider two rates: *coverage rate* P_c and *occurrence rate* P_o. We assume transmutation occurs only locally around the center of an original string Y. P_c represents the proportion of transmutation area to the whole Y, while P_o represents the probability of transmutation in the area. Thus, overall *transmutation rate* P_t can be evaluated as follows:

$$P_t = P_c \times P_o \tag{1}$$

Valid Transmutation

Simple transmutation will generate an invalid string, which means the string cannot be drawn through turtle graphics. To keep the transmutation valid, the numbers of left and right square brackets are monitored and controlled if necessary. That is, in the transmutation area the number $count_\ell$ of left square brackets should be larger than or equal to the number $count_r$ of right ones. Moreover, when the transmutation ends, we should assure $count_\ell = count_r$ by adding the right square brackets if necessary. Using such control, we get a valid transmuted string mY from the original string Y.

Enumerative Induction of L-system Grammar LGIC1

Nakano and Suzumura [7,8] proposed enumerative induction called LGIC1 with error correction capability. LGIC1 discovers L-system grammars from a transmuted string mY. LGIC1 works as follows. First, sets of parameter values are enumerated and those within the tolerable distance from mY are selected to form grammar candidates. Each grammar candidate is used to generate a string Z, and the similarity between mY and Z is calculated, and finally several best grammar candidates are selected as the final solutions.

Table 1 shows the success rates of LGIC1 with tolerable distance $tol\_diff = 150$ for r-type transmutation. When transmutation rate $P_t \leq 3/16$ ($= 0.188$), LGIC1 discovered the original grammar with probability 0.8 or 1. However, when $P_t = 0.25$ for $P_c = P_o = 0.5$, the success rate dropped to 0.2.

Table 2 shows the success rates of LGIC1 with $tol\_diff = 200$ for i-type transmutation. Here, tol_diff was increased up to 200 since LGIC1 did not work well for $tol\_diff = 100$ or 150. When $P_t = 1/16$ ($= 0.063$), LGIC1 discovered the original grammar with probability 1. When $P_t = 1/8$ ($= 0.125$), however, the success rate dropped to 0.8. Further, in all the combinations where $P_t \geq 3/16$ ($= 0.188$),

Table 1. Success rates of LGIC1 for r-type transmutation (tol_diff=150)

	$P_o = 0.25$	$P_o = 0.50$	$P_o = 0.75$	$P_o = 1.0$
$P_c = 0.25$	5/5	5/5	5/5	4/5
$P_c = 0.50$	5/5	1/5	0/5	(0/5)
$P_c = 0.75$	4/5	0/5	(0/5)	(0/5)

LGIC1 could not discover the original grammar at all. Compared with the results with $tol\_diff = 150$ for r-type transmutation, we see LGIC1 was rather weak on i-type even with $tol\_diff = 200$. This is because insertion-type transmutation drastically change occurrence frequencies of symbols, which prevents LGIC1 from selecting the right set of parameters. Deletion-type transmutation will work similarly as i-type.

Table 3 shows the average CPU time of LGIC1 with $tol\_diff = 200$ for i-type transmutation. For each P_o, average CPU time gets longer as P_c gets larger, but for each P_c, average CPU time does not always increase even if P_o gets larger.

Table 2. Success rates of LGIC1 for i-type transmutation (tol_diff=200)

	$P_o = 0.25$	$P_o = 0.50$	$P_o = 0.75$	$P_o = 1.0$
$P_c = 0.25$	5/5	4/5	0/5	(0/5)
$P_c = 0.50$	4/5	0/5	(0/5)	(0/5)
$P_c = 0.75$	0/5	(0/5)	(0/5)	(0/5)

Table 3. Average CPU time (sec) of LGIC1 for i-type transmutation together with the average number of LCS calculations in parentheses (tol_diff=200)

	$P_o = 0.25$	$P_o = 0.50$	$P_o = 0.75$	$P_o = 1.0$
$P_c = 0.25$	1050.9 (466)	869.9 (314)	797.9 (243)	n/a
$P_c = 0.50$	1806.4 (719)	2088.6 (666)	n.a	n/a
$P_c = 0.75$	3422.4 (1309)	n/a	n/a	n/a

3 LGIC2: Emergent Induction of L-system Grammar

An emergent induction method called LGIC2 (L-system Grammar Induction with error Correction, ver.2) is proposed. Given a transmuted string mY, LGIC2 generates grammar candidates, aiming at finding the original grammar.

Basic Framework

The basic framework of LGIC2 is very simple. Since a right side of each rule appears many times in mY, we extract frequently appearing substrings from mY to form rule candidates. Such substrings are extracted as if they emerge from mY. Then such a rule candidate is combined with its reasonable n, the number of rewritings, to form a grammar candidate.

The main drawback of the framework is the combinatorial growth in the number of grammar candidates. In our preliminary experiment it took ten days to finish two thirds of the processing for a string mY whose length is about 4,000. Thus, how to prune grammar candidates is vital for the method, and we introduce three pruning techniques as shown below.

Pruning by Frequency

Since the right side of each original rule appears many times in mY, we discard less frequent substrings whose occurrence frequency is less than the threshold min_frq. The threshold value may depend on the length of mY. In our experiments we use $min\_frq = 100$ *or* 50

The Number of Rewritings

Now we have a pair of rule candidates from the above extraction. Then we select n, the number of rewritings, as the largest integer satisfying $\text{len}(Z) \leq \text{len}(mY)$. Hereafter, $\text{len}(S)$ denotes the length of a string S.

Pruning by Goodness of Fit

The goodness of fit is a statistical measure which evaluates how well a model (mY) fits to observed data (Z). The goodness of fit can be evaluated by χ^2 values. Let the numbers of symbol occurrences in mY and Z be $\{y_i\}$ and $\{z_i\}$ respectively. Then calculate $\{p_i = y_i/\text{len}(mY)\}$, and we have the following χ^2 value. Here I is the number of all kinds of variables and constants.

$$\chi^2 = \sum_{i}^{I} (z_i - \text{len}(Z) \times p_i)^2 / (\text{len}(Z) \times p_i) \qquad (2)$$

We discard a grammar candidate if χ^2 is greater than the threshold max_chi2. In our experiments we use $max\_chi2 = 150$.

Similarity between Two Strings

As the similarity between two strings, LGIC2 employs the longest common subsequence (LCS) [1]. Let $\text{LCS}(S_1, S_2)$ denote LCS of two strings S_1 and S_2. For example, LCS(ABCDABC,BDCAB) is BDAB or BCAB. Given two strings we may have more than one LCSs, but the length of each LCS is the same. Note that LCS can cope with any type of transmutation. LCS can be found using dynamic programming [1]. Another reasonable measure is Levenshtein distance [5], which is defined as the minimum number of modifications required to transform one string into the other. We consider these two measures will result in much the same result.

Pruning by Contractive Embedding

In complex data, the cost of evaluating the distance of two objects is usually very high. Here we have to calculate the similarity LCS between two strings.

Our experiments show it takes about two seconds to calculate one LCS if each string length is around 4,000. Thus, the number of LCS calculations should be kept as small as possible. We can achieve this without false dismissals if we find suitable contractive embedding [4]. As such embedding we consider *ubLCS*, an upper bound of len(LCS), defined as below.

$$ubLCS(mY, Z) = \sum_i min(y_i, z_i) \tag{3}$$

This *ubLCS* can be used to prune grammar candidates further more. That is, we discard a grammar candidate if $ubLCS(mY, Z) < LCS(mY, bestZ)$, where $bestZ$ is the string generated by the best grammar found so far.

Procedure of LGIC2 Method

LGIC2 has four system parameters: maximum length of rule right side max_rsl, minimum frequency of rule right side occurrences min_frq, maximum χ^2 value max_chi2, and the number of final solutions *tops*. LGIC2 goes as below:

(step 1) Extract a substring rs_1 from mY as a right side candidate of rule A. Here the length of rs_1 should satisfy $2 \leq \text{len}(rs_1) \leq max\_rsl$, and the number of rs_1 occurrences in mY should be more than or equal to min_frq.
(step 2) Eliminate any occurrences of rs_1 from mY to get $mYrest$, and then extract a substring rs_2 from $mYrest$ as a right side candidate of rule B. Here the length of rs_2 should satisfy $2 \leq \text{len}(rs_2) \leq max\_rsl$, and the number of rs_2 occurrences in $mYrest$ should be more than or equal to min_frq.
(step 3) For each pair of rs_1 and rs_2 selected above, find the number of rewritings n to generate a string Z. Here n is selected to be the largest integer which satisfies $\text{len}(Z) \leq \text{len}(mY)$.
(step 4) Calculate χ^2 value using mY and Z, and then discard the grammar candidate as inappropriate if $\chi^2 > max\_chi2$.
(step 5) Calculate $ubLCS(mY, Z)$, the upper bound of $\text{LCS}(mY, Z)$, and then discard the grammar candidate if $ubLCS(mY, Z) < LCS(mY, bestZ)$. Here $bestZ$ is the string generated by the best grammar found so far.
(step 6) Calculate $\text{LCS}(mY, Z)$, and then keep the grammar candidate if the LCS is within best *tops* found so far. Go to step 1 if there is another candidate.
(step 7) Among the grammar candidates found so far, select *tops* candidates having the largest LCSs as the final solutions.

4 Experiments

The proposed method LGIC2 was evaluated using a transmuted plant model. A plant model ex05n, a variation of bracketed OL-system example [12], was used as a normal plant model in our experiments. Figure 1 shows ex05n whose string length is 4,243. PC with Xeon(R), 2.66GHz, dual was used in our experiments.

$$\begin{aligned}(\text{ex05n})\ \ n = 6,\ \ &\text{axiom} : X\\ &\text{rule} : X \rightarrow F[+X][-X]FX\\ &\text{rule} : F \rightarrow FF\end{aligned}$$

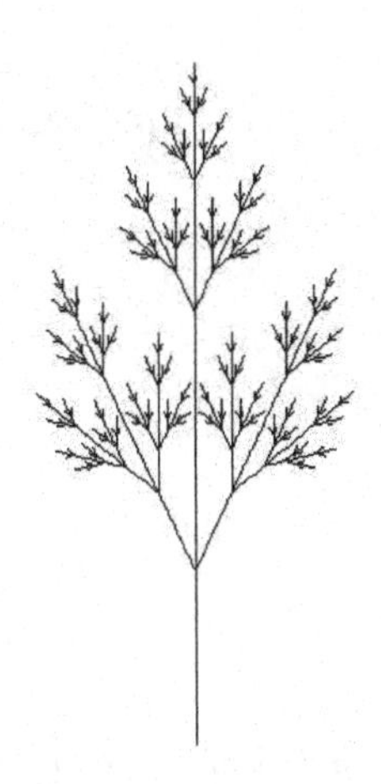

Fig. 1. Normal plant model ex05n

As is mentioned, we considered only insertion-type (i-type) transmutation. Here we considered combinations of three coverage rates P_c = 0.25, 0.5, 0.75 and four occurrence rates P_o = 0.25, 0.5, 0.75, 1.0. For each combination we transmuted ex05n five times changing a seed for random number generator.

Based on our early experiment, LGIC2 system parameters were set basically as follows: max_rsl = 15, min_frq = 100, max_chi2 = 150, and $tops$ = 50. However, LGIC2 did not work well for a wider range of transmutation whose $P_c = 0.75$; thus, $min\_frq = 50$ was adopted only for $P_c = 0.75$.

Table 4 shows the success rates of LGIC2 for i-type transmutation. LGIC2 discovered the original grammar for almost all cases except one. The single failure was caused by χ^2 pruning. Compared with Table 2, the difference between LGIC2 and LGIC1 is obvious. We verified that an emergent approach of LGIC2 together with pruning techniques works very well. Moreover, in LGIC2 the original grammar was rated as No.1 for most cases; otherwise was rated as No.2.

Figure 2 shows a plant model transmuted in i-type with P_c = 0.50 and P_o = 0.75. LGIC2 successfully discovered the original grammar from this rather heavily transmuted plant. Figure 3 shows a plant model transmuted in i-type with P_c = 0.75 and P_o = 1.00. Even from this very heavily transmuted plant, LGIC2 successfully discovered the original grammar.

Table 4. Success rates of LGIC2 for i-type transmutation

	$P_o = 0.25$	$P_o = 0.50$	$P_o = 0.75$	$P_o = 1.0$
$P_c = 0.25$	5/5	5/5	5/5	5/5
$P_c = 0.50$	5/5	5/5	5/5	5/5
$P_c = 0.75$	5/5	5/5	5/5	4/5

Fig. 2. Plant model transmuted in i-type ($P_c = 0.50$, $P_o = 0.75$)

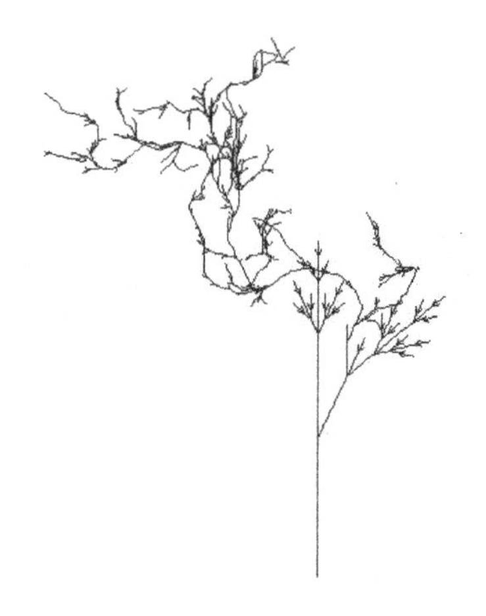

Fig. 3. Plant model transmuted in i-type ($P_c = 0.75$, $P_o = 1.00$)

Table 5 shows CPU time, the number of LCS calculations in parentheses, and χ^2 value for i-type transmutation in a left-to-right fashion. For each combination of P_c and P_o, we performed five runs. For each P_c, CPU time increases as P_o increases, and for each P_o, CPU time increases as P_c increases. This is because the number of grammar candidates became larger as the transmutation

Table 5. CPU time (sec) of LGIC2 for i-type transmutation together with the number of LCS calculations in parentheses, and χ^2 value at the rightmost position

	$P_o = 0.25$	$P_o = 0.50$	$P_o = 0.75$	$P_o = 1.0$
$P_c = 0.25$	266.6 (5) 2.93	329.0 (8) 11.68	389.7 (14) 21.78	452.2 (18) 34.33
	286.5 (5) 2.69	*348.1 (16) 7.93	*412.1 (22) 23.98	463.1 (21) 27.04
	282.2 (5) 3.99	322.5 (12) 13.15	393.5 (16) 20.29	499.5 (18) 35.57
	262.0 (4) 2.32	*338.6 (17) 11.73	394.7 (15) 27.02	512.0 (19) 34.46
	265.9 (4) 4.02	325.2 (12) 12.25	379.7 (12) 27.29	500.8 (17) 37.00
$P_c = 0.50$	447.2 (8) 10.80	611.7 (9) 37.71	767.0 (8) 67.84	875.5 (8) 85.92
	466.8 (10) 12.19	638.4 (10) 28.63	784.0 (9) 62.20	884.0 (9) 80.95
	481.7 (9) 12.76	629.4 (9) 34.09	760.6 (12) 65.83	798.8 (8) 103.21
	445.0 (8) 11.11	625.9 (9) 37.05	770.2 (9) 69.94	801.3 (12) 87.85
	460.2 (9) 10.06	613.1 (9) 29.22	743.1 (8) 70.41	803.4 (10) 90.65
$P_c = 0.75$	913.2 (18) 21.64	1186.2 (24) 66.63	1414.5 (15) 110.32	1729.0 (15) 138.81
	929.5 (20) 24.27	1223.9 (25) 52.06	1498.4 (18) 98.22	1675.0 (14) 134.39
	926.1 (21) 27.38	1197.8 (22) 64.43	1441.8 (18) 107.14	1796.1 (16) fail
	941.4 (21) 21.13	1259.3 (25) 64.53	1522.4 (17) 117.16	1713.2 (17) 145.76
	909.6 (22) 22.48	1152.3 (19) 60.72	1457.8 (20) 125.66	1756.1 (16) 147.35

Table 6. Average CPU time (sec) of LGIC2 for i-type transmutation together with the average number of LCS calculations in parentheses

	$P_o = 0.25$	$P_o = 0.50$	$P_o = 0.75$	$P_o = 1.0$
$P_c = 0.25$	272.6 (4.6)	332.7 (13.0)	393.9 (15.8)	485.5 (18.6)
$P_c = 0.50$	460.2 (8.8)	623.7 (9.2)	765.0 (9.2)	832.6 (9.4)
$P_c = 0.75$	924.0 (20.4)	1203.9 (23.0)	1467.0 (17.6)	1733.9 (15.6)

rate increased. The table also shows that the increase of P_c has greater impact on CPU time than that of P_o. The number of LCS calculations is very small due to a couple of pruning techniques. χ^2 values are about proportional to the transmutation rate P_t. Moreover, almost all χ^2 values are within 150.

Table 6 shows the average of Table 5. For each P_c, average CPU time gets longer as P_o gets larger, and for each P_o, average CPU time gets longer as P_c gets larger. Compared with Table 3, we see the average CPU time was reduced to be half or up to one-fourth. Moreover, the average number of LCS calculations was greatly reduced to be one-tenth or up to one-hundredth.

5 Conclusion

This paper proposes a noise-tolerant emergent induction method LGIC2 of deterministic context-free L-system grammar. LGIC2 induces L-system grammars from a transmuted string. The method extracts frequently appearing substrings from a given string to form grammar candidates. Since the number of grammar candidates gets huge, LGIC2 introduces three pruning techniques to make the method reasonably fast. Our experiments using insertion-type transmutation showed that LGIC2 discovered the original grammar for almost all cases in five minutes for the lowest transmutation and in 30 minutes for the heaviest transmutation. Thus, LGIC2 outperformed its former method LGIC1 in both the success rate and CPU time. In the future we plan to apply the method to other types of transmutation along with more improvement of its performance.

Acknowledgments. This work was supported by Grants-in-Aid for Scientific Research (C) 25330294 and Chubu University Grant 24IS27A.

References

1. Cormen, T.H., Leiserson, C.E., Rivest, R.L.: Introduction to algorithms. MIT Press (1990)
2. Damasevicius, R.: Structural analysis of regulatory DNA sequences using grammar inference and support vector machine. Neurocomputing 73, 633–638 (2010)
3. de la Higuera, C.: A bibliographical study of grammatical inference. Pattern Recognition 38, 1332–1348 (2005)
4. Hjaltason, G.R., Samet, H.: Contractive embedding methods for similarity searching in metric spaces. Technical Report CS-TR-4102, Univ. of Maryland (2000)
5. Levenshtein, V.: Binary codes capable of correcting deletions, insertions, and reversals. Soviet Physics Doklady 10(8), 707–710 (1966)
6. McCormack, J.: Interactive evolution of L-system grammars for computer graphics modelling. In: Complex Systems: From Biology to Computation, pp. 118–130. ISO Press, Amsterdam (1993)
7. Nakano, R.: Error correction of enumerative induction of deterministic context-free L-system grammar. IAENG Int. Journal of Computer Science 40(1), 47–52 (2013)
8. Nakano, R., Suzumura, S.: Grammatical induction with error correction for deterministic context-free L-systems. In: Proc. of the World Congress on Engineering and Computer Science 2012 (WCECS 2012), pp. 534–538 (2012)
9. Nakano, R., Yamada, N.: Number theory-based induction of deterministic context-free L-system grammar. In: Proc. Int. Joint Conf. on Knowledge Discovery, Knowledge Engineering and Knowledge Management 2010, pp. 194–199 (2010)
10. Nevill-Manning, C.G.: Inferring sequential structure. Technical Report Doctoral Thesis, Univ. of Waikato (1996)
11. Prusinkiewicz, P., Hanan, J.: Lindenmayer systems, fractals, and plants. Springer, New York (1989)
12. Prusinkiewicz, P., Lindenmayer, A.: The algorithmic beauty of plants. Springer, New York (1990)
13. Schlecht, J., Barnard, K., Springgs, E., Pryor, B.: Inferring grammar-based structure models from 3d microscopy data. In: Proc. of IEEE Conference on Computer Vision and Pattern Recognition, pp. 1–8 (2007)

Double Expert System for Monitoring and Re-adaptation of PID Controllers

Jana Nowaková and Miroslav Pokorný

VŠB-Technical University of Ostrava,
Faculty of Electrical Engineering and Computer Science,
Department of Cybernetics and Biomedical Engineering,
17. listopadu 15/2172, 708 33 Ostrava Poruba, Czech Republic
{jana.nowakova,miroslav.pokorny}@vsb.cz

Abstract. Finding of monitoring systems for deciding if or how re-adapt a PID controller in literature is not so complicated. These monitoring systems are also widely used in industry. But monitoring system which is based on non-conventional methods for deciding, takes into account the non-numeric terms and it is open for adding more rules, is not so common. Presented monitoring is designed for systems of second order and it is performed by the fuzzy expert system of Mamdani type with two inputs - settling time compared with the previous settling time (relative settling time) and overshoot. It is supplemented by using of non-conventional method for designing of classic PID controller. So it can be called as double expert system for monitoring and following re-adaptation of classical PID controller. The proof of efficiency of the proposed method and a numerical experiment is presented by the simulation in the software environment Matlab-Simulink.

Keywords: Monitoring, re-adaptation, expert system, knowledge base, PID controller, Ziegler-Nichols' combinated design methods, fuzzy system, feedback control.

1 Introduction

The paper is focused on adjustment of the monitoring system for deciding when re-adapt the classic PID controller. Presented monitoring system (Figure 1) has some common elements with [1]. It is created fuzzy expert system of Mamdani type (ES1) with two inputs - overshoot and settling time, but settling time is not defined as a classic time, but as the part or multiple of previous measured settling time (relative settling time) and one output - score. So there are monitored simply obtained process parameters. The score determines if is necessary to re-adapt the controller.

The following design of parameters of classic PID controller is done by the second fuzzy expert (ES2) system with a knowledge base is built on know-how obtained from the combination of the frequency response method and the step response Ziegler-Nichols design method [2].

A. Abraham et al. (eds.), *Innovations in Bio-inspired Computing and Applications*, Advances in Intelligent Systems and Computing 237,
DOI: 10.1007/978-3-319-01781-5_8,

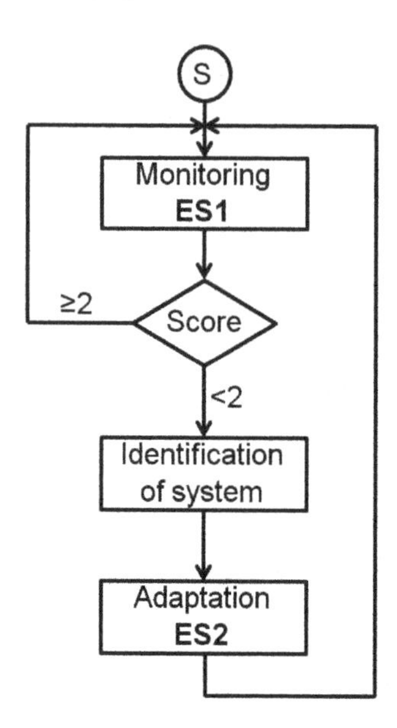

Fig. 1. The flowchart of monitoring and re-adaptation procedures

2 Monitoring System

The monitoring system is fuzzy expert system [3], [4], [5] of Mamdani type with two inputs, knowledge base with linguistic rules and one output and it has been created and the efficiency is proofed for controlled systems of the second order with transfer function in the form

$$G_S(s) = \frac{1}{a_2s^2 + a_1s + a_0}. \tag{1}$$

2.1 Inputs – Relative Overshoot and Relative Settling Time

The first input is the linguistic variable relative overshoot (RO) - the difference between the controlled value (CV) and the required value (RV) is rated relatively to the required value (2).

$$RO = \frac{|CV - RV|}{RV} \tag{2}$$

The maximal overshoot of the time response of the system is detected after the fast step change of timing, the fast step changes in timing could be caused by the e.g. change of the controlled system or change of the required value. As it is expressed in percentage, it is relative overshoot (RO). The value of the overshoot is stored in the memory and then used with the settling time for determining the score.

The second input is the linguistic variable relative settling time (RST). As the name says, it is not the classic settling time (ST_k), it is defined as the part or multiple of previous settling time $(ST_k - 1)(3)$. So the classic settling time is also stored in the memory as the overshoot, but the linguistic variable relative settling time is defined as the ratio of current settling time and previous settling time.

$$RST_k = \frac{ST_k}{ST_{k-1}} \tag{3}$$

For evaluation of the settling time the 3 % standard deviation from steady-state value [6]. The linguistic values of both linguistic variables are expressed by fuzzy sets, for each linguistic variable by three linguistic values (Figure 2,3).

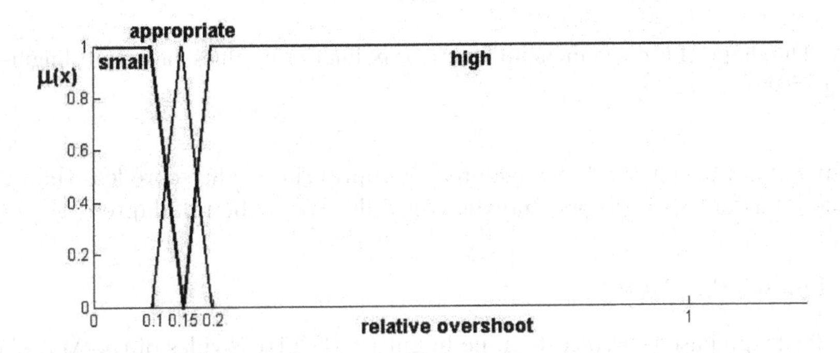

Fig. 2. The shape of the membership functions of linguistic values for input linguistic variable Relative Overshoot (RO)

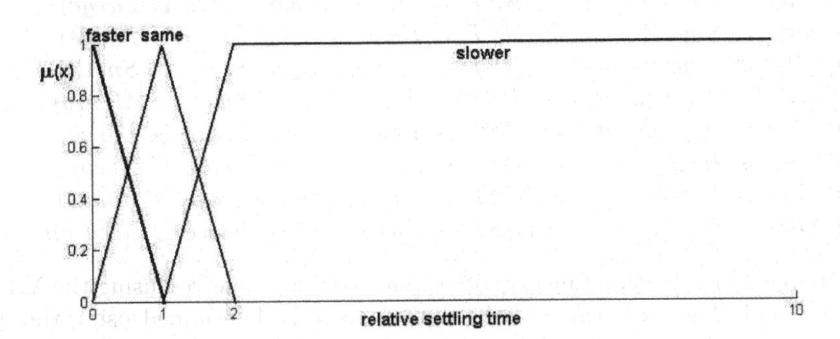

Fig. 3. The shape of the membership functions of linguistic values for input linguistic variable Relative Settling Time (RST)

2.2 Output – Score

The output of the monitoring fuzzy expert system is the score, which is also the linguistic variable and its linguistic values are expressed by fuzzy sets (Figure 4). As the fuzzy expert system of Mamdani type is used [7], the linguistic variable score must be defuzzificated. For defuzzification it is used the COA method (Center of Area) [8]. The score more than 2 means that the time response with

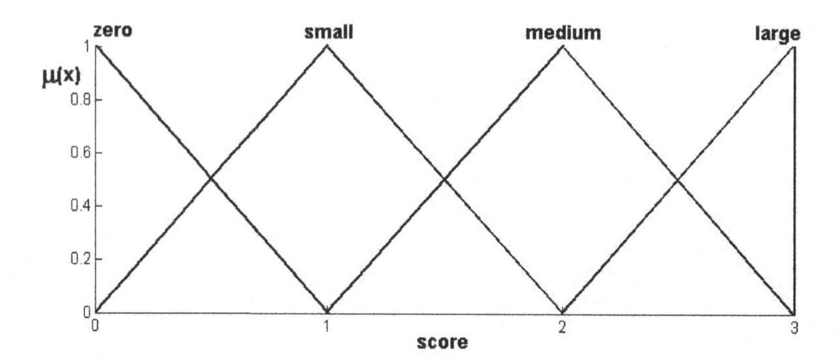

Fig. 4. The shape of the membership functions of linguistic values for output linguistic variable Score

the current controller can be considered as appropriate. The score less than 2 is considered as not satisfactory and the controller has to be re-adapted [1].

2.3 Knowledge Base

The knowledge base is formed by nine linguistic IF-THEN rules of the Mamdani type [8]:

1. If (*RO* is *Small*) & (*RST* is *Slower*) then (*Score* is *Medium*)
2. If (*RO* is *Small*) & (*RST* is *Same*) then (*Score* is *Large*)
3. If (*RO* is *Small*) & (*RST* is *Faster*) then (*Score* is *Large*)
4. If (*RO* is *Appropriate*) & (*RST* is *Slower*) then (*Score* is *Small*)
5. If (*RO* is *Appropriate*) & (*RST* is *Same*) then (*Score* is *Medium*)
6. If (*RO* is *Appropriate*) & (*RST* is *Faster*) then (*Score* is *Large*)
7. If (*RO* is *High*) & (*RST* is *Slower*) then (*Score* is *Zero*)
8. If (*RO* is *High*) & (*RST* is *Same*) then (*Score* is *Small*)
9. If (*RO* is *High*) & (*RST* is *Faster*) then (*Score* is *Medium*)

The shape of membership function of output variable is inferred using the Mamdani method. The crisp value of the output score is determined using the defuzzification method Center of Area [8].

3 PID Parameter Design System

As it was mentioned also for design parameters of conventional PID controller fuzzy expert system (ES2) is used [2]. It uses know-how obtained from the combination of the frequency response method and the step response Ziegler-Nichols design method. [6], [9]

The constant a_2, a_1 and a_0 from the denominator of the transfer function of the controlled system (1) are the inputs of the expert design systems (ES2) (Figure 5). The input parameters a_2, a_1 and a_0 are the linguistic variables expressed

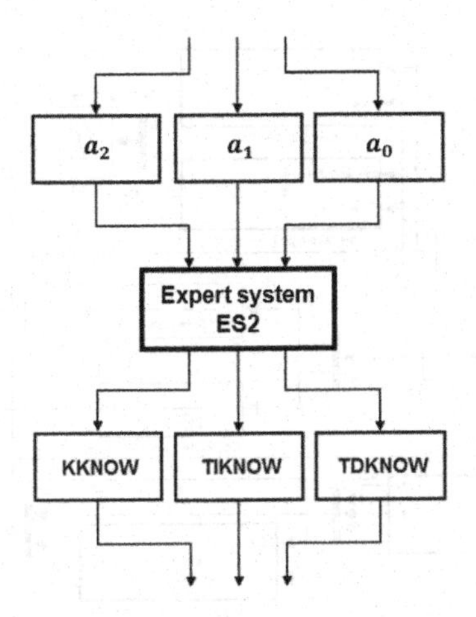

Fig. 5. Graphical representation of inputs and outputs of expert system ES2 [2]

by fuzzy sets for each linguistic variable by three linguistic values - small (S), medium (M) and large (L). It can be the constants from intervals a_2 - [0; 22], a_1 - [0; 20], a_0 - [0; 28].

The outputs *KKNOW*, *TIKNOW* and *TDKNOW* of ES2, which are also constants, are the parameters of conventional PID controller with the transfer function expressed as

$$G_R(s) = KKNOW\left(1 + \frac{1}{TIKNOW \cdot s} + TDKNOW \cdot s\right). \tag{4}$$

Expert design system is model of Takagi-Sugeno type [10] so it does not require defuzzification. The knowledge base of the ES2 is consisted of 27 linguistic rules. For detailed information see [2].

4 The Description of Implemented Algorithm

It is important to define the algorithm of monitoring and following re-adaptation of the controller (Figure 1). The relative overshoot is monitored and stored in memory after every step change of controlled value. The settling time is measured also after every step change of controlled value and is assessed to the previous settling time. If these two monitored parameters are obtained the score is assessed. According to the value of the score, the identification of the system starts and the controller is re-adapted (section 2.2). The simplified model in Matlab-Simulink [11] is depicted in Figure 6.

$$Score \begin{Bmatrix} \geq 2 & \text{do not re-adapt} \\ > 2 & \text{re-adapt} \end{Bmatrix}. \tag{5}$$

For identification the stochastic identification - ARMAX method is used [12]. The re-adaptation (change of parameters of controller) procedure ES2 is done only after the change of required value.

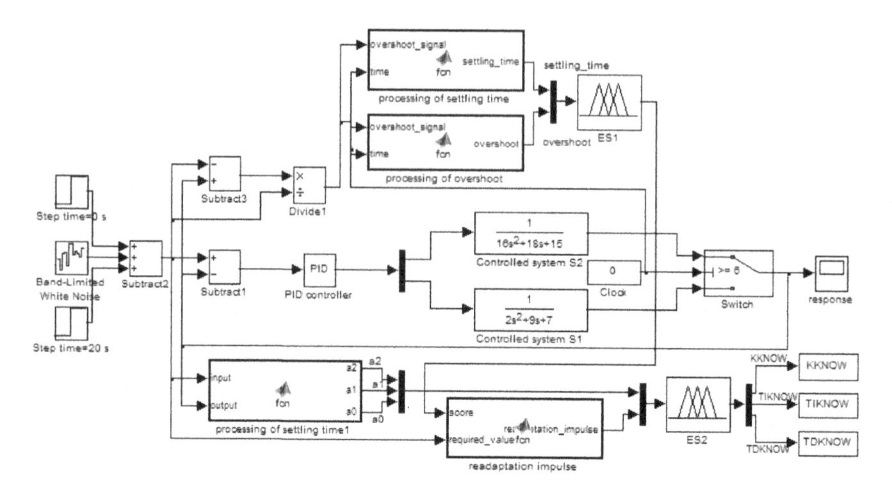

Fig. 6. Simplified model in Matlab-Simulink

5 Verification of Created System

The verification was done in Matlab-Simulink [11], the timing with description of important moments is depicted in Figure 7. Verification of the re-adaptation procedure proposed above (see Figure 1) is started using the controlled system (S1) with transfer function

$$G_{S1}(s) = \frac{1}{2s^2 + 9s + 7} \tag{6}$$

for which the controller with transfer function

$$G_{R1}(s) = 6.1\left(1 + \frac{1}{0.58s} + 0.14s\right) \tag{7}$$

designed through the identification system ES2 is used.

At the time t_A the unit step of required value is introduced. Therefore, the control process is carried out with 14%-overshoot and settling time $t_{st1} = 4.3\ sec.$ The appropriate calculated score by ES1 is

$$score_1 = 2.10 > 2, \tag{8}$$

which corresponds to the satisfactory control course.

At the time t_C a sudden change of the controlled system (to controlled system S2) is simulated from the transfer function $G_{S1}(s)$ to the transfer function

$$G_{S2}(s) = \frac{1}{16s^2 + 18s + 15}. \tag{9}$$

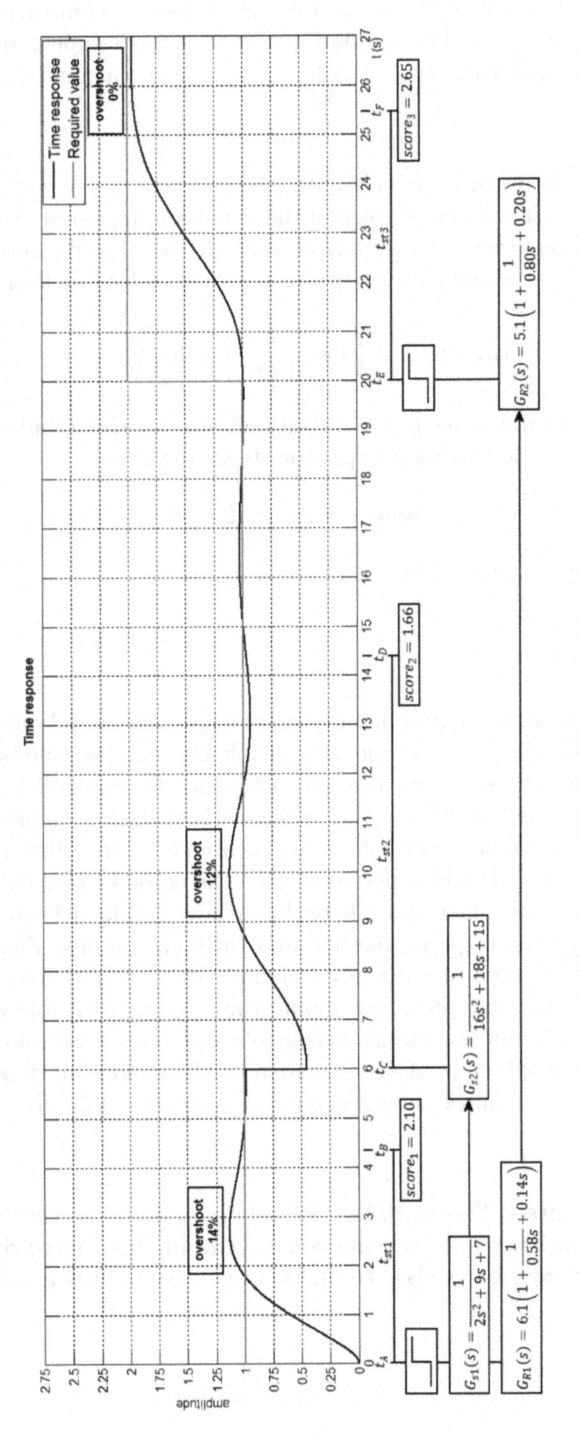

Fig. 7. Time response

Thus, a non-zero control deviation appeared which is compensated by the original controller $G_{R1}(s)$. The oscillating control course appeared with 12%-overshoot and settling time $t_{st2} = 8.4\ sec.$ Now, the calculated appropriate score is

$$score_2 = 1.66 < 2 \tag{10}$$

and insufficient control course is now indicated.

Therefore, when the nearest change of the deviation appeared at the time t_E (the unit step of reguired value is introduced) the re-adaptive process ES2 is initialized and it is designed a new controller with transfer function

$$G_{R1}(s) = 5.1\left(1 + \frac{1}{0.80s} + 0.20s\right). \tag{11}$$

Now, the control process is carried out without any overshoot and with settling time $t_{st3} = 5.5\ sec.$ The calculated appropriate score value is

$$score_3 = 2.65 > 2 \tag{12}$$

and the satisfactory control course is restored again.

6 Conclusion

The procedures of control quality monitoring and necessary re-adaptation of PID controller is solved using the fuzzy-logic principle through the rule-based expert systems. The first one concludes the initial impulse for controller adaptation. The rule base is created within two input linguistic variables - namely the relative settling time and relative overshoot are mentioned. The following design of parameters of classic PID controller is done by the second fuzzy expert system with a knowledge base which is built on know-how obtained from the combination of the frequency response method and the step response Ziegler-Nichols design method. The proof of efficiency was done using simulations in Matlab-Simulink. It is shown that presented monitoring system with following design of PID controller is useful for family of controlled systems of second order. The both described knowledge-based systems are open. Therefore, next time authors think of adding more monitored parameters and widening of family of controlled systems.

Acknowledgements. This work has been supported by Project SP2013/168, "Methods of Acquisition and Transmission of Data in Distributed Systems", of the Student Grant System, VŠB - Technical University of Ostrava.

References

1. Farsi, M., Karam, K.Z., Abdalla, H.H.: Intelligent multi-controller assessment using fuzzy logic. Fuzzy Sets and Systems 79, 25–41 (1996)
2. Nowaková, J., Pokorný, M.: On PID Controller Design Using Knowledge Based Fuzzy System. Advances in Electrical and Electronic Engineering 10(1), 18–27 (2012)
3. Jacskon, P.: Introduction To Expert Systems, 3rd edn. Addison Wesley (1998)
4. Siler, W., Buckley, J.J.: Fuzzy Expert Systems and Fuzzy Reasoning. Wiley-Interscience (2004)
5. Jones, C.H.: Knowledge Based Systems Methods: A Practitioners' Guide. Prentice Hall PTR (1995)
6. Astrom, K.J., Hagglund, T.: PID Controllers: theory, design, and tuning, USA (1995)
7. Novák, V., Perfilieva, I., Močkoř, J.: Mathematical Principles of Fuzzy Logic. Kluwer, Boston (1999)
8. Jager, R.: Fuzzy Logic in Control, Delft (1995)
9. Kilian, C.: Modern Control Technology. Thompson Delmar Learning (2005)
10. Klir, B.J., Yuan, B.: Fuzzy Sets and Fuzzy Logic: Theory and Application. Prentice Hall, New Jersey (1995)
11. MATLAB - The MathWorks-MATLAB and Simulink for Technical Computing, `http://www.mahworks.com` (cit. July 10, 2012)
12. Noskievič, P.: Modelování a identifikace systémů. Montanex, Ostrava (1999)

Multiplier System in the Tile Assembly Model with Reduced Tileset-Size

Xiwen Fang and Xuejia Lai*

Department of Computer Science and Engineering, Shanghai Jiao Tong University, Shanghai 200240, China
lai-xj@cs.sjtu.edu.cn

Abstract. Previously a 28-tile multiplier system which computes the product of two numbers was proposed by Brun. However the tileset-size is not optimal. In this paper we prove that multiplication can be carried out using less tile types while maintaining the same time efficiency: we propose two new tile assembly systems, both can deterministically compute $A * B$ for given A and B in constant time. Our first system requires 24 computational tile types while our second system requires 16 tile types, which achieve smaller constants than Brun's 28-tile multiplier system.

Keywords: tile assembly model, DNA computing, multiplier, tileset-size.

1 Introduction

1.1 Background and Related Work

Since Adleman's pioneering research which shows DNA could be used to solve Hamiltonian path problem [1], many researchers have explored the ability of biological molecules to perform computation [2,3,4]. The theory of tile self-assembly model which was developed by Winfree and Rothemund [5,6,7] provides a useful framework to study the self-assembly of DNA. This model has received much attention over the past few years. Researchers have demonstrated DNA implementations of several tile systems: Barish et al. [8] have demonstrated DNA implementations of copying and counting; Rothemund et al. [9] have demonstrated DNA implementation of *xor* tile system. Several systems solving satisfiability problem are also proposed [10,11,12].

The efficiency of a tile asssembly system involves two factors: the tileset-size and the assembly time. In [13], Brun proposed a multiplier system that computes the product of two numbers, which requires 28 distinct computational tile types besides the tiles constructing the seed configuration. The computation can be carried out in time linear in the input size. However, the tileset-size of Brun's system is not optimal, i.e. multiplier system can be implemented using less tile types.

* Corresponding author.

A. Abraham et al. (eds.), *Innovations in Bio-inspired Computing and Applications*, Advances in Intelligent Systems and Computing 237,
DOI: 10.1007/978-3-319-01781-5_9,

In this paper we present two new multiplier systems which achieve smaller constants for the multiplication problem than the previous 28-tile multiplier system proposed by Brun, while maintaining the same time efficiency. In our first system, we show that multiplication can be carried out using 24 tile types instead of 28 tiles types, then we propose a second multiplier system and show that the tileset-size can be further reduced to 16.

The remaining of this paper is organized as follow: in section 1.2 we briefly introduce the concept of tile assembly model to assist the reader. In section 1.3 we introduce the corresponding algorithms. Several subsystems are discussed in section 2. In section 3 we present two new multiplier systems and compare our system with existing system[13] in terms of tileset-size and assembly time. Our contributions are summarized in section 4.

1.2 Tile-Assembly Model

To assist the reader, in this section we briefly introduce the concept of tile assembly model. We refer to Σ as a finite alphabet of symbols called binding domains. We assume $null \in \Sigma$. Each tile has a binding domain on its north, east, south and west side. We represent the set of directions as $D = \{N, E, S, W\}$.

A tile over a set of binding domains is a 4-tuple. For a tile t, for $\langle \delta_N, \delta_E, \delta_S, \delta_W \rangle \in \Sigma^4$, we will refer to $bd_d(t)$ as the binding domain of tile t on d 's side.

A **strength function** $g : \Sigma \times \Sigma \to N$ denotes the strength of the binding domains. g is commutative and $\forall \delta \in \Sigma, g(\delta, null) = 0$. Let T be a set of tiles containing *empty*. A tile system S is a triple $< T, g, \epsilon >$. $\epsilon \in \mathbb{N}$ is the temperature. A tile can attach to a configuration only in empty positions if and only if the total strength of the appropriate binding domains on the tiles in neighboring positions meets or exceeds the temperature ϵ.

Given a set of tiles Γ, a **seed configuration** $S' : \mathbb{Z}^2 \to \Gamma$ and $S =< T, g, \epsilon >$, configurations could be produced by S on S'. If no more attachment is possible, then we have the **final configuration**.

The reader may refer to [5,6,7] for more discussion of the concept of tile assembly model.

Let $||$ be a special binding domain which connects the tiles constructing seed configuration. To simplify the discussion, for all of the systems involved in this paper, we define the strength funcion g as follow:

$$\forall \delta \in \Sigma, g(\delta, null) = 0; g(||, ||) = 2; \forall \delta \in \Sigma, \delta \neq ||, g(\delta, \delta) = 1. \quad (1)$$

1.3 Preliminary Algorithms

Intuitively, a multiplier system could be implemented by combining subsystems which compute $f(x) = 2x$ and $f(x, y) = 2x + y$ respectively.

Given n_A-bit binary integer A and n_B-bit binary integer B. We denote by A_i and B_i the i_{th} digit of A and B. $A = \sum_{i=0}^{n_A-1} 2^i A_i$, $B = \sum_{i=0}^{n_B-1} 2^i B_i$. We use Algorithm 1 to compute $A * B$.

Input: n_A-bit binary integer A , n_B-bit binary integer B ($n_B \geq 2$)
Output: $S = A * B$
1 $i \leftarrow n_B - 2$.
2 $S \leftarrow A$
3 **while** $i \geq 0$ **do**
4 | **if** ($B_i = 0$) **then**
5 | | $S \leftarrow 2S$
6 | **end**
7 | **else**
8 | | $S \leftarrow 2S + A$
9 | **end**
10 | $i \leftarrow i - 1$
11 **end**

Algorithm 1. Given binary integers A and B, compute $A * B$

2 Subsystems of Multiplier System

Our multiplier system is a combination of several subsystems. In this section we will define these subsystems respectively.

To simplify the definition of the seed configuration, for n_A-bit integer A, we denote by $A_i (0 \leq i \leq n_A - 1)$ the i_{th} digit of A. We also define $A_i (n_A \leq i \leq n - 1, n > n_A)$ as follow: we pad the n_A-bit binary integer A with n-n_A 0-bits and denote these 0-bits by $A_i (n_A \leq i \leq n - 1)$. Thus $A = \sum_{i=0}^{n-1} 2^i A_i$. The same definition holds for any other input variables in this paper.

2.1 Shifter Tile System

In this section we propose an 8-computational-tile system for computing $f(A, B) = (2A, B)$.

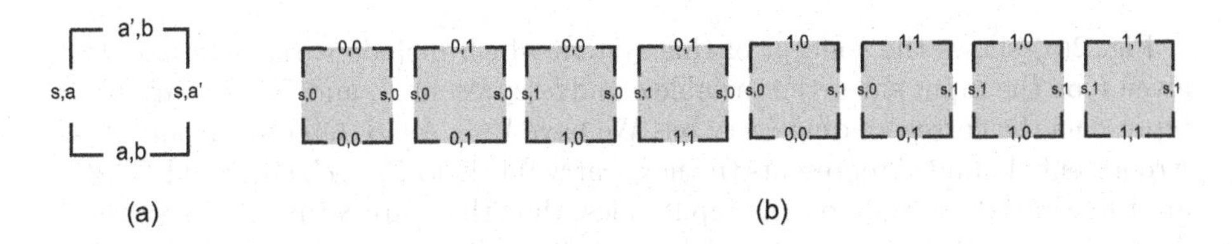

Fig. 1. Shifter tile system

Let Σ_S ={(0,0), (0,1), (1,0), (1,1), (s,0), (s,1)}, and T_s be a set of tiles over Σ_S defined as follow:

$$T_s = \{\langle (a', b), (s, a'), (a, b), (s, a) \rangle | a, b, a' \in \{0, 1\}\}. \tag{2}$$

Fig.1(a) shows the concept behind the system which includes variable a, b, a'. All of the tiles has two input sides (south and east) and two output sides (north and west), for input a' on the east side, we output the same value on the north side; for input a on the south side, we output the same value on the west side. We have $a, b, a' \in \{0, 1\}$, thus there are 8 tiles in the system (Fig.1(b)).

Let $\Gamma = \{\alpha_0\beta_0 = \langle(0,0), ||, null, ||\rangle,\ \alpha_0\beta_1 = \langle(0,1),\ ||,\ null,\ ||\rangle,\ \alpha_1\beta_0 = \langle(1,0),\ ||,\ null,\ ||\rangle,\ \alpha_1\beta_1 = \langle(1,1),\ ||,\ null,\ ||\rangle,\ \alpha\beta_{bound} = \langle(0,0),\ ||,\ null,\ null\rangle,\ \gamma_0 = \langle||,\ null,\ null,\ ||\rangle\}\ \gamma_1 = \langle null,\ null,\ ||,\ (s,0)\rangle\}$. Let $A_i, B_i \in \{0,1\}$, $n \geq max(n_A, n_B) + 1$. Let the seed configuration $S : \mathbb{Z}^2 \to \Gamma$ be such that

$$\begin{cases} S(1,1) = \gamma_1; S(1,0) = \gamma_0; \\ \forall i = 0, 1, \ldots, n-2 : S(-i,0) = \alpha_{A_i}\beta_{B_i}; S(-n+1,0) = \alpha\beta_{bound}; \\ \text{For all other } (x,y) \in \mathbb{Z}^2, S(x,y) = empty. \end{cases} \tag{3}$$

Theorem 1. *Let $\epsilon_S = 2$. Given a seed configuration S encoding A and B and the strength function g as defined in (3) and (1), the system $S_S =< T_S, g, \epsilon_S >$ computes the function $f(A,B) = (2A, B)$.*

2.2 Adder Tile System

In this section we propose a 8-tile system computing $f(A,B) = (A + B, B)$.

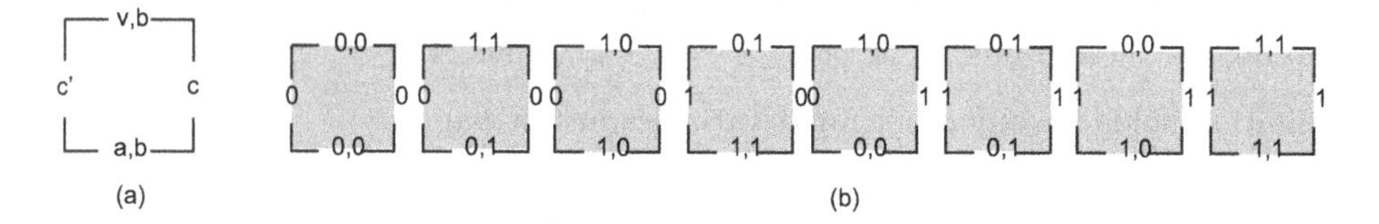

Fig. 2. Adder tile system (a)The tile has two input sides(south and east) and two output sides (north and west). The south side contains the value of the current bit of A and B; the east side is the carry bit. (b) There are 8 tiles in the system.

Fig. 2(a) shows the concept of this system which include variable a, b, c. For given i, on the input sides, the variable a and b represent A_i and B_i; the variable c represents the corresponding carry bit. We have $V = A+B$. On the output sides, v represents V_i and c' represents the next carry bit. Let $a, b, c, c', v \in \{0, 1\}$. There are three variables a, b, c on the input sides, thus there are 8 tiles in the system. According to addition rule, we define two functions $f_v(a, b, c)$ and $f_{c'}(a, b, c)$ which computes v and c' respectively:

$$\begin{cases} if(a+b+c=0), f_v(a,b,c) = 0, f_{c'}(a,b,c) = 0; \\ if(a+b+c=1), f_v(a,b,c) = 1, f_{c'}(a,b,c) = 0; \\ if(a+b+c=2), f_v(a,b,c) = 0, f_{c'}(a,b,c) = 1; \\ if(a+b+c=3), f_v(a,b,c) = 1, f_{c'}(a,b,c) = 1. \end{cases} \tag{4}$$

Let $\Sigma_+ = \{(0,0),(0,1),(1,0),(1,1),0,1\}$. We have $v = f_v(a,b,c), c' = f_{c'}(a,b,c)$. Let T_+ be a set of tiles over Σ_+ defined as follow:

$$T_+ = \{\langle (v,b), c, (a,b), c' \rangle\} \tag{5}$$

Fig.2(b) shows all of the eight tiles.

Let the seed configuration S be defined as it is in (3) except that $\gamma_1 = \langle null, null, ||, 0 \rangle$.

Theorem 2. *Let $\epsilon_+ = 2$, g defined in (1), Σ_+ defined as above, and T_+ be a set of tiles over Σ_+ as defined in (5). Given a seed configuration S encoding A and B which is defined above, the system $S_+ =< T_+, g, \epsilon_+ >$ computes the function $f(A,B) = (A+B,B)$*

2.3 Shifter-Adder Tile System

In this section we propose a 16-computational-tile system computing $f(A,B) = (2A+B,B)$. Fig. 3(a) shows the concept behind the tile system. For given i, on the input sides, the variable a and b represent A_i and B_i; a' represents A_{i-1}; the variable c represents the corresponding carry bit. Let $V = 2A+B$, thus for each bit we compute $A_{i-1} + B_i$. On the output sides, v represents V_i and c' represents the next carry bit. Let $a,b,c,a',c',v \in \{0,1\}$, we denote the function which computes v and c' by $f_v(a',b,c)$ and $f_{c'}(a',b,c)$ respectively. According to the addition rule, the functions f_v and $f_{c'}$ are defined as follow:

$$\begin{cases} if(a'+b+c=0), f_v(a',b,c)=0, f_{c'}(a',b,c)=0; \\ if(a'+b+c=1), f_v(a',b,c)=1, f_{c'}(a',b,c)=0; \\ if(a'+b+c=2), f_v(a',b,c)=0, f_{c'}(a',b,c)=1; \\ if(a'+b+c=3), f_v(a',b,c)=1, f_{c'}(a',b,c)=1. \end{cases} \tag{6}$$

Let $\Sigma_{S+} = \{(0,0),(0,1),(1,0),(1,1)\}$. We have $v = f_v(a',b,c)$ and $c' = f_{c'}(a',b,c)$. Let T_{S+} be a set of tiles over Σ_{S+} defined as follow:

$$T_+ = \{\langle (v,b), (c,a'), (a,b), (c',a) \rangle\} \tag{7}$$

The input sides involve 4 variables, i.e. a, b, a',c, thus there are 16 tiles in this system(Fig. 3(b)). Let the seed configuration S be defined as it is in (3) except that $\gamma_1 = \langle null, null, ||, (0,0) \rangle$.

Theorem 3. *Let $\epsilon_{S+} = 2$, g defined in (1), Given a seed configuration S encoding A and B as above, the system $S_{S+} =< T_{S+}, g, \epsilon_{S+} >$ computes the function $f(A,B) = (2A+B,B)$*

Fig. 4(a) shows a sample seed configuration which encodes $A = 0101011_2$, $B = 0100011_2$. Fig. 4(b) shows a sample execution of S_{S+} on the seed configuration. We could read the result from the top row of the final configuration, i.e. $2A+B = 1111001_2$.

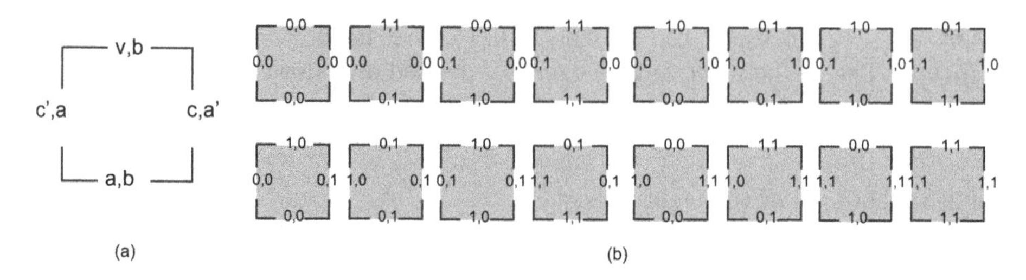

Fig. 3. Shifter-Adder tile system (a) The tile has two input sides(south and east) and two output sides (north and west). The south side contains the value of the current bit of A and B; the east side contains the value of the former bit of A and the carry bit. (b) There are 16 tiles in the system.

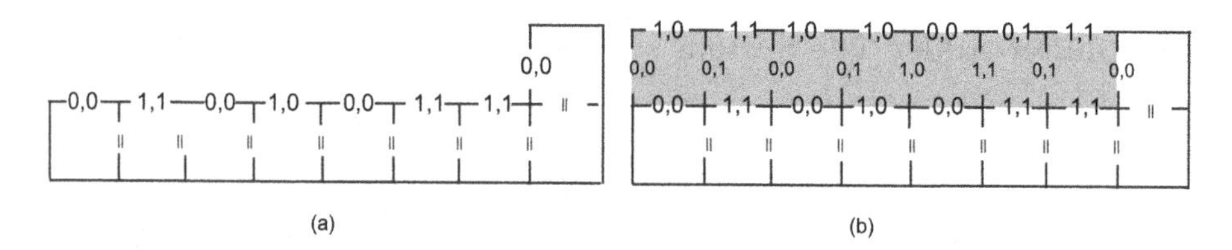

Fig. 4. (a) seed configuration: $A = 0101011_2$, $B = 0100011_2$.(b) The top row reads the solution: $2A + B = 1111001_2$.

3 Multiplier Tile System

3.1 Multiplier Tile System: Version 1

We use subsystems S_S and S_S+ to build the multiplier system. All of the tiles have two input sides(south and east) and two output sides (north and west).

The method of establishing the seed-configuration is described as follow: We use the two sides of the L-configuration to encode inputs. One of the input number, i.e. A, is encoded on the bottom row. On the same row there are also n_B extra tiles representing 0 in the most significant bit places because the product of the n_A-bit number A and the n_B-bit number B may be as large as $n_A + n_B$ bits.

Let $n = n_A + n_B$. Let $\Sigma_\times = \{(0,0),(0,1),(1,0),(1,1),(s,0),(s,1)\}$.

Let $\Gamma = \{\ \alpha_0 = \langle(0,0),||,null,||\rangle,\ \alpha_1 = \langle(1,1),||,null,||\rangle,\ \alpha_{bound} = \langle(0,0),||,\ null,null\rangle,\ \beta_0 = \langle||,null,||,(s,0)\rangle,\ \beta_1 = \langle||,null,||,(0,0)\rangle,\ \beta_{bound0} = \langle null,null,||,\ (s,0)\rangle,\ \beta_{bound1} = \langle null,null,||,(0,0)\rangle,\ \gamma_0 = \langle||,null,null,||\rangle\}$. Let the seed configuration $S : \mathbb{Z}^2 \to \Gamma$ be such that

$$\begin{cases} S(1,0) = \gamma_0 \\ \forall i = 0,1,\ldots,n-2 : S(-i,0) = \alpha_{A_i};\ S(-n+1,0) = \alpha_{bound} \\ \forall i, 1 \le i \le n_B - 2, S(1,n_B-i-1) = \beta_{B_i};\ S(1,n_B-1) = \beta_{boundB_0} \\ \text{For all other } (x,y) \in \mathbb{Z}^2, S(x,y) = empty. \end{cases} \tag{8}$$

The seed configuration is of length $n_A + n_B + 1$ and height n_B.

Theorem 4. *Let* $\epsilon_\times = 2$, g *defined in* (1), *and* $T_\times = T_S \cup T_{S+}$. *Given a seed configuration encoding* A *and* B *which is defined in* (8), *the system* $S_\times =< T_\times, g_\times, \epsilon_\times >$ *outputs* $(A * B, A)$.

Proof. We have $T_\times = T_S \cup T_{S+}$. Let $bd_{E,W}(T_S) = \{bd_E(t), bd_W(t)|t \in T_S\}$, $bd_{E,W}(T_{S+}) =\{bd_E(t), bd_W(t)|\ t \in T_{S+}\}$. These two sets are disjoint, thus S_S and S_{S+} work together without interfering.

For each tile $t(t \in T_\times)$, $bd_S(t)$ has two bits. Let $bd_Sl(t)$ be the first bit and $bd_Sr(t)$ be the second bit. Let $bd_S(F(i)) = (bd_SlF(i), bd_SrF(i))$. Let $bd_Sl(F(i)) = \sum_{j=0}^{n-1} 2^j * bd_Sl(F(-j,i))$, $bd_Sr(F(i)) = \sum_{j=0}^{n-1} 2^j * bd_Sr(F(-j,i)))$. The definition also holds for $bd_Nl(t)$, $bd_Nr(t)$, $bd_N(F(i))$, $bd_Nl(F(i))$ and $bd_Nr(F(i))$.

Let $b_k = \sum_{i=n_B-k}^{n_B-1} B_i * 2^{i-n_B+k}$. Thus $B = b_{n_B}$, $B_{n_B-1} = b_1 = 1$, $2 * B_{n_B-1} + B_{n_B-2} = b_2$. We are going to prove that given the seed configuration S encoding A and B which is defined in(8), for $1 \le i \le n_B - 1$, $bd_N(F(i)) = (A * b_{i+1}, A)$:

i) For $i = 1$, we have $bd_S(F(1)) = bd_N(F(0)) = (A, A)$.

If $B_{n_B-2} = 0$, i.e. $B_{n_B-i-1} = 0$,$b_2 = 10_2$, then $bd_W(F(1,1)) = (s, 0)$;

If $B_{n_B-2} = 1$, i.e. $B_{n_B-i-1} = 1$, $b_2 = 11_2$, then $bd_W(F(1,1)) = (0,0)$.

Thus according to Theorem 2 and Theorem 4, if $B_{n_B-2} = 0$, we have $bd_N(F(1)) = (2A, A) = (A * b_2, A)$; if $B_{n_B-2} = 1$, we have $bd_N(F(1)) = (2A + A, A) = (A * b_2, A)$.

Thus $bd_N(F(i)) = (A * b_{i+1}, A)$ holds for $i = 1$.

ii) Assume $bd_N(F(i)) = (A * b_{i+1}, A)$ holds for $i = k - 1$, i.e. $bd_N(F(k-1)) = (A * b_k, A)$.

For $i = k$, if $b_{(k+1)_0} = B_{n_B-k-1} = 0$, then $bd_W(F(1,i)) = (s, 0)$; If $b_{(k+1)_0} = B_{n_B-k-1} = 1$, then $bd_W(F(1,i)) = (0,0)$.

Thus according to Theorem 2 and Theorem 4, if $b_{(k+1)_0} = B_{n_B-k-1} = 0$, we have $bd_N(F(k)) = (2bd_Nl(F(k-1)), A) = (2 * A * b_k, A) = (A * b_{k+1}, A)$; if $b_{(k+1)_0} = B_{n_B-k-1} = 1$, we have $bd_N(F(k)) = (2bd_Nl(F(k-1)) + A, A) = (2 * (A * b_k) + A, A) = (A * b_{k+1}, A)$.

iii) Therefore $bd_N(F(i)) = (A * b_{i+1}, A)$ holds for $1 \le i \le k$. Let $k = n_B - 1$, we have $bd_N(F(n_B - 1)) = (A * b_{n_B}, A) = (A * B, A)$.

Thus we proved that given the seed configuration encoding A and B, the system outputs $(A * B, A)$. □

Fig. 5(a) shows a sample seed configuration which encodes $A = 1001_2$, $B = 1011_2$: A is encoded on the bottom row with 4 extra 0 tiles in the most significant bit place. We encode B from B_{n_B-2} to B_0. B_3 is the most significant bit, and in the seed configuration we discard this bit. $B_2 = 0$, $S(1,1) = \beta_0$; $B_1 = 1$, $S(1,2) = \beta_1$; $B_0 = 1$, $S(1,3) = \beta_{bound1}$. Fig. 5(b) shows a sample execution of the multiplier system on the seed configuration and the product could be read from the top row of the final configuration, i.e. $1001_2 * 1011_2 = 1100011_2$.

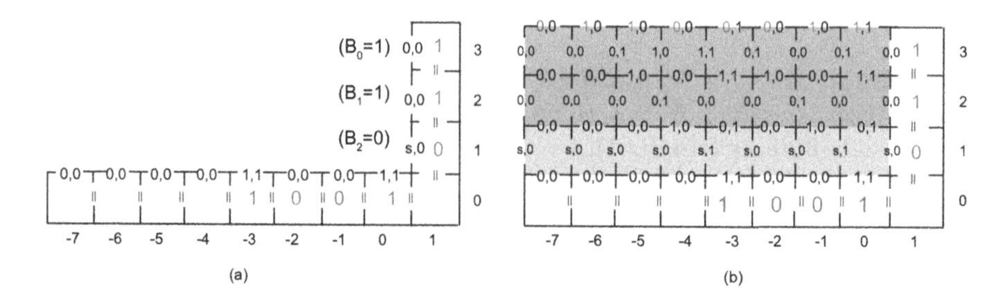

Fig. 5. Given a sample input of $A = 1001_2$, $B = 1011_2$, with A on bottom row and B on the right most column. (b) $1001_2 * 1011_2 = 1100011_2$.

3.2 Multiplier Tile System: Version 2

In this section we combine the shifter system S_S and the adder system S_+ to create another simplifier multiplier system which computes $f(A, B) = (A * B, A)$ and uses only 16 tiles.

We discuss how to establish the seed configuration for this system:

Let $\Gamma = \{\ \alpha_0 = \langle(0,0), ||, null, ||\rangle,\ \alpha_1 = \langle(1,1), ||, null, ||\rangle,\ \alpha_{bound} = \langle(0,0), ||, null, null\rangle,\ \beta_0 = \langle||, null, ||, (s,0)\rangle,\ \beta_1 = \langle||, null, ||, 0\rangle,\ \beta_{bound0} = \langle null, null, ||, (s,0)\rangle,\ \beta_{bound1} = \langle null, null, ||, 0\rangle,\ \gamma_0 = \langle||, null, null, ||\rangle\}$. Let $n = n_A + n_B$. Let the seed configuration $S : \mathbb{Z}^2 \to \Gamma$ be such that

$$\begin{cases} S(1,0) = \gamma_0 \\ \forall i = 0,1,\ldots,n-2, S(-i,0) = \alpha_{A_i}; S(-n+1,0) = \alpha_{bound} \\ j \leftarrow 1; k \leftarrow n_B - 2; \\ while(k \geq 1)\{ \\ if(B_k = 0)\{S(1,j) = \beta_0; j \leftarrow j+1;\} \\ else\{S(1,j) = \beta_0; S(1,j+1) = \beta_1; j \leftarrow j+2;\} \\ k \leftarrow k-1;\} \\ if(B_0 = 0), S(1,j) = \beta_{bound0} \\ else\{S(1,j) = \beta_{bound0}; S(1,j+1) = \beta_{bound1};\} \\ \text{For all other } (x,y) \in \mathbb{Z}^2, S(x,y) = empty. \end{cases} \tag{9}$$

Theorem 5. *Let $\epsilon_{\times 2} = 2$, g defined in (1), and $T_{\times 2} = T_S \cup T_+$. Given a seed configuration encoding A and B as defined in (9), the system $S_{\times 2} =< T_{\times 2}, g_{\times 2}, \epsilon_{\times 2} >$ outputs $(A * B, A)$.*

Proof. According to the establishment of the seed configuration, in this system we carry out the shift-addition in two steps: we shift the first input number by one bit, and then we add the second input number. Comparing with the first multiplier system presented in section 3.1, although the shift-addition is carried out in two steps, which takes two rows in the configuration, the algorithms of these two systems are just identical. Thus the correctness of this system follows directly from Theorem 4. □

3.3 Discussion

Lemma 6. *(Multiplier Assembly Time Lemma). For all $A \geq 1, B \geq 2$, the assembly time of the final configuration F produced by $S_\times$ or $S_{\times 2}$ on S that encodes A and B and pads A with n_B 0-tiles is $\Theta(n_A + n_B)$.*

Proof. For both of the two systems, the length of the final configuration is $n_A + n_B$, the height is $\Theta(n_B)$. According to Lemma 2.3 in[13], the assembly time of the final configuration F produced by $S_\times$ or $S_{\times 2}$ on S that encodes A and B and pads A with n_B 0-tiles is $\Theta(n_A + n_B)$. □

We use l and h to represent the length and the height of the final configuration.

For the first system, we use 24 distinct tile types. The final configuration is of height n_B. We reduce the tileset-size of the system to 16 in our second system. However the height of the final configuration is increased. The value of h depends on the number of 1-bits in B. Suppose there are x 1-bits in B, i.e. we have $n_B - x$ 0-bits in B. Each 1-bit(except the MSB 1-bit) takes two rows in the configuration while each 0-bit takes one row, thus $h = 1 + (n_B - x) + 2(x - 1) = n_B + x - 1$. We have $1 \leq x \leq n_B$,thus $n_B \leq h \leq 2n_B - 1$.

Along with Brun's scheme[13], we make comparison of these three systems in Table 1. We could see that the tileset-size of our second system is the least; however the height of the final configuration might be increased, depending on the number of 1-bits in B. The height of the final configuration of our first system is the least; the tileset-size is also less than Bruns. All of these three systems share the same assembly time, i.e. $\Theta(n_A + n_B)$.

While consider performing multiplication, one chould make a choice between these systems according to the actual demand. Intuitively, our first system could be choosed if one hopes to construct the seed configuration in a simpler way, while the second system is beneficial for those who want to perform multiplication with tile types as less as possible.

Table 1. Comparing the efficiency of multiplier systems

Scheme	version 1	version 2	Brun
Tileset-size	24	16	28
Assembly time	$\Theta(n_A + n_B)$	$\Theta(n_A + n_B)$	$\Theta(n_A + n_B)$
l	$n_A + n_B + 1$	$n_A + n_B + 1$	$n_A + n_B + 1$
h	n_B	$n_B \leq h \leq 2n_B - 1$	$n_B + 1$

4 Conclusion

In this paper we present two multiplication systems which compute $A * B$ for given A and B in constant time. Our first system is constructed using subsystems S_S and S_{S+}, which requiring 24 computational tile types; Further improvement is achieved in our second system: using subsystems S_S, S_+ and a different seed configuration, we show that multiplier system could be implemented using only 16 tile types. As a result, we achieved reduced tileset-size compared with that of Brun's 28-tile multiplier system.

Acknowledgements. This work was supported by the National Natural Science Foundation of China (No. 61272440, No. 61073149), Research Fund for the Doctoral Program of Higher Education of China (20090073110027), State Key Laboratory of ASIC & System (11KF0020), Key Lab of Information Network Security, Ministry of Public Security (C11603) and Scholarship Award for Excellent Doctoral Student granted by Ministry of Education.

References

1. Adleman, L.: Molecular computation of solutions to combinatorial problems. Science 266(5187), 1021–1024 (1994)
2. Lipton, R.: Using dna to solve np-complete problems. Science 268(4) (1995)
3. Liu, Q., Wang, L., Frutos, A., Condon, A., Corn, R., Smith, L., et al.: Dna computing on surfaces. Nature 403(6766), 175–179 (2000)
4. Ouyang, Q., Kaplan, P., Liu, S., Libchaber, A.: Dna solution of the maximal clique problem. Science 278(5337), 446–449 (1997)
5. Rothemund, P., Winfree, E.: The program-size complexity of self-assembled squares. In: Proceedings of the Thirty-Second Annual ACM Symposium on Theory of Computing, pp. 459–468. ACM (2000)
6. Winfree, E.: Algorithmic Self-Assembly of DNA. PhD thesis, California Institute of Technology (1998)
7. Winfree, E., Liu, F., Wenzler, L., Seeman, N., et al.: Design and self-assembly of two-dimensional dna crystals. Nature 394(6693), 539–544 (1998)
8. Barish, R., Rothemund, P., Winfree, E.: Two computational primitives for algorithmic self-assembly: Copying and counting. Nano Letters 5(12), 2586–2592 (2005)
9. Rothemund, P., Papadakis, N., Winfree, E.: Algorithmic self-assembly of dna sierpinski triangles. PLoS Biology 2(12), e424 (2004)
10. Brun, Y.: Solving np-complete problems in the tile assembly model. Theoretical Computer Science 395(1), 31–46 (2008)
11. Brun, Y.: Improving efficiency of 3-SAT-solving tile systems. In: Sakakibara, Y., Mi, Y. (eds.) DNA 16 2010. LNCS, vol. 6518, pp. 1–12. Springer, Heidelberg (2011)
12. Lagoudakis, M., LaBean, T.: 2-d dna self-assembly for satisfiability. In: DNA Based Computers V, vol. 54, pp. 141–154 (2000)
13. Brun, Y.: Arithmetic computation in the tile assembly model: Addition and multiplication. Theoretical Computer Science 378(1), 17–31 (2007)

A Bézier Curve-Based Approach for Path Planning in Robot Soccer

Jie Wu[1] and Václav Snášel[2]

[1] Department of Electrical Engineering,
School of Electrical and Electronic Engineering,
Hubei University of Technology, Wuhan 430068, China
defermat2008@hotmail.com
[2] Department of Computer Science,
Faculty of Electrical Engineering and Computer Science,
VŠB – Technical University of Ostrava, Ostrava 70032, Czech Republic
vaclav.snasel@vsb.cz

Abstract. This paper presents an efficient, Bézier curve-based path planning approach for robot soccer, which combines the function of path planning, obstacle avoidance, path smoothing and posture adjustment together. The locations of obstacles are considered as control points of Bézier curve, then according to the velocity and orientation of end points, a smooth curvilinear path can be planned in real time. For the sake of rapid reaching, it is necessary to decrease the turning radius. Therefore a new construction of curve is proposed to optimize the shape of Bézier path.

Keywords: Bézier curve, path planning, robot soccer.

1 Introduction

Robot soccer is a classic integration of robotics and artificial intelligence (AI). It almost covers the vast majority of important issues in these two fields. Robot soccer studies how mobile robots can be built and trained to play the game of soccer [9]. In the robot soccer game, robots stay in a dynamic environment all the time. Ball and two sides of robots move continually, which changes the situation of the game, while the teammate robots need to make decisions and take actions according to the changing situation, to gain ascendency and win the game by means of teamwork. In order to implement the team strategies, the robots need to plan a collision-free and time optimal path in real time. Therefore, path planning is the basis of robot movement, which is a kind of planning control in an unknown and dynamic environment. At present, there are two most important international competitions in the field of robot soccer, that is FIRA and RoboCup. In this paper, a Bézier curve-based method of path planning is presented for robot soccer game.

The paper is organized as follows. The path planning problem of robot soccer is discussed in Section 2. Section 3 presents a novel Bézier curve-based path

A. Abraham et al. (eds.), *Innovations in Bio-inspired Computing and Applications*,
Advances in Intelligent Systems and Computing 237,
DOI: 10.1007/978-3-319-01781-5_10,

planning approach for robot soccer. Section 4 describes the proposed path planning optimization techniques. Finally, Section 5 draws the conclusions of this work.

2 Path Planning in Robot Soccer

On the issue of path planning, a lot of approaches were presented, such as artificial potential field (APF) [2, 7, 10, 16], genetic algorithm (GA), fuzzy, artificial neural network (ANN), rapidly-exploring random tree (RRT) [1, 8, 15], and so on. All the approaches could be classified as AI methods [16,19] except APF and RRT. The AI methods are associated with optimization algorithms, resulting in optimal global path planning. However, optimization algorithms are relatively complex, time-consuming, and are not very useful for real-time applications. RRT is a universal algorithm, it can deal with all types of obstacles. However, it can not produce a smooth path.

APF [7,12,13,17] is a virtual force method, which has been studied extensively for autonomous mobile robot path planning [5]. There are many advantages of this method. However, there exist two problems that are inherent to the artificial potential field [11,14,20]. First, the robot is likely to be in a trap of local minima. The velocity and direction of the robot in the field depend on the magnitude and direction of the vector force, it means the robot can not move if the resultant force is zero, and that is why the robot could run into the trap of local minima. Secondly, in the area where obstacles are gathering, the planned path by APF presents a problem of oscillation. In this situation, although the robot can avoid obstacles and reach its destination point, the robot among the obstacles varies direction again and again, which may interfere with the movement fluency of the robot. Unless the robot moves slowly, the path does not meet the motion performance of robot nor the tactic of soccer game. If the robot move along the oscillating path, it will absolutely be under the heavy siege by opponents.

Jolly *et al.* [6] proposed a Bézier curve-based approach of path planning for a mobile robot in a multi-agent robot soccer system. Their work is applied to the MiroSot small league. The approach only considers a third degree Bézier curve, where the lengths of the polygon sides d_1 and d_2 are the key parameters, and therefore, no more complex situation can be considered. They proposed an iterative algorithm to determine d_1 and d_2, then a path can be planned by the lengths of d_1 and d_2. If the robot will intersect with an obstacle, the path is re-planned by changing parameters d_1 and d_2. The accuracy of the Bézier path depends on the accuracy of the pose estimation algorithm, and the optimization of the planned path is subjected to the parameters d_1 and d_2.

In robot soccer, path planning has two important functions, that is, obstacle avoidance and path smoothing. If there is no obstacle avoidance mechanism, collision will happen frequently in the game, which is going to stop the flow of the soccer game and damage the robot hardware. Furthermore, path planning must take into account the state of the initial and destination points, including the location and orientation, so that the robot can play the ball directly without posture adjustment when arriving the destination point. This requires the system to plan a smooth path, then the robot can adjust its posture through a smooth curvilinear movement in the process of moving to destination point. The presented approach in this paper combines path planning, obstacle avoidance, path smoothing and posture adjustment together.

3 Bézier Path Planning

Bézier Curves [3,4] were invented in 1962 by the French engineer Pierre Bézier for designing automobile bodies. Today Bézier Curves are widely used in computer graphics, computer aided geometric design and other related fields.

A Bézier curve is defined by a set of control points P_0 through P_n, where n is called its order ($n = 1$ for linear, 2 for quadratic, etc.). The first and last control point are always the end points of the curve; however, the intermediate control points (if any) generally do not lie on the curve.

Generally, we can define the Bézier curve as

$$P(t) = \sum_{i=0}^{n} b_{i,n}(t) P_i \, , \; t \in [0,1] \, , \tag{1}$$

where t is a normalized time variable, the points P_i are called *control points* for the Bézier curve, the polynomials

$$b_{i,n}(t) = C_n^i t^i (1 - t)^{n-i} \, , \; i = 0, 1, \cdots, n \, , \tag{2}$$

are known as *Bernstein basis functions* of degree n, and the binomial coefficient

$$C_n^i = \frac{n!}{i!(n-i)!} \, . \tag{3}$$

The soccer robot in MiroSot moves on two parallel wheels, with the centers of these wheels aligned, resulting in a common surface normal. For this wheel assembly, the robot turning is induced by the difference in the velocities of the two wheels. In other words, we control the robot through controlling the rotational speed of two independent wheels.

Let $P(t)$ be the location of robot. The generalized coordinates are defined in Fig. 1. Then we have

$$P(t) = [x \;\; y \;\; \theta]^T \, . \tag{4}$$

If the rotational speed of the left and the right wheels are ω_L and ω_R respectively, then assuming no slipping of the wheels, the wheel velocities at the contact point are respectively

$$V_L = r\,\omega_L \, , \tag{5}$$

$$V_R = r\,\omega_R \, , \tag{6}$$

where r is the radius of the wheel.

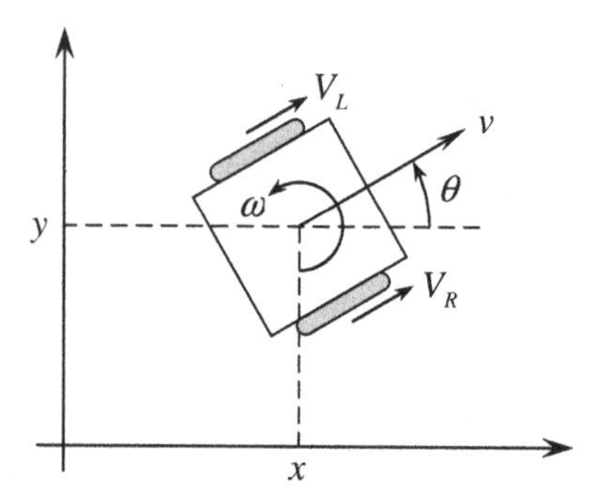

Fig. 1. Generalized Coordinates of MiroSot Robot

Let v be the tangential velocity of the robot at its center and ω the angular velocity of the robot. Then we have

$$v = \frac{V_R + V_L}{2} = r \cdot \frac{\omega_R + \omega_L}{2}, \tag{7}$$

$$\omega = \frac{V_R - V_L}{L} = r \cdot \frac{\omega_R - \omega_L}{L}, \tag{8}$$

that is

$$\begin{bmatrix} v \\ \omega \end{bmatrix} = \begin{bmatrix} \frac{r}{2} & \frac{r}{2} \\ \frac{r}{L} & -\frac{r}{L} \end{bmatrix} \begin{bmatrix} \omega_R \\ \omega_L \end{bmatrix}, \tag{9}$$

where L is the distance between the two wheels. Then we can solve the rotational speed of the wheels by

$$\begin{bmatrix} \omega_R \\ \omega_L \end{bmatrix} = -\frac{L}{r^2} \begin{bmatrix} -\frac{r}{L} & -\frac{r}{2} \\ -\frac{r}{L} & \frac{r}{2} \end{bmatrix} \begin{bmatrix} v \\ \omega \end{bmatrix} = \begin{bmatrix} \frac{1}{r} v + \frac{L}{2r} \omega \\ \frac{1}{r} v - \frac{L}{2r} \omega \end{bmatrix}. \tag{10}$$

The kinematics model of the robot is

$$\dot{P}(t) = \begin{bmatrix} \dot{x} \\ \dot{y} \\ \dot{\theta} \end{bmatrix} = \begin{bmatrix} \cos\theta & 0 \\ \sin\theta & 0 \\ 0 & 1 \end{bmatrix} \begin{bmatrix} v \\ \omega \end{bmatrix}. \tag{11}$$

Suppose that the robot can not slip in a lateral direction, hence

$$\dot{x}\sin\theta - \dot{y}\cos\theta = 0. \tag{12}$$

Eq. (12) is called the nonholonomic constraint of the robot.

In the following discussion, we would illustrate a fourth-order Bézier path of MiroSot robot (Fig. 2). For the sake of concision, we will omit the independent

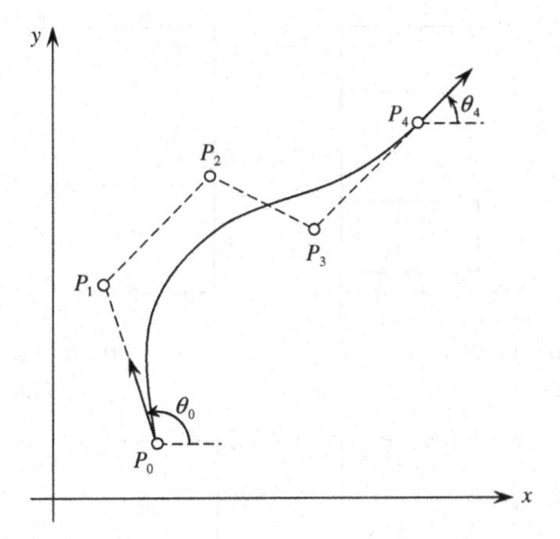

Fig. 2. Fourth-order Bézier Path of MiroSot Robot

variable and abbreviate $b_{i,n}(t)$ to $b_{i,n}$. Furthermore, we would use the notation b_{in} for $b_{i,n}$ whenever there is no confusion.

Suppose there are five control points, P_0, P_1, P_2, P_3 and P_4, which uniformly define the fourth-order Bézier curve (Fig. 2). The control points P_1 and P_3 are defined to fulfill the velocity and orientation requirements in the path. According to Eq. (1), the planned Bézier path vector could be expressed as

$$\begin{aligned} P(t) &= b_{04}P_0 + b_{14}P_1 + b_{24}P_2 + b_{34}P_3 + b_{44}P_4 \\ &= (1-t)^4 P_0 + 4t(1-t)^3 P_1 + 6t^2(1-t)^2 P_2 \\ &\quad + 4t^3(1-t)P_3 + t^4 P_4 . \end{aligned} \tag{13}$$

The tangential velocity of x-component and y-component, and angular velocity are defined as the derivative of the path vector,

$$\begin{aligned} [\,v_x(t) \;\; v_y(t) \;\; \omega\,]^T &= \frac{dP(t)}{dt} \\ &= 4b_{03}(P_1 - P_0) + 4b_{13}(P_2 - P_1) \\ &\quad + 4b_{23}(P_3 - P_2) + 4b_{33}(P_4 - P_3) . \end{aligned} \tag{14}$$

Generally, the velocity state of a n-order Bézier path is

$$[\,v_x(t) \;\; v_y(t) \;\; \omega\,]^T = \frac{dP(t)}{dt} = \sum_{i=0}^{n-1} n(P_{i+1} - P_i)\, b_{i,n-1} . \tag{15}$$

Additionally, the tangential velocity of the robot can be written as

$$v(t) = \sqrt{v_x^2(t) + v_y^2(t)} . \tag{16}$$

The v and ω can be obtained from joint solution of Eq. (15) and Eq. (16). Substituting v and ω in Eq. (10) gives the rotational speed of the wheels ω_R

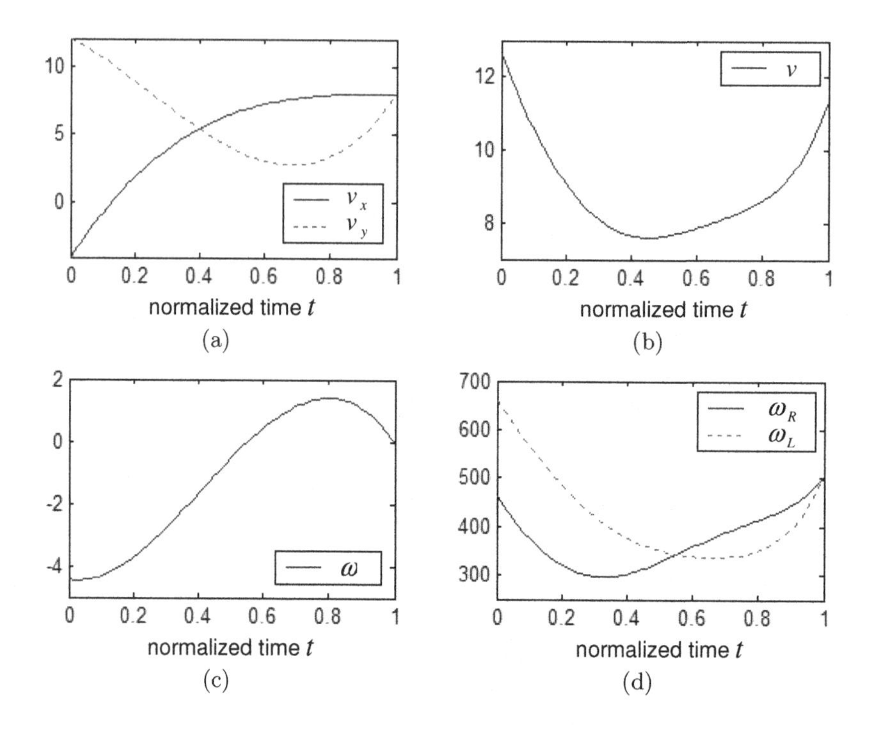

Fig. 3. Motion Control of MiroSot Robot (a) velocity of x-component and y-component; (b) tangential velocity of robot; (c) angular velocity of robot; (d) rotational speed of the robot wheels

and ω_L which are the final control variables of the motor in the MiroSot soccer robot. Then the robot can move along the planned Bézier path to its destination point.

Fig. 3 displays the waveforms of some control variables for the MiroSot robot on the fourth-order classical Bézier path shown in Fig. 2. Additionally, Fig. 3(d) shows the rotational speed of left and right robot wheel. We have known that the robot turning is induced by the difference in the velocities of the two wheels. Consequently, if we control the rotational speed of robot wheels in accordance with the calculated values in Fig. 3(d), the robot will move along the planned classical Bézier path as shown in Fig. 2.

4 Optimized Bézier Path Planning

Fig. 2 shows the planned Bézier path for the robot. In the domain, the robot current position is (300, 100), two obstacles stand at (400, 600) and (600, 500), the ball lies at (800, 700). There is a virtual control point at (200, 400). The robot is not able to turn sharply, then we set a virtual control point ahead of the direction of robot current velocity so that the robot can turn to its destination point smoothly. And we can as well set another virtual control point to control

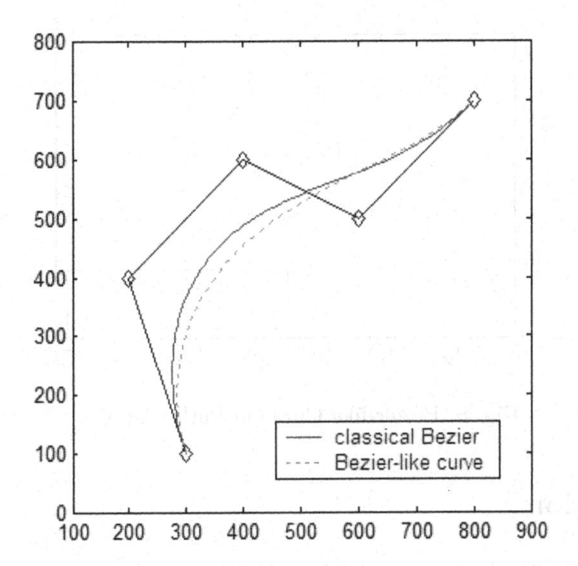

Fig. 4. Application of Bézier Curve in Path Planning

the robot hitting angle at the destination point if necessary. For the sake of rapid reaching, it is necessary to decrease the turning radius. Therefore we construct a new curve, which is on the basis of classical Bézier curve, to adjust the shape of Bézier path.

We have known that the Bézier curve is defined on the Bernstein basis functions, so we can adjust the shape of curve through introducing parameters into basis functions [18].

The *Bernstein-like basis functions* can be expressed explicitly as

$$b^*_{i,n}(t) = \left(1+\frac{3C^{i-1}_{n-2}+C^{i}_{n-1}-C^{i}_{n}}{C^{i}_{n}}\lambda - \frac{2C^{i}_{n-1}}{C^{i}_{n}}\lambda t+\lambda t^2\right)C^{i}_{n}t^i(1-t)^{n-i}, \tag{17}$$

where $\lambda \in [-1,1]$, $t \in [0,1]$, $i = 0,1,\cdots,n$ and $n \geq 2$, we set $C^p_q = 0$ in case of $p = -1$ or $p > q$. Then the Bézier-like path can be constructed by

$$P^*(t) = \sum_{i=0}^{n} b^*_{i,n}(t)P_i. \tag{18}$$

According to Eq. (17) and Eq. (18), we can reconstruct the Bézier path. The green dotted line in Fig. 4 displays a planned Bézier-like path in the case of $\lambda = -0.8$, its turning radius is smaller that that of classical Bézier path obviously. Consequently, the robot can reach its destination point more rapidly. In practice, the effect is more remarkable in the case of fewer obstacles. In Fig. 5, the green dashed Bézier-like path has much smaller turning radius than red solid classical Bézier path does.

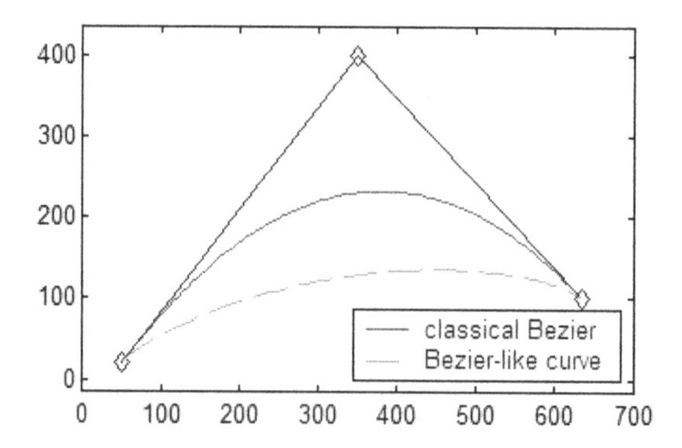

Fig. 5. Bézier-like Curve in Path Planning

5 Conclusion

An efficient and effective approach of path planning for soccer robots has been presented in this paper. In this approach, the current location of robot could be one of the curve's end points, another end point depends on the role of the robot. If a robot is assigned to be a striker, then the position of the ball is another end point of the planned path for the striker. All the robots are treated as obstacles. The end points and obstacle points are control points, by which a Bézier curve can be planned. In order to reduce the turning radius, a new construction of curve is proposed to optimize the shape of Bézier path. Because of the characteristic of Bézier curve, the planned path can avoid obstacles automatically. Furthermore, this approach combines the function of path smoothing and posture adjustment together.

References

[1] Bruce, J., Veloso, M.: Real-time randomized path planning for robot navigation. In: International Conference on Intelligent Robots and Systems, IROS 2002, Lausanne, Switzerland, vol. 3, pp. 2383–2388. IEEE (September 2002)

[2] Cheng, G., Gu, J., Bai, T., Majdalawieh, O.: A new efficient control algorithm using potential field: extension to robot path tracking. In: Electrical and Computer Engineering, CCECE 2004, Niagara Falls, Canada, vol. 4, pp. 2035–2040. IEEE (May 2004)

[3] Farin, G., Hoschek, J., Kim, M.S.: Handbook of Computer Aided Geometric Design, 1st edn. North Holland (August 2002)

[4] Foley, J.D., van Dam, A., Feiner, S.K., Hughes, J.F.: Computer Graphics: Principles and Practice, 2nd edn. Addison-Wesley (September 2004)

[5] Ge, S.S., Cui, Y.J.: New potential functions for mobile robot path planning. IEEE Transactions on Robotics and Automation 16(5), 615–620 (2000)

[6] Jolly, K.G., Sreerama Kumar, R., Vijayakumar, R.: A bezier curve based path planning in a multi-agent robot soccer system without violating the acceleration limits. Robotics and Autonomous Systems 57, 23–33 (2009)

[7] Khatib, O.: Real-time obstacle avoidance for manipulators and mobile robots. The International Journal of Robotics Research 5(1), 90–98 (1986)
[8] Kim, J., Ostrowski, J.P.: Motion planning a aerial robot using rapidly-exploring random trees with dynamic constraints. In: IEEE International Conference on Robotics and Automation, ICRA 2003, Taipei, China, vol. 2, pp. 2200–2205. IEEE (September 2003)
[9] Kim, J.H., Kim, D.H., Kim, Y.J., Seow, K.T.: Soccer Robotics, vol. 11. Springer (2004)
[10] Kim, J.H., Kim, K.C., Kim, D.H., Kim, Y.J., Vadakkepat, P.: Path planning and role selection mechanism for soccer robots. In: IEEE International Conference on Robotics and Automation, ICRA 1998, Leuven, Belgium, vol. 4, pp. 3216–3221. IEEE (May 1998)
[11] Koren, Y., Borenstein, J.: Potential field methods and their inherent limitations for mobile robot navigation. In: IEEE International Conference on Robotics and Automation, Sacramento, CA, USA, pp. 1398–1404. IEEE (April 1991)
[12] Park, M.G., Jeon, J.H., Lee, M.C.: Obstacle avoidance for mobile robots using artificial potential field approach with simulated annealing. In: IEEE International Symposium on Industrial Electronics, ISIE 2001, Pusan, Korea, vol. 3, pp. 1530–1535. IEEE (June 2001)
[13] Park, M.G., Lee, M.C.: Artificial potential field based path planning for mobile robots using a virtual obstacle concept. In: International Conference on Advanced Intelligent Mechatronics, AIM 2003, Kobe, Japan, vol. 2, pp. 735–740. IEEE (July 2003)
[14] Rimon, E., Koditschek, D.E.: Exact robot navigation using artificial potential functions. IEEE Transactions on Robotics and Automation 8(5), 501–518 (1992)
[15] Rodriguez, S., Tang, X., Lien, J.M., Amato, N.M.: An obstacle-based rapidly-exploring random tree. In: IEEE International Conference on Robotics and Automation, ICRA 2006, Orlando, USA, pp. 895–900. IEEE (May 2006)
[16] Vadakkepat, P., Tan, K.C., Wang, M.L.: Evolutionary artificial potential fields and their application in real time robot path planning. In: Evolutionary Computation, CEC 2000, La Jolla, CA, USA, vol. 1, pp. 256–263. IEEE (July 2000)
[17] Warren, C.W.: Multiple robot path coordination using artificial potential fields. In: IEEE International Conference on Robotics and Automation, ICRA 1990, Cincinnati, Ohio, USA, vol. 1, pp. 500–505. IEEE (May 1990)
[18] Yan, L., Liang, J.: An extension of the bézier model. Applied Mathematics and Computation 218(6), 2863–2879 (2011)
[19] Yi, Z., Heng, P.A., Vadakkepat, P.: Absolute periodicity and absolute stability of delayed neural networks. IEEE Transactions on Circuits and Systems 49(2), 256–261 (2002)
[20] Yun, X., Tan, K.C.: A wall-following method for escaping local minima in potential field based motion planning. In: International Conference on Advanced Robotics, ICAR 1997, Monterey, Canada, pp. 421–426. IEEE (July 1997)

Adaptively Nearest Feature Point Classifier for Face Recognition

Qingxiang Feng, Jeng-Shyang Pan, Lijun Yan, and Tien-Szu Pan

Shenzhen Graduate School
Harbin Institute of Technology
Shenzhen, China
jengshyangpan@gmail.com

Abstract. In this paper, an improved classifier based on the concept of feature line space, called as adaptively nearest feature point classifier (ANFP) is proposed for face recognition. ANFP classifier uses the new metric, called as adaptively feature point metric, which is different from metrics of NFL and the other classifiers. ANFP gain better performance than NFL classifier and some others classifiers based on feature line space, which is proved by the experiment result on Yale face database.

Keywords: Nearest Feature Line; Nearest feature centre; Nearest Neighbor; Face Recognition.

1 Introduction

The procedure of face recognition contains two steps. The first step is feature extraction. For instance PCA [1], LDA [2], ICA [3] and other methods [4-6]. The second step is classification. One of the most popular classifiers is the nearest neighbor (NN) classifier [7]. However, the performance of NN is limited by the available prototypes in each class. To overcome this drawback, nearest feature line (NFL) [8] was proposed by Stan Z. Li. NFL was originally used in face recognition, and later began to be used in many other applications.

NFL attempts to enhance the representational capacity of a sample set of limited size by using the lines through each pair of the samples belonging to the same class. NFL shows good performance in many applications, including face recognition [9-12], audio retrieval [13], speaker identification [14], image classification [15], object recognition [16] and pattern classification [17]. The authors of NFL explain that the feature line can give information about the possible linear variants of the corresponding two samples very well.

Though successful in improving the classification ability, there are still some drawbacks in NFL that limit their further application in practice, which can be summarized as two main points. Firstly, NFL will have a large computation complexity problem when there are many samples in each class. Secondly, NFL may

A. Abraham et al. (eds.), *Innovations in Bio-inspired Computing and Applications*,
Advances in Intelligent Systems and Computing 237,
DOI: 10.1007/978-3-319-01781-5_11, © Springer International Publishing Switzerland 2014

fail when the prototypes in NFL are far away from the query sample, which is called as extrapolation inaccuracy of NFL.

To solve the above problems, extended nearest feature line [18] (ENFL), nearest feature mid-point [19] (NFM), shortest feature line segment [20] (SFLS) and nearest feature centre [21] (NFC) are proposed. They gains better performance in some situation. However, they are not so good in other situation.

In this paper, an improved classifier based on the concept of feature line space, called as adaptively nearest feature point classifier (ANFP) is proposed for face recognition. ANFP classifier uses the new metric, called as adaptively feature point metric, which is different from metrics of NFL and the other classifiers. ANFP gain better performance than NFL classifier and some others classifiers based on feature line space. A large number of experiments are executed on Yale face database. Detailed comparison result is given.

2 Background

In this section, we will introduce nearest feature line classifier, extended nearest feature line classifier and nearest feature centre classifier. Suppose that $Y=\{y_i^c, c=1,2,\cdots,M, i=1,2,\cdots,N_c\}\subset R^D$ denote the prototype set, where y_i^c is the *i*th prototype belonging to *c*-class, M is the number of class, and N_c is the number of prototypes belonging to the *c*-class.

2.1 Nearest Feature Line

The core of NFL is the feature line metric. As is shown in Fig. 1, the NFL classifier doesn't compute the distance of query sample y and y_i^c; doesn't calculate the distance of y and y_j^c, while NFL classifier calculates the feature line distance between query sample y and the feature line $\overline{y_i^c y_j^c}$. The feature line distance between point y and feature line $\overline{y_i^c y_j^c}$ is defined as:

$$\mathrm{d}(y, \overline{y_i^c y_j^c}) = \| y - y_p^{ij,c} \| \tag{1}$$

where $y_p^{ij,c}$ is the projection point of y on the feature line $\overline{y_i^c y_j^c}$, $\|.\|$ means the L2-norm.

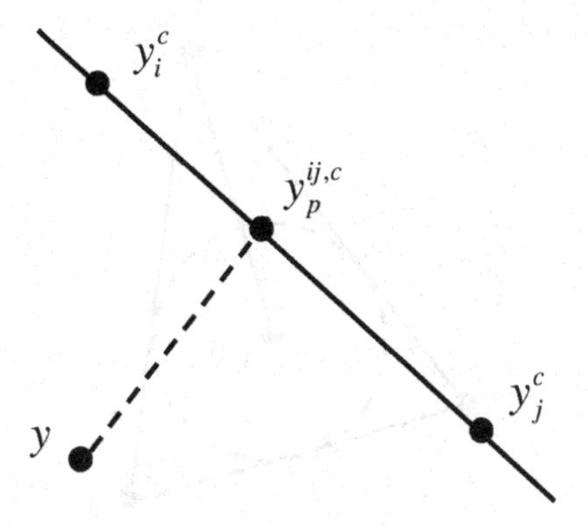

Fig. 1. The metric of NFL

The projection point $y_p^{ij,c}$ is calculated by $y_p^{ij,c} = y_i^c + t(y_j^c - y_i^c)$ where $t \in R$, which is the positional parameters. After simple deformation, we can see that the location parameter can be computed as follows.

$$t = \frac{(y - y_i^c)^T (y_j^c - y_i^c)}{(y_j^c - y_i^c)^T (y_j^c - y_i^c)} \tag{2}$$

2.2 Nearest Feature Centre

Show in the Fig. 2, NFC uses the feature center metric, which is defined as the Euclidean distance between query sample y and the feature center $y_o^{ij,c}$, which is $d_{NFC}(y, \overline{y_i^c y_j^c}) = \| y - y_o^{ij,c} \|$, where $y_o^{ij,c}$ is the center of inscribed circle of the triangle $\Delta y y_i^c y_j^c$. The feature center $y_o^{ij,c}$ can be calculated as follows.

$$y_o^{ij,c} = \frac{b_{ij}^c \times y + b_{yi}^c \times y_j^c + b_{yj}^c \times y_i^c}{b_{ij}^c + b_{yi}^c + b_{yj}^c} \tag{3}$$

where $b_{yi}^c = \| y - y_i^c \|$, $b_{yj}^c = \| y - y_j^c \|$ and $b_{ij}^c = \| y_i^c - y_j^c \|$.

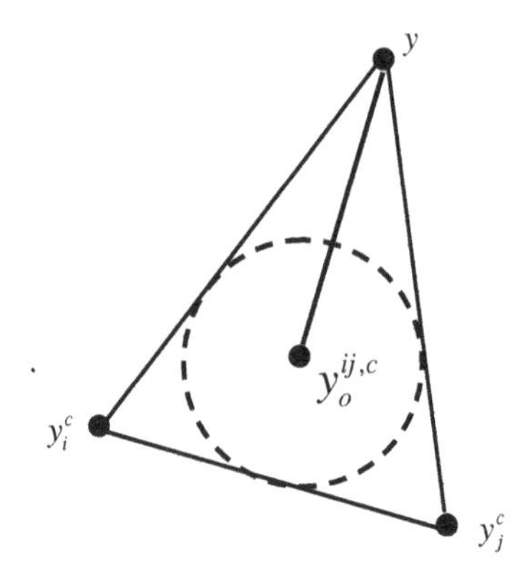

Fig. 2. The metric of NFC

2.3 Extended Nearest Feature Line

Borrowing the concept of feature line spaces from the NFL method, the extended nearest feature line (ENFL) is proposed in 2004. However, the distance metric of ENFL is different from the feature line distance of NFL.

ENFL does not calculate the distance between the query sample and the feature line. Instead, ENFL calculates the product of the distances between query sample and two prototype samples. Then the result is divided by the distance between the two prototype samples. As shown in Fig. 3. The new distance metric of ENFL is described as

$$d_{ENFL}(y, \overline{y_i^c y_j^c}) = \frac{\| y - y_i^c \| \times \| y - y_j^c \|}{\| y_i^c - y_j^c \|} \tag{4}$$

The distance between the pair of prototype samples can strengthen the effect when the distance between them is large.

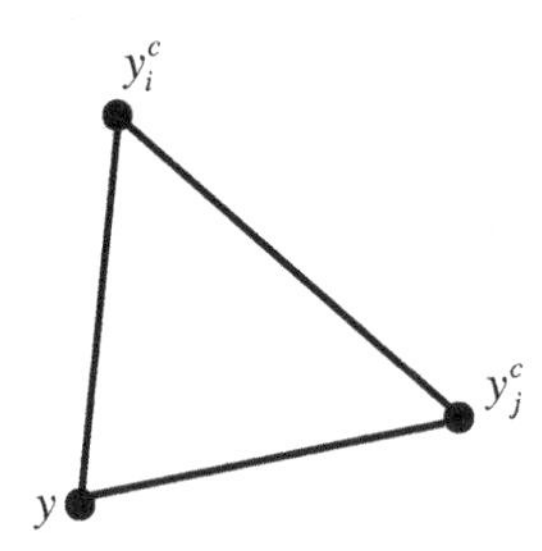

Fig. 3. The metric of ENFL

3 The Proposed Methods

Similar to NFL, ENFL and NFC, the adaptively nearest feature point (ANFP) classifier supposes that at least two prototype samples are available for each class. However, the metric is different. A better distance metric, called as adaptively feature metric, is proposed in this section. The basic idea is shown as follows. The adaptively feature point $y_o^{ij,c}$ can be computed by formula (5)-(8).

$$s_{i,j}^c = \frac{1}{b_{yi}^c \times b_{yi}^c + b_{yj}^c \times b_{yj}^c} \tag{5}$$

$$s_i^c = \frac{1}{b_{yi}^c \times b_{yi}^c} \times s_{i,j}^c \tag{6}$$

$$s_j^c = \frac{1}{b_{yj}^c \times b_{yj}^c} \times s_{i,j}^c \tag{7}$$

$$y_o^{ij,c} = \frac{s_i^c \times y_i^c + s_j^c \times y_j^c}{s_i^c + s_j^c} \tag{8}$$

where $b_{yi}^c = \| y - y_i^c \|$, $b_{yj}^c = \| y - y_j^c \|$ and $b_{ij}^c = \| y_i^c - y_j^c \|$.

After the point $y_o^{ij,c}$ being got, the adaptively feature metric is computed as in formula (9).

$$d_{ANFP}(y, \overline{y_i^c y_j^c}) = \| y - y_o^{ij,c} \| \tag{9}$$

The detailed classification procedure of ANFP is described as follows. Firstly, the adaptively feature distance between the query sample y and each pair prototypes y_i^c and y_j^c is computed, which produces a number of distances. Secondly, the distances are sorted in ascending order, each of which is marked with a class identifier and two prototypes. Then, the ANFP distance can be determined as the first rank distance shown in the following formula (10).

$$d(y, \overline{y_{i^*}^{c^*} y_{j^*}^{c^*}}) = \min_{1 \le c \le L, 1 \le i < j \le N_c} d(y, \overline{y_i^c y_j^c}) \tag{10}$$

The first rank gives the best matched class c* and the two best matched prototypes i* and j* of the class. The query sample y will be classified into the class c*.

4 Experimental Results

The classification performance of the proposed classifiers is compared with NN, NFL, ENFL, and NFC classification approach. The experiments are executed on Yale face database.

Yale [22] face database contains 165 greyscale images in GIF format of 15 persons. There are 11 images per people, one per different facial expression or configuration: center-light, w/glasses, happy, left-light, w/no glasses, normal, right-light, sad, sleepy, surprised, and wink. All images are copped in 100×100.

"Randomly-chose-*N*" scheme is taken for comparison: *N* images per person are randomly chosen from the Yale face database as prototype set. The rest images of Yale face database are used for testing. The whole system runs 20 times. To test the robustness of new algorithms, the average recognition rate (ARR) and the average running time are used to assess the performance of new algorithms.

In the experimental, the "randomly-choose-*N*" scheme is adopt on Yale face database. The average recognition rates (ARR) of NN, NFL, ENFL, NFC and ANFP are shown in Fig. 4 and Table 1. From the Fig. 4, we can know that the recognition rate of the proposed is better than the recognition rate of NN, NFL, ENFL and NFC on Yale face database.

Table 1. The ARR of several classifiers using "randomly-choose-*7*" scheme

Classifier	RR
NN	84.33%
NFL	95.50%
ANFP	86.50%
NFC	85.00%
ENFL	84.83%

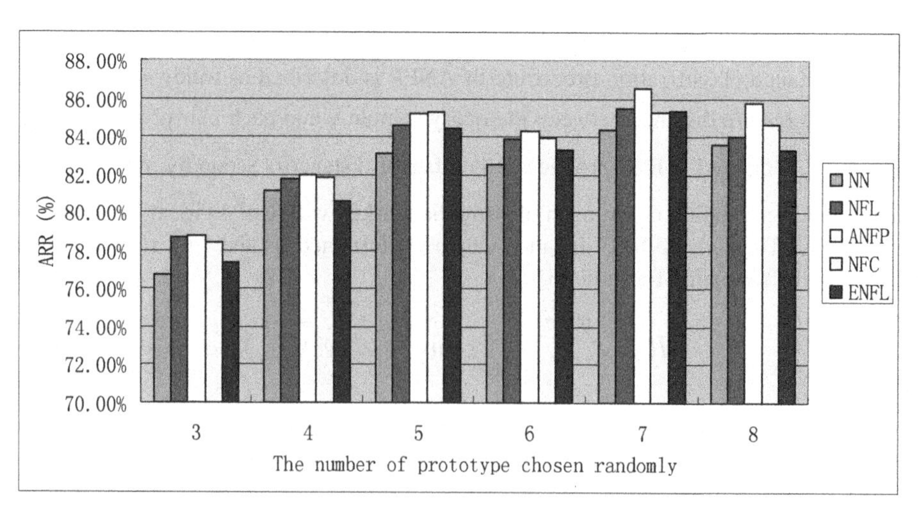

Fig. 4. The recognition rate of several classifiers using "randomly-choose-*N*" scheme on Yale face database

5 Conclusion

In this paper, a new classifier based on feature line space, called as adaptively nearest feature point classifier, is proposed. The proposed classifier uses the new metric, called as adaptively feature point metric. The average recognition rate of new classifier surpasses the other classifiers based on feature line space. Experimental result on Yale face database affirms the performance of the new classifier.

References

1. Turk, M., Pentland, A.: Eigenfaces for recognition. Journal of Cognitive Neuroscience 3(1), 71–86 (1991)
2. Belhumeur, P., Hespanha, J., Kriegman, D.: Eigenfaces vs. fisherfaces: recognition using class specific linear projection. IEEE Transactions on Pattern Analysis and Machine Intelligence 19(7), 711–720 (1997)
3. Bartlett, M.S., Movellan, J.R., Sejnowski, T.J.: Face recognition by independent component analysis. IEEE Transactions on Neural Networks 13(6), 1450–1464 (2002)
4. He, X., Yan, S., Hu, Y., Niyogi, P., Zhang, H.J.: Face recognition using laplacianfaces. IEEE Transactions on Pattern Analysis and Machine Intelligence 27(3), 1–13 (2005)
5. Kekre, H.B., Shah, K.: Performance Comparison of Kekre's Transform with PCA and Other Conventional Orthogonal Transforms for Face Recognition. Journal of Information Hiding and Multimedia Signal Processing 3(3), 240–247 (2012)
6. Zhou, X., Nie, Z., Li, Y.: Statistical analysis of human facial expressions. Journal of Information Hiding and Multimedia Signal Processing 1(3), 241–260 (2010)
7. Cover, T.M., Hart, P.E.: Nearest neighbor pattern classification. IEEE Trans. Inform. Theory 13(1), 21–27 (1967)
8. Li, S.Z., Lu, J.: Face Recognition Using the Nearest Feature Line Method. IEEE Transactionson Neural Networks 10(2), 439–443 (1999)
9. Chien, J.T., Wu, C.C.: Discriminant waveletfaces and nearest feature classifiers for face recognition. IEEE Trans. Pattern Anal. Machine Intell. 24(12), 1644–1649 (2002)
10. Lu, J., Tan, Y.-P.: Uncorrelated discriminant nearest feature line analysis for face recognition. IEEE Signal Process. Lett. 17(2), 185–188 (2010)
11. Feng, Q., Pan, J.-S., Yan, L.: Restricted Nearest Feature Line with Ellipse for Face Recognition. Journal of Information Hiding and Multimedia Signal Processing 3(3), 297–305 (2012)
12. Feng, Q., Huang, C.-T., Yan, L.: Resprentation-based Nearest Feature Plane for Pattern Recognition. Journal of Information Hiding and Multimedia Signal Processing 4(3), 178–191 (2013)
13. Li, S.Z.: Content-based audio classification and retrieval using the nearest feature line method. IEEE Trans. Speech Audio Process. 8(5), 619–625 (2000)
14. Chen, K., Wu, T.Y., Zhang, H.J.: On the use of nearest feature line for speaker identification. Pattern Recognition Lett. 23(14), 1735–1746 (2002)
15. Li, S.Z., Chan, K.L., Wang, C.L.: Performance evaluation of the nearest feature line method in image classification and retrieval. IEEE Trans. Pattern Anal. Machine Intell. 22(11), 1335–1339 (2000)
16. Chen, J.H., Chen, C.S.: Object recognition based on image sequences by using inter-feature-line consistencies. Pattern Recognition 37(9), 1913–1923 (2004)

17. Zheng, W., Zhao, L., Zou, C.: Locally nearest neighbor classifiers for pattern classification. Pattern Recognition 37(6), 1307–1309 (2004)
18. Zhou, Y., Zhang, C., Wang, J.: Extended nearest feature line classifier. In: Zhang, C., Guesgen, H.W., Yeap, W.K. (eds.) PRICAI 2004. LNCS (LNAI), vol. 3157, pp. 183–190. Springer, Heidelberg (2004)
19. Zhou, Z., Kwoh, C.K.: The pattern classification based on the nearest feature midpoints. International Conference on Pattern Recognition 3, 446–449 (2000)
20. Han, D.Q., Han, C.Z., Yang, Y.: A Novel Classifier based on Shortest Feature Line Segment. Pattern Recognition Letters 32(3), 485–493 (2011)
21. Feng, Q., Pan, J.S., Yan, L.: Nearest feature centre classifier for face recognition. Electronics Letters 48(18), 1120–1122 (2012)
22. The Yale faces database (2001), http://cvc.yale.edu/projects/yalefaces/yalefaces.html

Comparison of Classification Algorithms for Physical Activity Recognition*

Tomáš Peterek[1,2], Marek Penhaker[1,2], Petr Gajdoš[1,3], and Pavel Dohnálek[1,3]

[1] IT4 Innovations, VSB-Technical University of Ostrava, Czech Republic
[2] Department of Cybernetics and Biomedical Engineering, VSB-Technical University of Ostrava, Czech Republic
[3] Department of Computer Science, VSB-Technical University of Ostrava, Czech Republic
tomas.peterek@vsb.cz

Abstract. The main aim of this work is to compare different algorithms for human physical activity recognition from accelerometric and gyroscopic data which are recorded by a smartphone. Three classification algorithms were compared: the Linear Discriminant Analysis, the Random Forest, and the K-Nearest Neighbours. For better classification performance, two feature extraction methods were tested: the Correlation Subset Evaluation Method and the Principal Component Analysis. The results of experiment were expressed by confusion matrixes.

1 Introduction

Nowadays, cell phones do not serve us only as communication medium, they are equipped by powerful CPU and a GPU therefore it opens new opportunities for this field. The cell phones together with external sensors are possible to use as devices for control e.g. home appliances or they can be used for recording and pre-processing biomedical signals (ECG, EEG). Modern smartphones are equipped by embedded accelerometers which do not be necessarly used only for rotating the screen or playing games but they can be useful as a sensor of regular accelerometric data, which be used for many purposes. One of these purposes can be monitoring of physical activity. The smartphones allow preprocesing of data and due to powerful CPU perform sophisticated classification. The Smartphones with applications for physical activity recognition can be used in homes for elderly people, for recording circadian rhythms, or as fall detectors.

* This article has been elaborated in the framework of the IT4Innovations Centre of Excellence project, reg. no. CZ.1.05/1.1.00/02.0070 supported by Operational Programme Research and Development for Innovations funded by Structural Funds of the European Union and state budget of the Czech Republic. This work was also supported by the Bio-Inspired Methods: research, development and knowledge transfer project, reg. no. CZ.1.07/2.3.00/20.0073 funded by Operational Programme Education for Competitiveness, co-financed by ESF and state budget of the Czech Republic.

A. Abraham et al. (eds.), *Innovations in Bio-inspired Computing and Applications*, Advances in Intelligent Systems and Computing 237,
DOI: 10.1007/978-3-319-01781-5_12, © Springer International Publishing Switzerland 2014

2 Previous Work

The issue of physical activity recognition was solved many times by different classification approaches. e.g: the Support Vector Machine [1]. the Hidden Markov Model [2] [3], classification based on sparse representation [4], Bayes aproaches [5], the Neural Network [6][7], or the Linear Discriminant Analysis [8],[9].

3 Dataset

The dataset was downloaded from the UCI learning repository. The data were recorded by a smartphone Samsung Galaxy SII. 30 volunteers within ages of 19-48 years participated at this experiment. Each participant alternated six types of movement (walking, walking upstairs, walking downstairs, sitting standing, lying).The smartphones were placed on the volunteer's waists. Data were captured by 3-axis embedded accelerometer and 3 axial angular velocity sensor. The sampling frequency was set to 50 Hz. The data from accelerometers and gyroscope were filtered against noise and consequently they were divided segments into 2.56 seconds length with 50 % overlapping. Jerk signals were derived from these segments and magnitudes were calculated, and consequently The Fast Fourier Transform was applied on some these signals. Using this procedure ten different signals in time domain and four signals in frequency domain were obtained. The Jerks signals are derivative signals of regular accelerometric and gyroscopic data and they are used in many useful applications: dynamic motion aerial vehicle trace measurement, earthquake-resistant mechanisms of structures, mechanisms of high speed auto-control of machines, human responses in high speed moving vehicle and high-speed elevators. From each segment a few basic parameters were counted: (mean value, standard deviation, median of absolute deviation, max and min values, energy of segment, interquartile range, entropy, autorregresion coefficients, correlation coefficient between two signals, index of the frequency component with largest magnitude, weighted average of the frequency components to obtain a mean frequency, skewness of the frequency domain signal, kurtosis of the frequency domain signal, energy of a frequency interval within the 64 bins of the FFT of each window, angle between to vectors). By these statistical operations 561 features were calculated, which were used for classification. [1]

4 Classification Method

4.1 Random Forest

The Random Forest (RF) is very popular classification and regression algorithm. The RF algorithm belongs to ensemble learning methods. Likewise a regular forest consists of a number of trees, the RF algorithm consists of a number of classification or regression trees (CART). The algorithm does not use all features for the CART construction but only a few of them. The RF was designed by Leo Breiman in 2001.[12] In this paper, Breiman himself compare the RF with other ensemble techniques and mentioned that this method has higher accuracy than e.g. Adaboost method. Since that time the RF is used in bioinformatics, medical informatics and so on. The algorithm is resistant to outliers, missing values, or noise. The RF stands out especially in simplicity of parameter tuning but the main problem is his interpretability. For optimal settings of the algorithm only two parameters have to be set: the number of trees in the forest and the number of variables in trees.

4.2 K-Nearest Neighbours

The K-nearest neighbours (KNN)is a non-parametric algorithm for classification. The KNN is one of the easiest method for data classification. The training set T and the testing sample x_i are given. The KNN classifier tries to find sample x_r from the training set T, with a minimal Euclidian distance to the testing sample. Better results are achieved if more than one sample from the training set are founded. This algorithm achieves satisfactory results but is not suitable for solving difficult tasks.

4.3 Linear Discriminant Analysis

Th Linear discriminant analysis (LDA) is a parametric classification technique. The LDA has a lot in common with the Principal Component Analysis (PCA) but with different that PCA does more for feature separation and LDA does more feature classification. The main aim of this method is to find linear combinations of features, which provide the best separation between classes. These combinations are called discriminant functions. The algorithm was developed by Fischer in 1931 but in his original form the algorithm is able to classify only to two classes. In 1988, the algorithm was improved for multiclass classification problem. There are n classes. The intra-class matrix can be calculated by

$$\hat{\Sigma}_w = S1 + ... + S_n = \sum_{i=1}^{n} \sum_{x \in c_i} (x - \hat{x}_i)(x - \hat{x}_i)' \tag{1}$$

and inter-class matrix by an equation

$$\hat{\Sigma}_b = \sum_{i=1}^{n} m_i (\bar{x}_i - \bar{x})(\bar{x}_i - \bar{x})' \tag{2}$$

where m_i is a number of samples in each class from the training set and x_i is the mean for each class and $\hat{x}$ is the total mean vector. Now, a linear transformation Φ should be suggested, in order to maximize Rayleigh coefficient, which is the ratio of determinants of inter-class and intra-class scatter matrixes.

$$J(\Phi) = \frac{\left|\Phi^T \hat{\Sigma}_b \Phi\right|}{\left|\Phi^T \hat{\Sigma}_w \Phi\right|} \tag{3}$$

The linear transformation Φ can be counted by solving equation

$$\hat{\Sigma}_b \Phi = \lambda \hat{\Sigma}_w \Phi \tag{4}$$

The last step is classification itself. The basic principle is measurement of metric or cosine distances between a new instances and centroid of classes. The new instances are classified acording to expression:

$$arg\, min\, d\left(z\Phi, \bar{x}_k \Phi\right) \tag{5}$$

5 Feature Extraction

All mentioned methods are sensitive to irrelevant features. Therefore the number of features has to be reduced so that only relevant features are preserved. There are three basic approaches:

- Feature transform: The basic idea of feature transform method is to transform the original feature space to a new feature space with smaller dimensions. There are algorithms such as: The Singular Value Decomposition, Non-negative Matrix Factorization (NNMF) or Principal Component Analysis (PCA), which are able to solve this task. The last named method showed promising results in similar classification problems, therefore it was also tested in this task.
- Feature filter carry out the feature selection process as a pre-processing step with no induction algorithm. The featerue filter are faster than wrapper and have better results in generalization.
- Feature wrapper achieves the best results but it is very time consuming especially if the number of features is high. The wrappers search for the best result in the whole space of possibly solution. It means that the wrapper methods have to design the classifier for every possible set of feature combination. In this case it is 2^{561} possible solutions. Therefore only first two approaches were tested.

5.1 Feature Filter – Correlation Feature Selection Method

The idea of feature filters is based on a hypothesis: *Good feature subsets contain features highly correlated with the classification, yet uncorrelated to each other.* The Feature filters try to decide which features should be included in the final subset and which should be ignored. The Correlation Feature Selection is typical feature filter method, which evaluates each subsets according to a hearuastic function:

$$M_s = \frac{k_{\overline{rcf}}}{\sqrt{k + k\left(k-1\right) k_{\overline{rff}}}} \tag{6}$$

where M_s is the merit of a subset S, which contains k features. Parameters $k_{\overline{rcc}}$ and $k_{\overline{rcf}}$ expres corelation between a feature and class respectively a feature and a feature. The acceptance of a feature depends on the extent to which it predicts classes in areas of the instance space not already predicted by other features. The numerator of equation provides information how a set of features is predictive of the class and the denominator provides informations how much redudancy there is among the features. [10]

5.2 Principle Components Analysis

The PCA tries to transform a set of observations of possibly correlated variables to a set of new uncorrelated variables called principal components. The PCA components can be counted by

$$X = YP \tag{7}$$

where X is a centering matrix and Y is an input matrix, P is a matrix of the eigenvector of the covariance vector matrix C, expressed by equation

$$C_x = P\Delta P^T \tag{8}$$

where P is orhonormal and Δ is a diagonal matrix of eigenvalues. The number of components which will be used for classification should cover more than 80 % of original variables dispersion. [11]

6 Testing and Results

The data were devided into a training part (70%) and a testing part (30%). Firstly, the full dataset was classified by the three algorithms. In the second part the dataset were transformed by the PCA and lastly the dataset was reduced by the CFS algorithm. The results were expressed by the confusion matrixes and precisions and recalls were counted for each class. For the CFS method the evolutionary search strategy was used. The parameters of the search strategy were set up on values which are shown in the Table 1. Using the CFS algorithm, 250 of features were screened out and the rest of them were used for classification. The PCA counted 10 principal components which represents over 90 % of original variables dispersion. Covererage of the original dispersion for each component are shown in the Fig. 1. These 10 principal components were used as a new feature vector. The results are expresed by the confusion matrixes and by values of the precision and the recall. The results for all classifiers and for both types of feature selection are shown in the Table 2, Table 3 and Table 4.

Table 1. The parameters of the evolutionary search strategy

Parameter	Value
Crossover probability	0.6
Generations	100
Mutation probability	0.01
Population Size	50

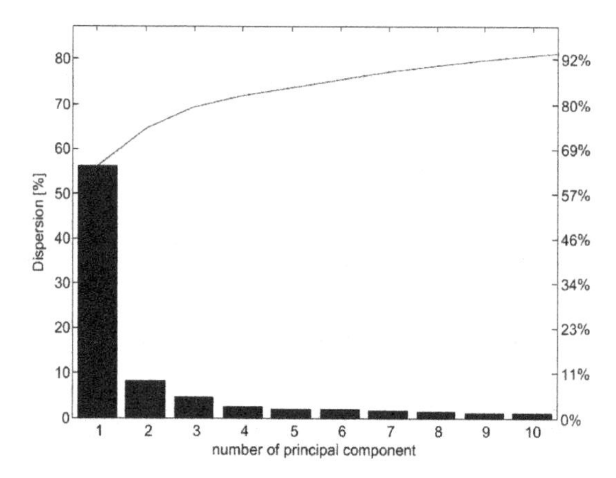

Fig. 1. The coverage of the original variable dispersion for the first ten principal components

Table 2. The confusion matrixes for the LDA

LDA	Walking	Upstairs	Downstairs	Sitting	Standing	Laying	Precision [%]
Walking	490	6	0	0	0	0	98.79
Upstairs	11	460	0	0	0	0	97.66
Downstairs	1	15	404	0	0	0	96.16
Sitting	0	1	0	435	55	0	88.59
Standing	0	0	0	22	510	0	95.86
Laying	0	0	0	0	0	537	100
Recall [%]	97.61	95.44	100	95.19	90.27	100	
LDA+CFSsubset	Walking	Upstairs	Downstairs	Sitting	Standing	Laying	Precision [%]
Walking	494	2	0	0	0	0	99.6
Upstairs	10	461	0	0	0	0	97.88
Downstairs	8	20	392	0	0	0	93.33
Sitting	0	1	0	396	94	0	80.65
Standing	0	0	0	65	467	0	87.78
Laying	0	0	0	0	0	537	100
Recall [%]	96.48	95.25	100	85.9	83.24	100	

Table 3. The confusion matrixes for the KNN

KNN	Walking	Upstairs	Downstairs	Sitting	Standing	Lying	Precision [%]
Walking	473	8	15	0	0	0	95.36
Upstairs	31	422	18	0	0	0	89.60
Downstairs	53	46	321	0	0	0	76.43
Sitting	0	2	0	389	99	1	79.23
Standing	0	0	0	88	451	0	84.77
Lying	0	0	0	3	1	533	99.26
Recall [%]	84.92	88.28	90.68	82.24	81.85	99.81	
KNN+CFSsubset	**Walking**	**Upstairs**	**Downstairs**	**Sitting**	**Standing**	**Lying**	**Precision [%]**
Walking	460	10	26	0	0	0	92.74
Upstairs	46	420	5	0	0	0	89.17
Downstairs	32	62	326	0	0	0	77.62
Sitting	0	1	0	378	112	0	76.99
Standing	0	0	0	41	491	0	92.29
Lying	0	0	0	0	0	537	100
Recall [%]	85.50	85.19	91.32	90.21	81.43	100	
KNN+PCA	**Walking**	**Upstairs**	**Downstairs**	**Sitting**	**Standing**	**Lying**	**Precision [%]**
Walking	110	108	43	96	66	73	22.18
Upstairs	128	56	66	100	57	64	11.89
Downstairs	75	52	67	77	50	99	15.95
Sitting	81	73	55	102	66	114	20.77
Standing	91	79	67	108	89	98	16.73
Lying	119	110	68	82	59	99	18.44
Recall [%]	18.21	11.72	18.31	18.05	23.0	18.1	

Table 4. The confusion matrixes for the RF

Random Forest	Walking	Upstairs	Downstairs	Sitting	Standing	Lying	Precision [%]
Walking	482	6	8	0	0	0	97.18
Upstairs	36	430	5	0	0	0	91.30
Downstairs	14	38	368	0	0	0	89.76
Sitting	0	0	0	422	69	0	85.95
Standing	0	0	0	54	478	0	89.85
Lying	0	0	0	0	0	537	100
Recall [%]	90.6	92.67	96.59	88.66	87.39	100	
Random Forest+CFS	**Walking**	**Upstairs**	**Downstairs**	**Sitting**	**Standing**	**Lying**	**Precision [%]**
Walking	478	6	12	0	0	0	96.37
Upstairs	56	402	13	0	0	0	85.35
Downstairs	7	41	372	0	0	0	88.57
Sitting	0	0	0	395	96	0	80.45
Standing	0	0	0	65	467	0	87.78
Lying	0	0	0	0	0	537	100
Recall [%]	88.35	89.53	93.7	85.87	82.95	100	
Random Forest+PCA	**Walking**	**Upstairs**	**Downstairs**	**Sitting**	**Standing**	**Lying**	**Precision [%]**
Walking	97	70	9	120	82	118	19.56
Upstairs	102	57	20	124	65	103	12.10
Downstairs	57	38	15	71	60	179	3.57
Sitting	80	59	33	69	82	168	14.05
Standing	95	108	174	4	3	148	15.79
Lying	73	77	28	109	84	161	27.19
Recall [%]	17.9	13.97	12.3	11.73	20.39	16.69	

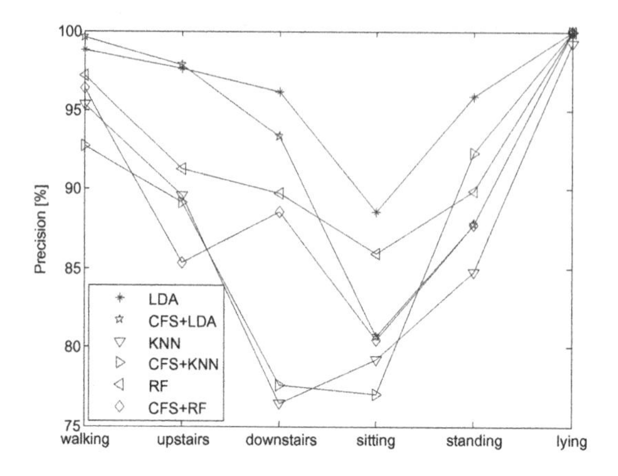

Fig. 2. The achieved values of precision for all classifiers

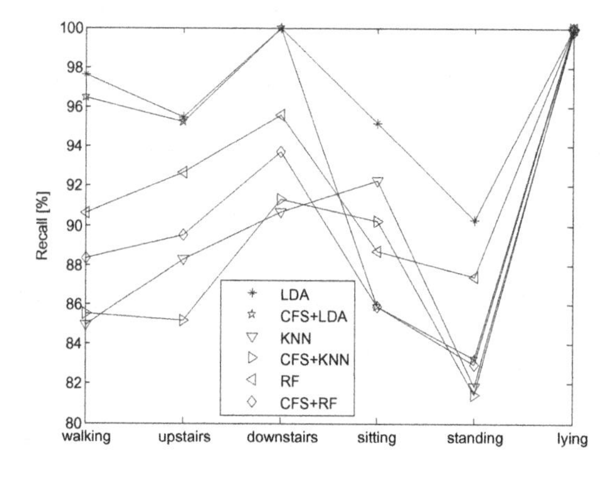

Fig. 3. The achieved values of recall for all classifiers

7 Conclusion

In this work three classification methods were tested for human physical activity recognition. The best classification performance was achived by the Linear Discriminant Analysis. The Random Forest and the K-NN methods were not able to achieve comparable results with the LDA. For simplification of classifier and possible performance classification improvement, two feature selection method were tested: the Corelation Feature Selection Method and the Principal Component Analysis. Reduction of the dataset by the CFS increased the precision of walking and walking upstairs but decreased the precision of the rest states. The most easist state for recognition was lying. The LDA and the RF achieved 100 % of accuracy. The PCA method completely failed because averagy precission is about 20 %. It possible to say, that this method is not suitable for this task. Generaly, the LDA seems like very promising method for humans activity recognition , what is claimed in others works too.

References

1. Anguita, D., Ghio, A., Oneto, L., Parra, X., Reyes-Ortiz, J.L.: Human Activity Recognition on Smartphones using a Multiclass Hardware-Friendly Support Vector Machine. In: Bravo, J., Hervás, R., Rodríguez, M. (eds.) IWAAL 2012. LNCS, vol. 7657, pp. 216–223. Springer, Heidelberg (2012)
2. Li, A., Ji, L., Wang, S., Wu, J.: Physical activity classification using a single triaxial accelerometer based on HMM. In: IET International Conference on Wireless Sensor Network, IET-WSN, November 15-17, pp. 155–160 (2010), doi:10.1049/cp.2010.1045
3. Wu, J.-K., Dong, L., Xiao, W.: Real-time Physical Activity classification and tracking using wearble sensors. In: 2007 6th International Conference on Information, Communications & Signal Processing, December 10-13, pp. 1–6 (2007), doi:10.1109/ICICS.2007.4449890
4. Liu, S., Gao, R.X., John, D., Staudenmayer, J., Freedson, P.S.: Classification of physical activities based on sparse representation. In: 2012 Annual International Conference of the IEEE Engineering in Medicine and Biology Society (EMBC), August 28-September 1, pp. 6200–6203 (2012), doi:10.1109/EMBC.2012.6347410
5. Madabhushi, A., Aggarwal, J.K.: A Bayesian approach to human activity recognition. In: Second IEEE Workshop on Visual Surveillance (VS 1999), pp. 25–32 (July 1999), doi:10.1109/VS.1999.780265
6. Fang, H., He, L.: BP Neural Network for Human Activity Recognition in Smart Home. In: 2012 International Conference on Computer Science & Service System (CSSS), August 11-13, pp. 1034–1037 (2012), doi:10.1109/CSSS.2012.262
7. Khan, A.M., Lee, Y.-K., Kim, T.-S.: Accelerometer signal-based human activity recognition using augmented autoregressive model coefficients and artificial neural nets. In: 30th Annual International Conference of the IEEE Engineering in Medicine and Biology Society, EMBS 2008, August 20-25, pp. 5172–5175 (2008), doi:10.1109/IEMBS.2008.4650379
8. Uddin, M.Z., Lee, J.J., Kim, T.-S.: Independent Component feature-based human activity recognition via Linear Discriminant Analysis and Hidden Markov Model. In: 30th Annual International Conference of the IEEE Engineering in Medicine and Biology Society, EMBS 2008, August 20-25, pp. 5168–5171 (2008), doi:10.1109/IEMBS.2008.4650378
9. Abidine, M.B., Fergani, B.: Evaluating C-SVM, CRF and LDA classification for daily activity recognition. In: 2012 International Conference on Multimedia Computing and Systems (ICMCS), May 10-12, pp. 272–277 (2012), doi:10.1109/ICMCS.2012.6320300
10. Hall, M.A.: Correlation-based feature selection for machine learning. The University of Waikato (1999)
11. Peterek, T., Krohova, J., Smondrk, M., Penhaker, M.: Principal component analysis and fuzzy clustering of SA HRV during the Orthostatic challenge. In: 2012 35th International Conference on Telecommunications and Signal Processing (TSP), July 3-4, pp. 596–599 (2012), doi:10.1109/TSP.2012.6256366
12. Breiman, L.: Random Forests. Mach. Learn. 45(1), 5–32 (2001), http://dx.doi.org/10.1023/A:1010933404324, doi:10.1023/A:1010933404324

Modular Algorithm in Tile Self-assembly Model

Xiwen Fang and Xuejia Lai*

Department of Computer Science and Engineering, Shanghai Jiao Tong University,
Shanghai 200240, China
lai-xj@cs.sjtu.edu.cn

Abstract. In this paper we propose a system computing $A \bmod B$ for given n_A-bit binary integer A and n_B-bit binary integer B, which is the first system directly solving the modulus problem in tile assembly model. The worst-case assembly time of our system is $\Theta(n_A(n_A - n_B))$ and the best-case assembly time is $\Theta(n_A)$.

Although the pre-existing division system which computes A/B can also be used to compute $A \bmod B$, the assembly time of this system is not ideal in some cases. Compared with the pre-existing division system, we achieved improved time complexity in our system. Our advantage is more significant if n_A is much greater than n_B.

Keywords: tile assembly model, DNA computing, modulus problem, assembly time.

1 Introduction

1.1 Background and Related Work

The tile assembly model theory [1,2] provides a useful framework to study the self-assembly of DNA. Researchers have demonstrated DNA implementations of several tile systems: Barish et al. [3] have demonstrated DNA implementations of copying and counting; Rothemund et al. [4] have demonstrated DNA implementation of xor tile system. Several systems solving satisfiability problem are also proposed [5,6].

The efficiency of a tile assembly system involves two factors: the tileset-size and the assembly time. In [7], a division system which computes A/B for given n_A-bit binary integer A and n_B-bit binary integer B was proposed. The assembly time is always $\Theta(n_A(n_A - n_B))$. The system can also be used to compute A mod B by computing the remainder of A/B. However, the time complexity of this system is not ideal in some cases:

Assume we have $A = 110000001_2$, $B = 11_2$, we can get the result of A mod B by computing $A - B * 10000000_2$. Such expression can be computed using tile assembly system which requires linear time. Our goal in this paper is to construct a system which directly solves the modulus problem, thus the assembly time can be improved.

* Corresponding author.

A. Abraham et al. (eds.), *Innovations in Bio-inspired Computing and Applications*, Advances in Intelligent Systems and Computing 237,
DOI: 10.1007/978-3-319-01781-5_13, © Springer International Publishing Switzerland 2014

The remaining of this paper is organized as follow: in section 1.2 we briefly introduce the concept of tile assembly model to assist the reader. In section 1.3 we introduce the corresponding algorithms. Several subsystems are discussed in section 2. In section 3 we present a system solving modulus problem and compare our system with existing system [7] in terms of time complexity. Our contributions are summarized in section 4.

1.2 Tile-Assembly Model

To assist the reader, in this section we briefly introduce the concept of tile assembly model. We refer to Σ as a finite alphabet of symbols called binding domains. We assume $null \in \Sigma$. Each tile has a binding domain on its north, east, south and west side. We represent the set of directions as $D = \{N, E, S, W\}$.

A tile over a set of binding domains is a 4-tuple. For a tile t, for $\langle \delta_N, \delta_E, \delta_S, \delta_W \rangle \in \Sigma^4$, we will refer to $bd_d(t)$ as the binding domain of tile t on d 's side.

A **strength function** $g : \Sigma \times \Sigma \to N$ denotes the strength of the binding domains. g is commutative and $\forall \delta \in \Sigma, g(\delta, null) = 0$. Let T be a set of tiles containing *empty*. A tile system S is a triple $< T, g, \epsilon >$. $\epsilon \in \mathbb{N}$ is the temperature. A tile can attach to a configuration only in empty positions if and only if the total strength of the appropriate binding domains on the tiles in neighbouring positions meets or exceeds the temperature ϵ.

Given a set of tiles Γ, a **seed configuration** $S' : \mathbb{Z}^2 \to \Gamma$ and $S =< T, g, \epsilon >$, configurations could be produced by S on S'. If no more attachment is possible, then we have the **final configuration.**

The reader may refer to [1,2,8] for more discussion of the concept of tile assembly model.

1.3 Preliminary Algorithms

To illustrate the basic idea of the modular tile system, we introduce Algorithm 1 which computes $A \bmod B$ without proof. According to the algorithm, a subsystem which is able to compare two numbers and a subsystem which performs subtraction according to the compare result are required.

Given n_A-bit binary integer A and n_B-bit binary integer B. We denote by A_i and B_i the i_{th} digit of A and B. $A = \sum_{i=0}^{n_A-1} 2^i A_i$, $B = \sum_{i=0}^{n_B-1} 2^i B_i$.

Let $A \geq B$. In order to analysis the assembly time of our modular system in later sections, we define the best-case of Algorithm 1 as follow: for the first time we compute $A - B * 2^{n_A - n_B}$, we get the difference and the difference is less than B, i.e. for the best-case the loop body will be executed only once. A typical example is $A = B * 2^m + C$, $C < B$. In this case we have $A - B * 2^{n_A - n_B} < B$, thus the loop body will be executed only once.

We define the worst-case as follow: each time we perform the subtraction, the length of A is decreased only by 1. Let $Q = A/B$, the loop body will be executed repeatedly n_Q times. A typical example is $A = 2^{n_A} - 1, B = 2^{n_B - 1}$. The loop body will be executed repeatedly $n_A - n_B + 1$ times.

Input: A , B
Output: $S = A \bmod B$
1 **while** $A \geq B$ **do**
2 **if** $A \geq B * 2^{n_A - n_B}$ **then**
3   $A = A - B * 2^{n_A - n_B}$.
4 **end**
5 **else**
6   $A = A - B * 2^{n_A - n_B - 1}$
7 **end**
8 **end**
9 $S = A$
10 return S

Algorithm 1. Modular arithmetic

2 Subsystems of Modular System

2.1 Connector System

Fig. 1 shows the tile set of the connector system that will encode the modulus B and connect B to the configuration encoding A. This system will also be used to connect the extended-subtractor system and the shift-comparer system, which will be discussed in later sections. The tile set consists of two parts: the 9-computational-tile set(Fig. 1(a)) and the $3(n_B - 1) + 1$-inputting-tile set(Fig. 1(b)). The inputting tile set includes $< (1, 1), con_{n_B} - 1, 1, pre >$ which encodes the most significant bit(MSB) of B and connects it to the MSB of A. For the remaining bits of B, according to the inputs on the west side and south side, the encoding rule is shown in Fig. 1(b).

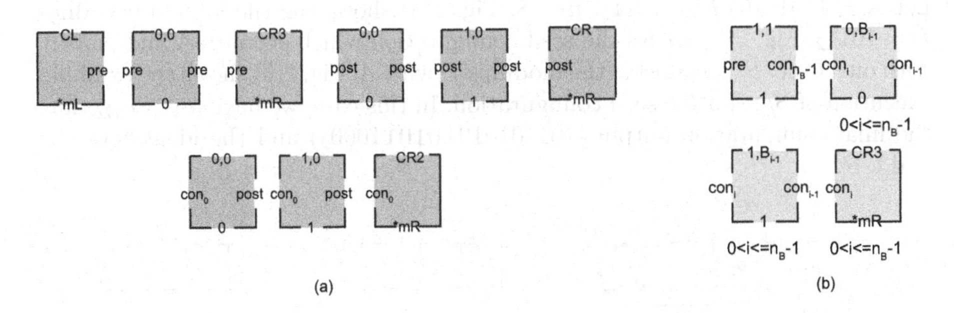

Fig. 1. The tiles have two input sides(south and west) and two output sides (north and east). (a) the computational tile set (b) the inputting tile set.

Let Σ_{con} $=\{||, pre, post, *mL, *mR, CL, CR, CR2, CR3,\ 0, 1,\ (0,0),\ (0,1),$ $(1,0),\ (1,1)\} \cup \{con_i | 0 \le i \le n_B - 1\}$, the strength function g_{con} is defined as follow:

$$\begin{cases} \forall \delta \in \Sigma_{con}, g_{con}(\sigma, null) = 0 \\ g_{con}(CL, CL) = 2; g_{con}(*mL, *mL) = 2; g_{con}(||, ||) = 2 \\ \forall \sigma \in \Sigma_{con} \backslash \{null, *mL, CL\}, g_{con}(\sigma, \sigma) = 1 \\ \forall \sigma, \sigma' \in \Sigma_{con} \backslash \{null\}, \text{if} \sigma \neq \sigma', \text{then} g_{con}(\sigma, \sigma') = 0 \end{cases} \tag{1}$$

Let $A_i, B_i \in \{0,1\}$, $n \ge n_A$. Let $\Gamma = \{\alpha_0 = \langle 0, ||, null, || \rangle, \alpha_1 = \langle 1, ||, null, || \rangle,$ $\gamma_h = \langle *mL, ||, null, null \rangle, \gamma_t = \langle *mR, null, null, || \rangle\}$. Let the seed configuration $S : \mathbb{Z}^2 \to \Gamma$ be such that

$$\begin{cases} S(1,0) = \gamma_t; S(-n,0) = \gamma_h \\ \forall i = 0, 1, \ldots, n-1 : S(-i,0) = \alpha_{A_i} \\ \text{For all other } (x,y) \in \mathbb{Z}^2, S(x,y) = empty. \end{cases} \tag{2}$$

Lemma 1. *Let* $\epsilon_{con} = 2$, *and* T_{con} *be a set of tiles over* Σ_{con}*(Fig. 1). Given a seed configuration encoding* A *which is defined in*(2), $S_{con} =< T_{con}, g_{con}, \epsilon_{con} >$ *connects* B *with* A*:*

- *If* $n_A > n_B$, *pads* B *with* $n_A - n_B$ *0-bits in the least significant bit place and* $n - n_A$ *0-bits in the most significant bit place, outputs the identifiers* CL *and* CR*;*
- *If* $n_A = n_B$, *pads* B *with* $n - n_B$ *0-bits in the most significant bit place, outputs the identifiers* CL *and* $CR2$*;*
- *If* $n_A < n_B$, *pads* B *with* $n - n_A$ *0-bits in the most significant bit place, outputs the identifiers* CL *and* $CR3$

Let $A = 1101010_2$,$B = 1011_2$, $n = 8$. Fig. 2(a) shows the tile system encoding $B = 1011_2$. Fig. 2(b) shows the seed configuration which encodes A and pads it with one 0-tile. S_{con} connects the modulus B with A. Fig. 2(c) shows the example execution of S_{con} on the seed configuration. In this case, we have $n_A > n_B$, thus the final configuration outputs (01101010_2,0**1011**000_2) and the identifiers CL and CR.

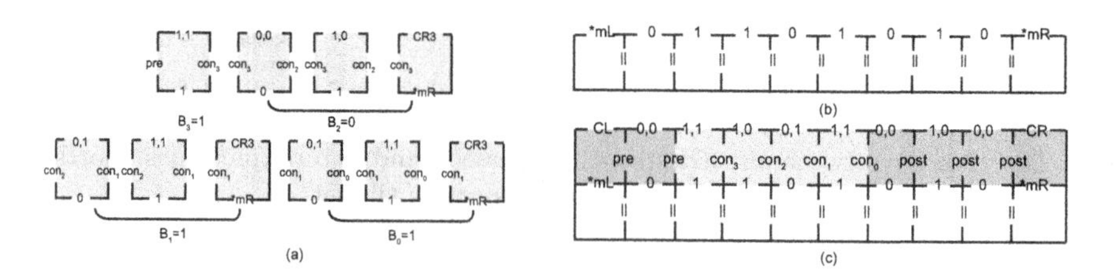

Fig. 2. (a) Inputting tile set encoding $B = 1011_2$ (b) seed configuration encoding $A = 1101010_2$ (c) S_{con} connects B with A

2.2 Shift-Comparer System

In this section we present a system with two jobs: to compare A with B and to perform right-shift on B. Fig. 3(a) shows the concepts behind the tiles in T_C, which have two input sides (west and south) and two output sides(north and east). The concepts includes variables a, b, c, r, r'. We have $a, b, c \in \{0, 1\}$ and $r, r' \in \{>, =, <\}$. The input sides includes variables a, b, c, r, thus there are 24 tiles(Fig. 3 (c)). The rule of comparing A with B is shown in Fig. 3 (b). We compare A_i with $B_i(i = n-1, n-2, ..., 0)$. The assumption $A = B$ holds until we find $A_i \neq B_i$: if $A_i \geq B_i$, then we have $A \geq B$; else if $A_i < B_i$, we have $A < B$(Fig. 3(b)). Fig. 3(d) shows the corresponding boundary tiles.

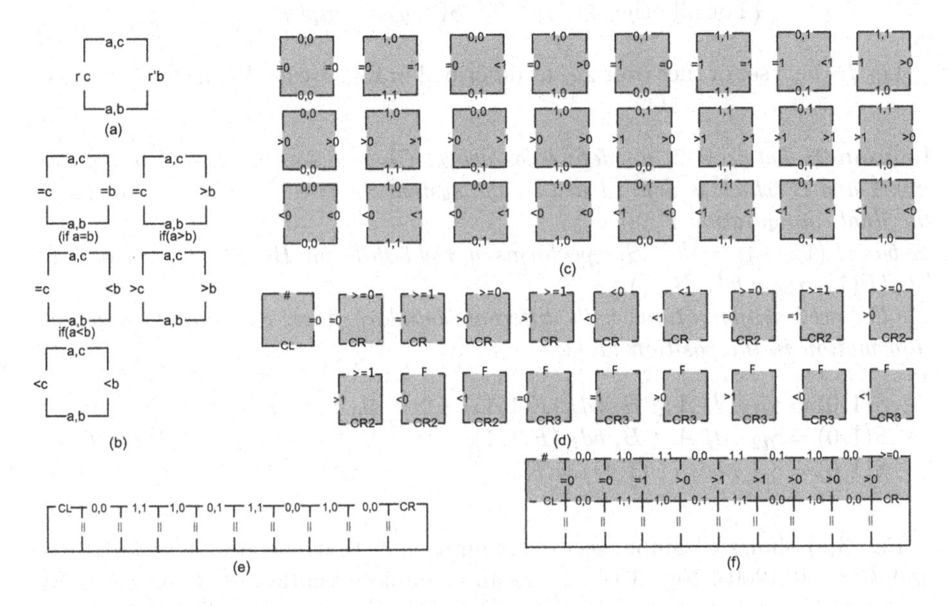

Fig. 3. (a) concepts behind the comparer system (b) rules of comparing A and B (c) computational tiles (d) boundary tiles (e) $A = 1101010_2$ and $B = 1011000_2$. (f)$A \geq B$, $B_0 = 0$. F outputs (1101010_2,101100_2) and the boundary flags (#, >= 0).

Let $\Sigma_C = \{||, (0,0), (0,1), (1,0), (1,1), =0, =1, <0, <1, >0, >1, >=0, >=1, CL, CR, CR2, CR3, F, \#\}$. Let $n \geq \max(n_A, n_B)$. Let $\Gamma = \{\alpha_0\beta_0 = \langle(0,0), ||, null, ||\rangle, \alpha_0\beta_1 = \langle(0,1), ||, null, ||\rangle, \alpha_1\beta_0 = \langle(1,0), ||, null, ||\rangle, \alpha_1\beta_1 = \langle(1,1), ||, null, ||\rangle, \gamma_h = \langle CL, ||, null, null\rangle, \gamma_{t1} = \langle CR, null, null, ||\rangle\}$, $\gamma_{t2} = \langle CR2, null, null, ||\rangle$, $\gamma_{t3} = \langle CR3, null, null, ||\rangle\}$. The strength function g_C is defined as follow:

$$\begin{cases} \forall \sigma \in \Sigma_C, g_C(\sigma, null) = 0 \\ g_C(CL, CL) = 2; g_C(>= 0, >= 0) = 2 \\ g_C(>= 1, >= 1) = 2; g_C(<, <) = 2; g_C(||, ||) = 2 \\ \forall \sigma \in \Sigma_C' \backslash \{null, CL, >= 0, >= 1, <, ||\}, g_C'(\sigma, \sigma) = 1 \\ \forall \sigma, \sigma' \in \Sigma_C' \backslash \{null\}, \sigma \neq \sigma', g_C(\sigma, \sigma') = 0 \end{cases} \tag{3}$$

Let $n \geq \max(n_A, n_B)$. Let the seed configuration $S : \mathbb{Z}^2 \to \Gamma$ be such that

$$\begin{cases} S(1,0) = \gamma_t, \gamma_t \in \{\gamma_{t1}, \gamma_{t2}, \gamma_{t3}\}; S(-n, 0) = \gamma_h \\ \forall i = 0, 1, \ldots, n-1 : S(-i, 0) = \alpha_{A_i} \beta_{B_i} \\ \text{For all other } (x, y) \in \mathbb{Z}^2, S(x, y) = empty. \end{cases} \tag{4}$$

Let T_C be a set of tiles over Σ_C as described in Fig. 3(c,d). We have $|T_C| = 43$.

Lemma 2. *Let $\epsilon_C = 2$, g_C defined as above. Given a seed configuration encoding A and B which is defined in* (4)*, the system $S_C =< T_C, g_C, \epsilon_C >$ produces the final configuration F:*
I) $bd_N(F(1, -n))$ = '#'. S_C' performs a right-shift on B: $\forall 0 \leq i \leq n-1$, $bd_N(F(1, -i)) = (A_i, B_{i+1})$.
II) For seed configurations with different boundary flags, S_C' outputs different information in the position (1,1):

- $S(1,0) = \gamma_{t1}$ *: If $A \geq B$, $bd_N(F(1,1))$ = '>= B_0'; else $bd_N(F(1,1))$ = '<'.*
- $S(1,0) = \gamma_{t2}$ *: If $A \geq B$, $bd_N(F(1,1))$ = '>= B_0'; else $bd_N(F(1,1))$ = 'F'.*
- $S(1,0) = \gamma_{t3}$ *: $bd_N(F(1,1))$ = 'F'.*

Fig. 3(e) shows a sample seed configuration S that encodes $A = 1101010_2$ and $B = 1011000_2$. Fig. 3 (f) shows an example execution of S_c on the seed configuration and the top row reads the right-shift solution of B, i.e. 101100_2; the rightmost boundary flag '>= 0' indicates that $A \geq B$ and the initial LSB of B is 0.

2.3 Extended Subtractor System

In this section we propose a extended subsystem that performs subtraction and shift-subtraction. The concepts behind the system are showed in Fig. 4(a). The concepts include variables a, b, c, b', c', and s. each of which can take on as values the elements of the set $\{0, 1\}$. The tiles have two input sides (east, south) and two output sides (west, north). For the subset that performs subtraction, on the input sides there are 3 variables, thus there are 8 tile types in this system; for the subset that performs shifter-subtraction, on the input sides there are 4 variables, thus there are 16 actual tile types in this system. Therefore in this system we have totally 24 tile types(Fig. 4(b)). Fig. 4(c) shows 5 extra boundary tiles.

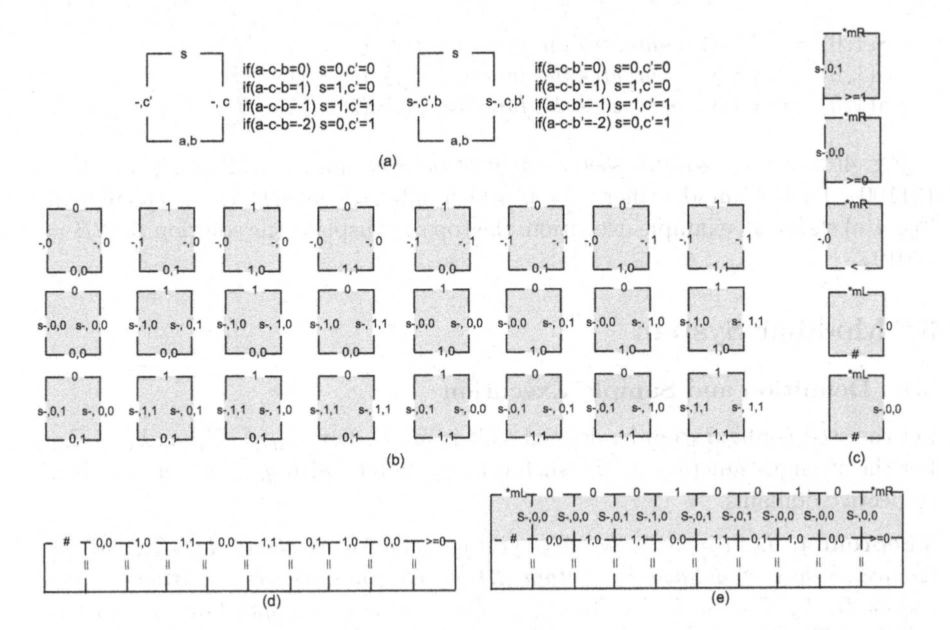

Fig. 4. (a) the concept behind the system computing $f(A,B) = A - B$ and $f(A,B) = A - 2B - t$ (b) the 24-computational-tile types (c) the 5-boundary-tile types (d)$A = 1101010_2$ and $B = 101100_2$ (e)$A - 2 * B = 10010_2$

Let $\Sigma_- = \{||, (0,0), (0,1), (1,0), (1,1), (-,0), (-,1), 0{,}1, (s-,0,0), (s-,0,1), (s-,1,0), (s-,1,1), \#, >= 0, >= 1, <, *mL, *mR\}$, the strength function g_- is defined as follow:

$$\begin{cases} \forall \sigma \in \Sigma_-, g_-(\sigma, null) = 0 \\ g_-(*mL, *mL) = 2; g_-(>= 0, >= 0) = 2 \\ g_-(>= 1, >= 1) = 2; g_-(<, <) = 2; g_-(||, ||) = 2 \\ \forall \sigma \in \Sigma_- \backslash \{null, *ml, >= 0, >= 1, <, ||\}, g_-(\sigma, \sigma) = 1 \\ \forall \sigma, \sigma' \in \Sigma_- \backslash \{null\}, \text{if} \sigma \neq \sigma', \text{then} g_-(\sigma, \sigma') = 0 \end{cases} \tag{5}$$

Let $n \geq \max(n_A, n_B)$, $A \geq B$. Let $\Gamma = \{\alpha_0\beta_0 = \langle (0,0), ||, null, || \rangle, \alpha_0\beta_1 = \langle (0,1), ||, null, || \rangle, \alpha_1\beta_0 = \langle (1,0), ||, null, || \rangle, \alpha_1\beta_1 = \langle (1,1), ||, null, || \rangle, \gamma_h = \langle \#, ||, null, null \rangle, \gamma_{t1} = \langle >= 0, null, null, || \rangle, \gamma_{t2} = \langle >= 1, null, null, || \rangle, \gamma_{t3} = \langle <, null, null, || \rangle\}$. Let the seed configuration $S : \mathbb{Z}^2 \to \Gamma$ be such that

$$\begin{cases} S(1,0) = \gamma_t, \gamma_t \in \{\gamma_{t1}, \gamma_{t2}, \gamma_{t3}\}; S(-n,0) = \gamma_h \\ \forall i = 0, 1, .., n-1 : S(-i, 0) = \alpha_{A_i}\beta_{B_i} \\ \text{For all other } (x,y) \in \mathbb{Z}^2, S(x,y) = empty. \end{cases} \tag{6}$$

Lemma 3. *Let $\epsilon_- = 2$, let Σ_-, g_- defined as above, and T_- be a set of tiles over Σ_- (see Fig. 4(b,c)). Given a seed configuration encoding A and B which is defined in* (6)*, the system $S_- =< T_-, g_-, \epsilon_- >$ computes different functions:*

- $S(1,0) = \gamma_{t1}$: *S_- computes the function $f(A,B) = A - 2B$*
- $S(1,0) = \gamma_{t2}$: *S_- computes the function $f(A,B) = A - 2B - 1$*
- $S(1,0) = \gamma_{t3}$: *S_- computes the function $f(A,B) = A - B$*

Fig. 4(d) shows a sample seed configuration encoding $A = 1101010_2$ and $B = 101100_2$. There is an identifier '$>= 0'$ which indicates $A - 2B$ will be calculated. Fig. 4(e) shows an example execution: the top row displays the solution $A - 2B = 10010_2$.

3 Modular System

3.1 Definition and Sample Execution

Let the seed configuration be defined as it is in (2). Let $\Sigma_M = \Sigma_{con} \cup \Sigma_C \cup \Sigma_-$. Let the strength function g_M be such that g_M agrees with g_{con}, g_C, g_- on their respective domains.

Theorem 4. *Let $\epsilon_M = 2$. Let Σ_M and g_M defined as above. Let T_{con} be the connector system defined in section 2.1 which contains an inputting set encoding B. Let $T_M = T_{con} \cup T_C \cup T_-$. Given a seed configuration encoding A which is defined in* (2)*, the system $S_M =< T_M, g_M, \epsilon_M >$ computes the function $f(A,B) = A \bmod B$.*

Due to limitations of space, we are not able to give the proof in detail. A brief explanation of Theorem 4 is given as follow:

According to Lemma 1, Lemma 2 and Lemma 3, S_C could be executed on the final configuration of S_{con}, S_{con} could be executed on the seed configuration of S_M and the final configuration of S_-, and S_- could be executed on the final configuration of S_C. Thus at the very beginning, S_{con} will be executed. Then S_C will be executed according to the output of S_{con}, which is $(A,\ B * 2^{n_A - n_B})$.

According to Algorithm 1, the intuition behind the modular system computing $A \bmod B$ is that comparing A with B:

- If $A < B$, then A itself is the solution;
- Else if $A \geq B * 2^{n_A - n_B}$, S_C outputs the binding domain '$>= B_0$' which indicates the comparison result; later S_- will calculate $A = A - B * 2^{n_A - n_B}$;
- Else if $A < B * 2^{n_A - n_B}$, S_C outputs the binding domain '$<$', and S_- will calculate $A = A - B * 2^{n_A - n_B - 1}$.

Then for A with the updated value, $A \bmod B$ will be computed again.The process will be repeated until we have $A < B$. The system outputs A as the final solution.

Fig. 5 shows an example execution of S_M with $A = 1101010_2$ and $B = 1011_2$. The seed configuration encoding $A = 1101010_2$ is shown in Fig. 5(a). The modulus $B = 1011_2$, $n_B = 4$, thus $B_3 = 1$, $B_2 = 0$, $B_1 = 1$ and $B_0 = 1$. Fig. 5(b) shows the corresponding inputting tile set encoding $B = 1011_2$. Fig. 5(c) shows the final configuration F_M. On the north side of row 8, S_C outputs $(111_2, 10_2)$ and the final identifier 'F', which indicates 111_2 is the solution, i.e. $1101010_2 \bmod 1011_2 = 111_2$.

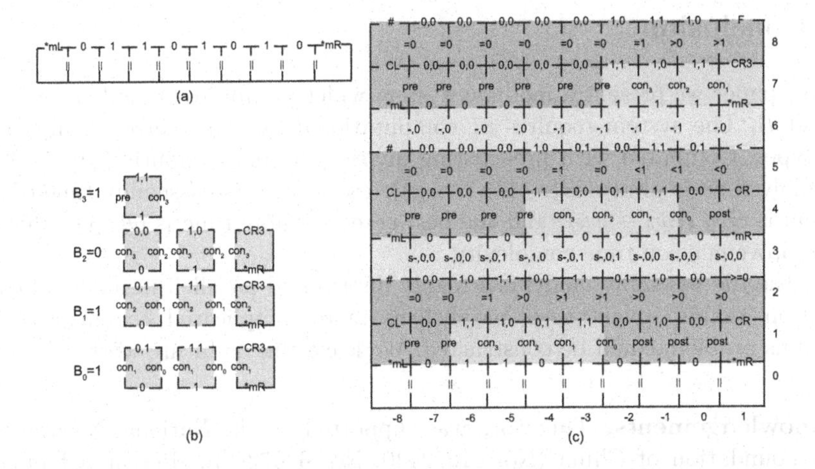

Fig. 5. (a) a sample seed configuration S that encodes $A = 1101010_2$, $n_A = 7$, $n = 8$. (b) the inputting tile set encoding the modulus $B = 1011_2$ (c) the modular system performs the computation 1101010_2 mod 1011_2 and produces a final configuration F_M on S. Along the top row F_M encodes the solution, i.e. 00000111_2.

3.2 Discussion

Lemma 5. *(Modular Assembly Time Lemma). Given the seed configuration defined in* (2) *which is of length* n. *For all* A, B, *if* n *and* n_A *have the same order of magnitude, the worst-case assembly time of* S_M *computing* $A \bmod B$ *is* $\Theta(n_A(n_A - n_B))$ *and the best-case assembly time is* $\Theta(n_A)$.

Although the division system proposed in [7] which computes A/B can also be used to compute $A \bmod B$, our scheme enjoys advantage in aspect of assembly time: the overall assembly time of our scheme is less than that of scheme[7]. In [7], to compute the remainder, the assembly time is always identical to that of the worst-case in our scheme. Assume we have $A = 110000001_2$, $B = 11_2$, $B' = 110000000_2$, i.e. B padded with seven 0-bits. To compute the remainder of A/B, in [7] it costs seven cycles to reduce the length of B' until it is identical to n_B; while in our scheme, such operation is not required. The fact that $A - B'$ is exactly the final solution of $A \bmod B$ could be verified directly after $A - B'$ is computed. Thus to compute $A \bmod B$, $A > B$, in our scheme, the best-case assembly time is $\Theta(n_A)$ while in [7] the assembly time is always $\Theta(n_A * (n_A - n_B))$, which is identical to our worst-case assembly time. Our advantage is more significant if n_A is much greater than n_B.

Lemma 6. *(Modular Tile-set Size Lemma). To compute* $A \bmod B$, T_M *requires 81 computational tile types and* $3n_B - 2$ *inputting tile types.*

4 Conclusion

In this paper we present a modular system which computes A mod B for given A and B. The system requires 81 computational tiles and $3n_B - 2$ inputting tile types. Compared with pre-existing division system computing A/B whose assembly time is always $\Theta(n_A(n_A - n_B))$, the worst-case assembly time of our system is $\Theta(n_A(n_A - n_B))$ and the best-case assembly time is $\Theta(n_A)$, which is an improvement on time complexity.

As the tile assembly model is a highly distributed parallelism model of computation, taking advantage of parallelism, more complicated system involving modulus problem could be constructed. We leave this as future work.

Acknowledgements. This work was supported by the National Natural Science Foundation of China (No. 61272440, No. 61073149), Research Fund for the Doctoral Program of Higher Education of China (20090073110027), State Key Laboratory of ASIC & System (11KF0020), Key Lab of Information Network Security, Ministry of Public Security (C11603) and Scholarship Award for Excellent Doctoral Student granted by Ministry of Education.

References

1. Rothemund, P., Winfree, E.: The program-size complexity of self-assembled squares. In: Proceedings of the Thirty-Second Annual ACM Symposium on Theory of Computing, pp. 459–468. ACM (2000)
2. Winfree, E.: Algorithmic Self-Assembly of DNA. PhD thesis, California Institute of Technology (1998)
3. Barish, R., Rothemund, P., Winfree, E.: Two computational primitives for algorithmic self-assembly: Copying and counting. Nano Letters 5(12), 2586–2592 (2005)
4. Rothemund, P., Papadakis, N., Winfree, E.: Algorithmic self-assembly of dna sierpinski triangles. PLoS Biology 2(12), e424 (2004)
5. Brun, Y.: Solving np-complete problems in the tile assembly model. Theoretical Computer Science 395(1), 31–46 (2008)
6. Brun, Y.: Improving efficiency of 3-SAT-solving tile systems. In: Sakakibara, Y., Mi, Y. (eds.) DNA 16 2010. LNCS, vol. 6518, pp. 1–12. Springer, Heidelberg (2011)
7. Zhang, X., Wang, Y., Chen, Z., Xu, J., Cui, G.: Arithmetic computation using self-assembly of dna tiles: subtraction and division. Progress in Natural Science 19(3), 377–388 (2009)
8. Winfree, E., Liu, F., Wenzler, L., Seeman, N., et al.: Design and self-assembly of two-dimensional dna crystals. Nature 394(6693), 539–544 (1998)

LLLA: New Efficient Channel Assignment Method in Wireless Mesh Networks

Mohammad Shojafar[1], Zahra Pooranian[2], Mahdi Shojafar[3], and Ajith Abraham[4]

[1] Department of Information Engineering, Electronics (DIET), Sapienza University of Rome, Rome, Italy
Shojafar@diet.uniroml.it

[2] Department of Computer Engineering, Dezful Islamic Azad University, Dezful, Iran
zahra.pooranian@gmail.com

[3] Department of Electrical Engineering, Islamic Azad University of Noor, Noor, Iran
mahdishojafar@yahoo.com

[4] Machine Intelligence Research Labs (MIR Labs), WA, USA
ajith.abraham@ieee.org

Abstract. Wireless mesh networks (WMNs) have emerged as a promising technology for providing ubiquitous access to mobile users, and quick and easy extension of local area networks into a wide area. Channel assignment problem is proven to be an NP-complete problem in WMNs. This paper aims proposing a new method to solve channel assignment problem in multi-radio, multichannel wireless mesh networks for improving the quality of communications in the network. Here, a new hybrid state channel assignment method is employed. This paper proposes a Link-Layard Protocol and Learning Automata (LLLA) to achieve a smart method for suitable assignment. Simulation results show that the proposed algorithm has better results compared to AODV method. E.g., it reduces the packet drop considerably without degrading.

Keywords: Wireless mesh network (WMN); Channel Assignment (CA); Learning Automata (LA); Network Throughput.

1 Introduction

Recent improvements in Micro-Electro-Mechanical-Systems (MEMS), wireless telecommunication and also digital electronic have made possible manufacturing small, low energy consuming and cost effective nodes that are able to have wireless connection [1]. Generally networks are classified as wired and wireless. Wireless networks include infrastructure based wireless networks and infrastructure less wireless networks. The first class of wireless networks have central controller and service providers called access points that have the same duty as routers in wired networks and nodes connect to each other through access points. But in infrastructure-less wireless networks there is no central controller and access point and every node

A. Abraham et al. (eds.), *Innovations in Bio-inspired Computing and Applications*, Advances in Intelligent Systems and Computing 237,
DOI: 10.1007/978-3-319-01781-5_14, © Springer International Publishing Switzerland 2014

acts as final node and router for other nodes in the network. Infrastructure-less wireless networks include mobile ad-hoc networks (MANET), wireless sensor network (WSN) and wireless mesh network (WMN). WMNs are of connection systems that their connection to clients is high speed and wide band. These networks are completely wireless and self-organized and guide traffic to internet or from internet in multi-hop and ad-hoc method. Wireless mesh networks consist of some nodes that are stable and static. These nodes usually have one or more radio or network interface. These networks use available channels that are supported by 802.11 protocols to reach to maximum capacity. In these networks every node connects to its neighboring nodes or nodes in its transmission range if these two nodes have at least one radio that uses a common channel. It is necessary that each of these two nodes has a radio and these radios be adjusted to a common channel. Because nodes can have connection to each other only if this connection is made through a common channel [2, 3].

One of the problems in wireless mesh networks is optimum Channel Assignment (CA) for nodes and interference. In order to decrease interference, channels may be dedicated in such a way that nodes have the least number of common channels. This means decreasing nodes connections. Therefore, high level of connection in the network and low interference are not possible simultaneously. In other words, there is a trade-off between connection in network and network interference. Interference in these networks is inevitable and is the factor that limits capacity. So CA in WMN methods tries to decrease interference. So in this paper, we want to propose a new method to improve CA and decrease interference in multi-channel wireless mesh networks [4, 5].

The purpose of this study is proposing a new method for solving CA problem with better performance than mentioned methods. We compare our algorithm with AODV [6, 18] method based on interference of channels and simultaneous connection of network nodes and will show that our proposed method yields better result. In section 2 related works in CA and routing in WMN will be mentioned. In section 3 the proposed idea that is based on learning automata is presented in details. Section 4 includes tables and graphs of simulation and efficiency estimation of the proposed idea and other methods. Finally, conclusion and future works are presented in the last section.

2 Related Research

In this section studies on recent channel assignment protocols and reliable multi-section protocols are reviewed. Prior work on channel assignment schemes can be broadly classified into three categories: static assignment, dynamic assignment and hybrid assignment. *Static assignment or Constant* strategies assign a channel to each interface for permanent use. *Dynamic assignment* forces nodes to switch their interfaces from one channel to another between successive data transmission dynamically and *Hybrid approaches* apply a static or semi-dynamic assignment to the fixed interfaces and a dynamic assignment to the switching interfaces [7].

There are several works that have been done in all three categories. In the first method in which assignment is dynamic, Pediaditaki has presented a method named LCAP in [8]. In this method algorithm is learning automata such that every node can do CA based on its neighbor's conditions. Also neighboring nodes are able to automatically find their considered neighbors in information transmission in the channel by considering information of each channel. This algorithm is similar to Asymmetric and Distributed graph Coloring algorithm or ADC. This method leads to increase of channel efficiency at high traffic. In the second method [9], channel selection algorithm that is presented below is used. The purpose of this algorithm is heuristic specification of a channel. Here the aim is decreasing work of interferers in different channels and decreasing surplus generation in a channel by entering one or more new nodes in data transmission. This idea not only tolerates various errors but also adapts to changes caused by external interferers effective on channel performance. In order to make adaptability, intelligent LA algorithm is used. Here a new architecture is presented by its details for WMN and CA. The proposed algorithm is added to network layer as an effective algorithm in CA like [9-13]. Specifically, there are several meta-heuristic works have done on Wireless mesh networks in routing [12-14], Channel assignments [15] that belongs to the hybrid types of Channel Assignments. Also, authors in [16] specify a heuristic method to optimize channel allocation like graph coloring in the mesh graph of nodes.

Das et al. [17] proposed a Carrier sense multiple access (CSMA) algorithm with RTS/CTS/ACK in MAC layers of WMNs, which dedicates channel constantly. This method is of constant CA methods. Here, the problem of CA for Multi-Channel Multi-Radio in wireless mesh networks is studied. This paper focuses on static mesh wireless networks in which multiple channels are available for each wireless medium without interference. Also the purpose is finding a constant CA that has the most mutual links that can be activated simultaneously by considering limitation of interferers. This paper has proposed two Integral Linear Programming (ILP) models to solve problem of constant CA with multiple radios. Numerical results show that much benefit is obtained through increasing number of radios in each node and channels in the network. Pirzada et al. [18] assessed performance on Ad hoc On-Demand Distance Vector (AODV) in some radio wireless mesh networks. Simulation results show that in high load traffic, multiple-radio (AODV-MR) is able to use extended spectrum in the best way and it is established that this is much better than single-radio AODV.

3 The Proposed Method

In this section we want to present an algorithm that can mitigate this problem by learning automata (LA). In the following we use one of the famous wireless network protocols: AODV. Also for CA link-layer protocols (LLP) is applied that is of hybrid assignment in CA section, so we called our algorithm (LLLA means mix of LA in LLP). Finally learning automata is used to complete the algorithm to get an intelligent method for proper assignment.

The purpose of this intelligent algorithm is directing to the best states that can meet mentioned criteria. Consider the following figure 1 (Initial Step). It is supposed that each node has at least 3 radio Medias and can transmit information by Time-division multiplexing (TDM) and frequency-division multiplexing (FDM) methods simultaneously. One of the most important suppositions is that medium radio cards are the same. It means that all of them can exchange information in X to X+10000 frequency ranges. Of course this supposition is not impossible because many medium cards made by different companies use the same standards, as figure 1.

In fact we want to create an intelligent method by LLP method and Learning Automata (LA) and TDM and FDM multiplexing. Intelligence of this method is related to the learning automata when it transforms media from static to dynamic state or conversely according to environment feedback. Each node has two functions. These functions are related to the media types. In fact we want to present an algorithm that adapts itself by network requirements, network topology and application in the network. At first, medium type is chosen according to data volume. Then according to the feedback got from network, medium type changes its state. This state change occurs when medium probability reaches to an amount less than 10 percent of its original state.

In this algorithm always one third of nodes are considered static and the remainder is considered dynamic at first. It is supposed that the red node has 3 Medias. So one of its radios is considered as static medium and the other two radios are considered as dynamic media.

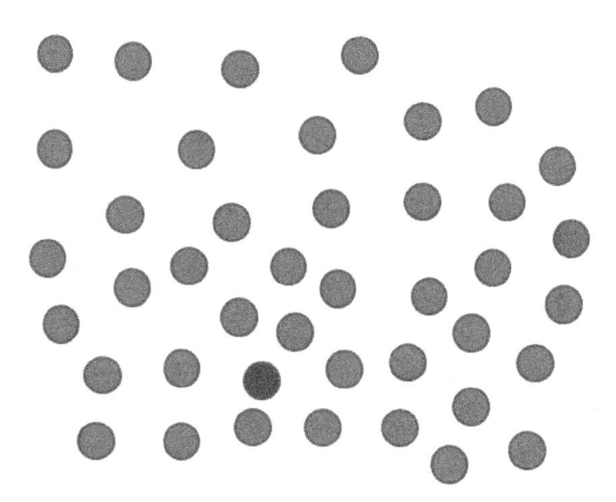

Fig. 1. Proposed method in the initial step

Also in implementation, static media are considered with TDM multiplex and dynamic media are considered with FDM multiplex. In the first stage, transmitter node goes through the same stages of AODV protocol. This node transmits a packet titled "Hello" to its neighbors. Neighbors who have received "Hello" packet send its response. In the next stage the second phase of AODV protocol is implemented. In this stage some packets are sent by REQUEST title and neighbors who have received the message send list of their neighbors as response to the transmitter (echo). It is supposed that red nod should have data transmission with five of its neighbors simultaneously.

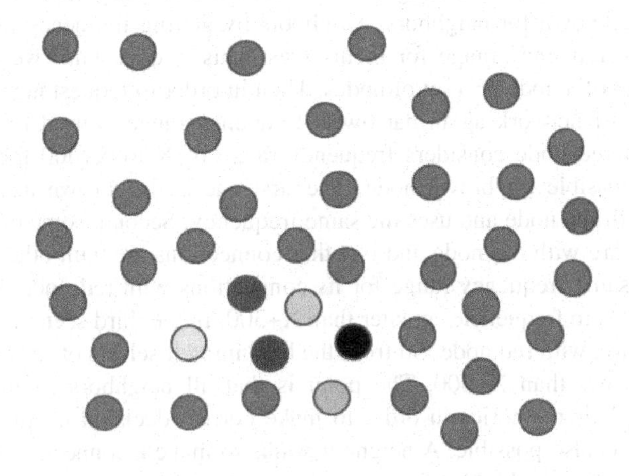

Fig. 2. Proposed Method Transmission Step

As it was considered (as figure 2) red node has 3 radio Medias that one of them is static and the other two nodes are in dynamic state. Then red node should have five connections with 3 radio Medias (Transmission Step). All the nodes in transmission stage should send volume of their packets to their neighbors. Also nodes that are media of transmission should receive packet volume from transmitter and send it to the next node. In this way all nodes will be aware of volume of packet that is to be transmitted. Now it is supposed that gray, black and yellow nodes carry small packets and orange and brown nodes carry large packets. The considered learning automata is standard automata[19]. Also since desired and not desired states of criteria are considered, our environment is considered P environment (we have different LAs, P-Type or P environment is one of this types). In this environment feedback from environment determines if the environment is desired or not desired (undesired). For example, if end to end delay time, packet drop and etc. are desired, it means desired performance. But if the considered criteria do not have desired performance, the environment is considered undesired. According to LLLA algorithm, red node has 5 functions. Red node makes a connection channel for each of neighboring nodes. According to the original definition, large packets use static media and small packets use dynamic media. In the next stage, probability of connection type selection is determined by receiving routing performance. For example, at first brown node sends a packet to its destination. This node uses a static medium to make connection with red node. If red node has received acceptable time from transmission of packet of brown node to receiving acceptance of packet accuracy, probability of connection with static media for brown node is increased and this connection is maintained for some time.

A very important note is interference area among radio media. For example consider neighboring nodes that one of them has 3 Medias, the other has 4 Medias and the other one has 2 radio Medias. If each of these nodes is others circle, at first glance interference is inevitable. This problem is solvable by considering some notes. The first one is that before making any connection, frequency range of connecting node should be known for neighbors. Neighbors by getting frequency range of node consider wider frequency range for themselves. This is done until we get far from frequency range of a node or a set of nodes. Then in order to request next connections in other points of network again narrowest frequency range is used for connection. For example if red node considers frequency range of X to X+300 for itself, three scenarios are possible for brown node. The first one is that brown node has all its connections with red node and uses the same frequency. Second, some of connections of brown node are with red node and its other connections are with other nodes. This node uses the same frequency range for its connections with red node but makes its other connections in frequencies greater than X+300. In the third scenario brown node has no connection with red node. So from the beginning it selects other connections in frequencies greater than X+300. The point is that all neighbors should save used frequencies in their memories in order to make correct decision in this regard. The following state is also possible. A neighbor wants to make a connection and wants to say this. In this state the highest frequency range is used for transmitting information so that no interference with existing connections is created.

4 Experimental Results

For algorithm implementation AODV code along with OPNET 14 [20] software package was used. In LLLA algorithm implementation at first LLP concept was entered in the second layer and learning automata algorithm was considered on nodes.

4.1 Simulation Scenario

In order to study accuracy and quality of the proposed algorithm some scenarios were considered and each of them was studied separately based on different criterion or criteria. Also in all states of standard learning automata reward rate was considered 0.1 and punishment rate was 0.05.

4.2 Network Environment

In this state nodes of network are considered almost static and some nodes transmit large packets and some other nodes transmit light packets in the network. To simulate this scenario a 100*100 (m^2) environment with 70 nodes is considered; these nodes are distributed randomly in the network environment. Simulation running time was considered 10 minutes and the purpose was transmitting data from 10 nodes as offset and reaching to the considered destination by AODV algorithm. Each of nodes starts random transmission of information from 30th second. Of these 10 nodes, 3 nodes create high traffic and7 nodes create light traffic. This scenario was run 5 times and the following results were obtained in different states of learning automata. In the following we will describe different QoS parameters that are tested in the proposed method with AODV.

4.2.1 Criterion of Packet Drop with Various Learning Automata

According to figure 3 it can be noted that there is no interference among data except in some exceptional situations because each of media are active in their own frequency range and this greatly decreases interference. Almost 3 times meaning 300 percent decrease in interference is obtained and accordingly dropped packets are decreased.

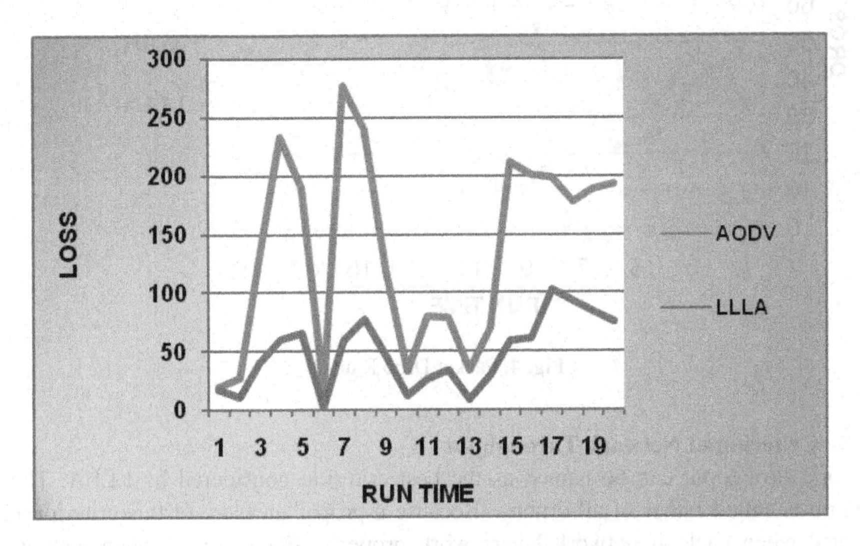

Fig. 3. Loss Rate based on Run Time

4.2.2 Criterion of Drop of Packets

Drop measure of packets depends on buffer memory of each node. If memory of buffers is less than required for packet buffering, nodes have should drop the received data and request data in another time. Of course this increases network traffic. On one side control messages should flow in the network and on the other side the same packet should flow in the network again and this increases network congestion. Generally buffer fullness occurs when there is mobility in the route and destination node or medium nodes have moved or there is failure and current node can`t direct data forward and its buffer is full so it should drop received data.

In the following method drop occurs less frequently due to less failure and mobility, and compared to standard AODV algorithm it has about 70% improvement in this regard. The reason is that when there are less interferences, packet failure is less so packet buffering is less required. Usually one of the reasons for packet buffering by nodes is uncertainty of receiving packets without failure because if medium buffers don't do buffering, routing should be done from offset to destination even for a small packet. In 4 states of learning automata, LRI learning automata has the best performance and it can be said that this is because of more adaptability with network due to less mobility, as figure 4.

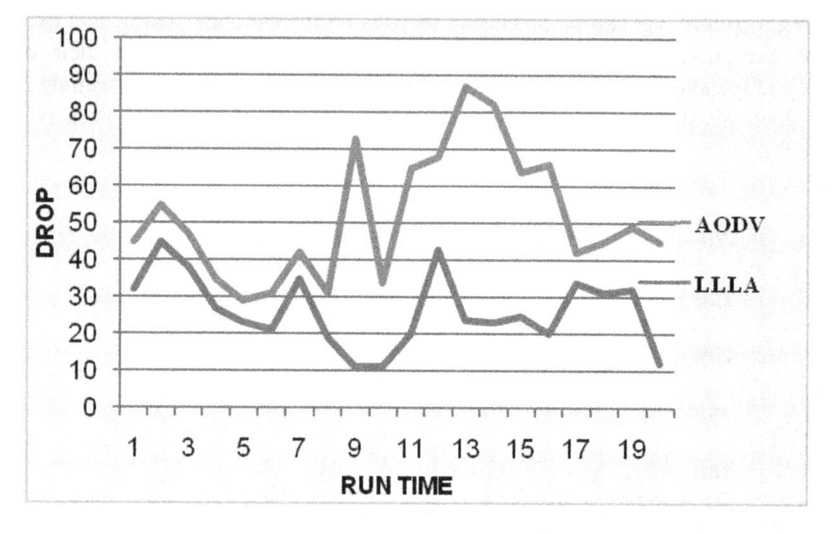

Fig. 4. Packet Drop Rate

4.2.3 Criterion of Network Throughput

Network throughput can be named as the best criterion considered by LLLA. This criterion is called father of all criteria. Because a proper measure of this criterion is obtained when each of network layers work properly. For example when network delay is decreased it means that more data can be handled in a time period or when encounters are decreased, more data traffic can be exchanged in the same time period. The purpose of the proposed LLLA algorithm has been more network throughput compared to AODV standard algorithm and about 650 percent improvement has been obtained in simulation. The reason is that: first, in the AODV standard algorithm there is just one radio medium but in the proposed algorithm a number of radio media are used and this certainly makes performing speed much more. Second. In the proposed algorithm a state is considered in which traffics are differentiated but in the standard AODV algorithm there is not such differentiation (LLP statement). Third, network adapts itself with traffic type and traffic handling type and this has significant effect on data transmission that is learning automata algorithm usage. In the studies it was shown that learning automata algorithm of LRI type has the best performance because of constant state from mobility and failure point of view (figure 5).

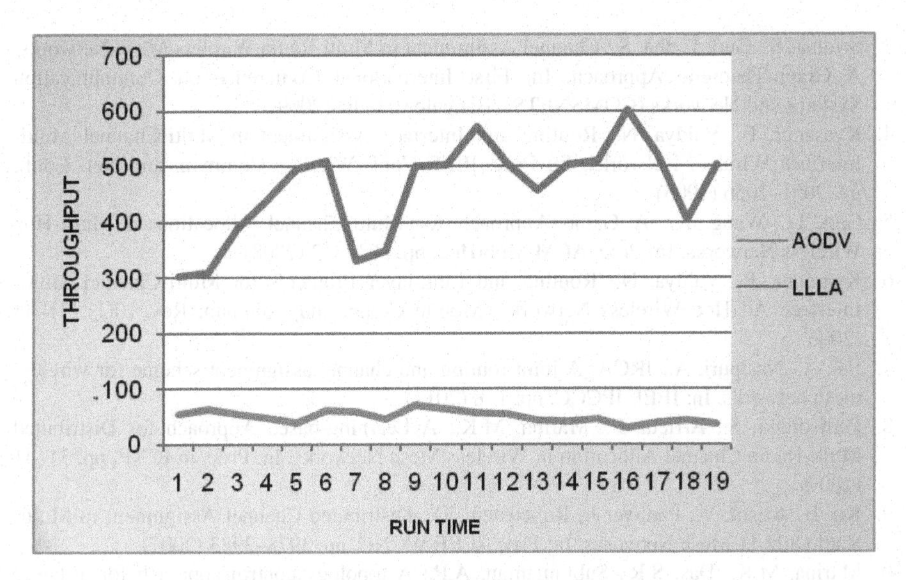

Fig. 5. Network Throughput based on Run Time

5 Conclusion and the Future Work

As was discussed in this paper, loss rate, packet drop, and network throughput are studied and compared with AODV. What is resulted from simulation study is that if algorithms implemented in the network can adapt themselves with network and function based on it, network throughput would increase significantly and our proposed algorithm is an example of such a case. This could be applied in VANET that needs high throughputs and availability but we applied our work in the special cases that mentioned without considering the routing jointly.

In future works we plan to study the effect of using the proposed algorithm in directional antenna to reduce the co-channel interference between some of the neighboring links in MC-WMNs. Of course this simulation is just a proposition. Here network nodes are considered almost dynamic and some nodes transmit large packets in the network and some other transmits light packets. This method has more intelligent performance than simple TDM. Also failure states can be considered and network with special applications can be studied.

References

1. Xu, S., Saadawi, T.: Does the IEEE 802.11MAC Protocol Work Well in Multi hop Wireless Adhoc Networks. IEEE Communications Magazine 39, 130–137 (2001)
2. Wanli, D., Kun, B., Lei, Z.: Distributed Channel Assignment Algorithm for Multi-Channel Wireless Mesh Networks. In: International Colloquium on Computing, Communication, Control, and Management (CCCM), vol. 2, pp. 444–448 (2008)

3. Sridhar, S., Guo, J., Jha, S.: Channel Assignment in Multi-Radio Wireless Mesh Networks: A Graph-Theoretic Approach. In: First International Conference on Communication Systems and Networks (COMSNETS), Bangalore, India (2009)
4. Kyasanur, P., Vaidya, N.: Routing and Interface Assignment in Multi-Channel Multi-Interface Wireless Networks. In: Proc. IEEE Conf. Wireless Commun. and Net. Conf., pp. 2051–2056 (2005)
5. Gao, L., Wang, X.: A Game Approach for Multi-Channel Allocation in Multi-Hop Wireless Networks. In: Proc. ACM MobiHoc, pp. 303–312 (2008)
6. Kyasanur, P., Vaidya, N.: Routing and Link-layer Protocols for Multi-Channel Multi-Interface Ad Hoc Wireless Networks. Mobile Comp. and Commun. Rev. 10(1), 31–43 (2006)
7. Pal, A., Nasipuri, A.: JRCA: A joint routing and channel assignment scheme for wireless mesh networks. In: IEEE IPCCC, pp. 1–8 (2011)
8. Pediaditaki, S., Arrieta, P., Marina, M.K.: A Learning-based Approach for Distributed Multi-Radio Channel Allocation in Wireless Mesh Networks. In: Proc. in ICNP, pp. 31–41 (2009)
9. Ko, B., Misra, V., Padhye, J., Rubenstein, D.: Distributed Channel Assignment in Multi-Radio 802.11 Mesh Networks. In: Proc. IEEE WCNC, pp. 3978–3983 (2007)
10. Marina, M.K., Das, S.R., Subramanian, A.P.: A topology control approach for utilizing multiple channels in multi-radio wireless mesh networks. Computer Networks 54, 241–256 (2010)
11. Si, W., Selvakennedy, S., Zomaya, A.Y.: An overview of channel assignment methods for multi-radio multi-channel wireless mesh networks. Journal of Parallel and Distributed Computing 70(5), 505–524 (2010)
12. Sayyad, A., Shojafar, M., Delkhah, Z., Ahamadi, A.: Region Directed diffusion in Sensor Network Using Learning Automata: RDDLA. Journal of Advances in Computer Research, 71–83 (2011)
13. Sayyad, A., Ahmadi, A., Shojafar, M., Meybodi, M.R.: Improvement Multiplicity of Routs in Directed Diffusion by Learning Automata New Approach in Directed Diffusion. In: International Conference on Computer Technology and Development, pp. 195–200 (2010)
14. Sayyad, A., Shojafar, M., Delkhah, Z., Meybodi, M.R.: Improving Directed Diffusion in sensor network using learning automata: DDLA new approach in Directed Diffusion. In: IEEE ICCTD 2010, pp. 189–194 (2010)
15. Sridhar, S., Guo, J., Jha, S.: Channel assignment in multi-radio wireless mesh networks - A graph-theoretic approach. In: Communication Systems and Networks and Workshops, COMSNETS 2009 (2009)
16. Omranpour, H., Ebadzadeh, M., Barzegar, S., Shojafar, M.: Distributed coloring of the graph edges. In: Proc. IEEE Int. Conf. on Cybernetic Intelligent Systems (CIS 2008), pp. 1–5 (2008)
17. Das, A.K., Alazemi, H.M.K., Vijayakumar, R., Roy, S.: Optimization Models for Fixed Channel Assignment in Wireless Mesh Networks with Multiple Radios. In: SECON, pp. 463–474 (2005)
18. Pirzada, A.A., Portmann, M., Indulska, J.: Evaluation of multi-radio extensions to AODV for wireless mesh networks. In: Proceedings of the 4th ACM International Workshop on Mobility Management and Wireless Access, pp. 45–51 (2006)
19. Thathachar, M.A.L., Sastry, P.S.: Varieties of learning automata: An overview. IEEE Transactions on Systems, Man, and Cybernetics 32 (2002)
20. http://www.opnet.com/

Binary Matrix Pseudo-division and Its Applications

Aleš Keprt

Department of Computer Science and Applied Mathematics
Moravian University College Olomouc, Czech Republic
Ales.Keprt@mvso.cz

Abstract. The benefit of each key algorithm also depends on many additional supporting algorithms it uses. It turned out that class of problems related to dimensionality reduction of binary spaces used for statistical analysis of binary data, e.g. binary (Boolean) factor analysis, is dependent on the possibility and ability of performing pseudo-division of binary matrices. The paper presents novelty computation approach to it, giving an algorithm for reasonably fast exact solution.

Keywords: Binary Factor Analysis, BFA, matrix, division, factor.

1 Introduction

The benefit of each key algorithm depends also on many additional supporting algorithms it uses. It turned out that class of problems related to dimensionality reduction of binary spaces used for statistical analysis of binary data is dependent on the possibility and ability of performing pseudo-division of binary matrices. The algorithm presented in this paper is a result of research in the field of Binary Factor Analysis (BFA), which is a statistical analysis method used to express binary matrix data by hidden factors, i.e. as a product of two much smaller binary matrices, thus reducing the dimensionality of problem space. BFA can also be used for compression, or feature and similarity based searching in binary data.

Binary spaces are the analogy of classic linear (vector) spaces, but they are nonlinear by their nature, so it isn't possible to use linear algebra to study or work with them. The goal of BFA is to find the hidden relationships among binary representations of objects and their attributes. The process of binary factorization is somewhat similar to a clustering of attributes, i.e. a binary factor is something like a cluster. The difference is that binary factors can arbitrarily overlap, and inclusion of an attribute in a binary factor has no influence of the likelihood of its inclusion in other binary factors. In contrast, the clustering of attributes usually assigns each one of them to a single cluster, and inclusion of an attribute in a particular cluster also reduce the likelihood that the same attribute will be included in another cluster. (Read more on BFA in [1],[2],[3],[6].)

BFA is a pure binary method using Boolean algebra, i.e. a set of operations expressible by a combination of Boolean sum, product and negation (or triplet of computer

A. Abraham et al. (eds.), *Innovations in Bio-inspired Computing and Applications*,
Advances in Intelligent Systems and Computing 237,
DOI: 10.1007/978-3-319-01781-5_15, © Springer International Publishing Switzerland 2014

operations and, or, not). It is nonlinear, that's why it can't use classic factorization methods based on linear algebra.

Solving of BFA problems is computationally intensive. It consists of large number of operations on binary numbers, binary vectors and binary matrices. All the former BFA algorithms tried to avoid the pseudo-division of binary matrices at any cost, as it is known to be the most expensive one. A novelty computational approach to the problem of pseudo-division is presented here.

2 Binary Matrix Pseudo-division

Binary matrix pseudo-division is a generic operation on binary matrices. In BFA it is used to evaluate a candidate solution of matrix decomposition into Boolean product of two smaller matrices. The ability of fast pseudo-division opens the possibility of using algorithms like genetic ones where the evaluation of candidate solutions is very important.

The term pseudo-division is also well known from a classical linear algebra. In there the division is based on the orthogonality (and orthonormality) of vectors. There is no orthogonality in binary spaces, so we can't use classical division algorithms either.

2.1 Notation

The following text is related to binary algebra. All matrices are binary, i.e. contain just 0s and 1s. All the matrix, vector, and scalar operations are pure binary (Boolean). Operators $\otimes$ and $\oplus$ denote Boolean product and sum, and are equivalent to computer operations and and or respectively.

When limited to binary numbers, we can see just one single difference to classic math: $1 \oplus 1 = 1$. See table 1.

Table 1. Two-element Boolean algebra operators

$\odot$	0	1
0	0	0
1	0	1

$\oplus$	0	1
0	0	1
1	1	1

	neg
0	1
1	0

In order to make the formulas and algorithms clearer, we use the simplified notation in place where ambiguity doesn't threaten. Binary matrices are marked upper case (e.g. A), their fields are called bits and marked lower case with row and column indices a_{ij}. Rows (row vectors) are marked lower case with one index (e.g. a_i), column vectors aren't used. All indices are 1-based. We use also classic math operations whenever it's advantageous and clearly understandable. Obviously, the simplified notation is omitted at places where confusion could occur (like the difference between classic sum $\sum$ based on +, and binary sum $\bigvee$ based on binary $\oplus$).

2.2 Problem Definition

Definition 1. (Binary matrix pseudo-division)
Let X be a binary matrix of size $n \times p$ and A be a binary matrix of size $m \times p$. Binary matrix pseudo-division means to find a binary matrix F of type $n \times m$ which approximately fulfils the formula

$$\mathrm{X}_{[n \times p]} \approx \mathrm{F}_{[n \times m]} \odot \mathrm{A}_{[m \times p]} \tag{1}$$

■

We search for such F whose discrepancy d is the lowest possible in respect to formula 1.

$$\hat{\mathrm{X}} = \mathrm{F} \odot \mathrm{A}$$

$$d = \sum_{i=1}^{n} \sum_{j=1}^{p} |\hat{x}_{ij} - x_{ij}|$$

Note that we always search for the first coefficient (multiplicand) while knowing the second one (multiplicator). This is important because the matrix product $\mathrm{F} \odot \mathrm{A}$ is noncommutative.

3 Algorithms

3.1 Algorithm #1: Simple Approach

The simplest algorithm is straightforward: We scan through all possible combinations for F, and find the one with the lowest discrepancy d. In the past, the computation of BFA really used this slow algorithm.

Time complexity of this algorithm is $O(2^{mn})$, which limits its practical usability to small matrices only.

3.2 Algorithm #2: Row-by-Row

Row-by-row computation was firstly presented in the paper [2]. We start with unwinding the matrix product formula $\mathrm{X} = \mathrm{F} \odot \mathrm{A}$.

$$x_{ij} = \bigvee_{k=1}^{m} (f_{ik} \odot a_{kj}) \tag{2}$$

When we express matrices using row vectors $\mathrm{X} = [x_1, \ldots, x_n]$ and $A = [a_1, \ldots, a_m]$, we can transcribe equation 2 this way

$$x_i = \bigvee_{k=1}^{m} (f_{ik} \odot a_k) \tag{3}$$

Similarly, we can express total discrepancy d as a sum of row discrepancies

$$d = \sum_{i=1}^{n} d_i$$

Equation 3 shows that to compute i-th row of matrix X we need just i-th row of F. Obviously, this applies reversely as well: We can search for F row-by-row using just a corresponding single row of X for each single row of F. Thus, we scan through all possible combinations for each particular row f_i and use the one with lowest row discrepancy d_i.

Time complexity of this algorithm is $O(n2^m) = O(2^m)$.

3.3 Algorithm #3: Classification of Candidates

From now on, we will focus to the computation of a single row vector f_i, what is then repeated for each $i \in [1, n]$. The key to success is a suitable interpretation of equation 3. We must understand each row x_i as a binary vector sum of a subset of rows of matrix A. The presence or absence of a particular row a_k in this subset is determined by bit f_{ik}, and we say: "Row a_k is included in x_i iff $f_{ik} = 1$." Our goal is to find the mentioned subset as fast as possible, which gives us the whole wanted row vector f_i.

First of all we count number of bit discrepancies for each row a_k, i.e. number of bits of a_k which generate positive or negative discrepancy when a_k is included in x_i.
Positive discrepancy: $a_{kj} = 0$ a $x_{ij} = 1$
Negative discrepancy: $a_{kj} = 1$ a $x_{ij} = 0$
Row positive and negative discrepancy is computed as the sum of bit discrepancies in the particular row (x_i or a_k).

Especially unwelcome are all negative bit discrepancies; it is a situation where there are 1s where 0s should be instead. These 1s can't be removed by adding more rows of A to our subset, because in Boolean algebra the result of 1 + any number of any values is always 1.

Positive bit discrepancy is not such a big problem, because these *missing 1s* can be supplemented by adding another row of A into the selected subset.

It is also wise to use parallel Boolean instructions, which are natively supported on all microprocessors, even in higher programming languages. They are classic operations like and, or, not and xor working over integer data types (like 32bit int/dword). When the matrices are stored in the memory on row-by-row basis, and successive bits of particular rows are stored on successive places in memory, the computation can be significantly accelerated. (Note that the order or rows in memory isn't important, we need just the bits of each row to be together.) For example common x86 CPUs can directly operate on 32bits at a time. So, the optimum dividing algorithm works with 32bit row vectors. Some bits may be left unused, or we can also use multiple times 32 bits if needed. The source code in pseudo-language uses classical computer notation of row vectors (X[i] means x_i, A[k] means a_k, etc.) and operations on them.

```
for all rows A[k] {
  match = A[k] AND X[i]
  if match = 0 {
    //rows A[k] and X[i] have got no matching bits, so we can
    //simply throw the row A[k] away (don't need to remember)
  } else {
    //match now contains the matching 1s
    //next we are going to compute discrepancy of remaining bits
    goodbits = nzb(match)
    negative_d = nzb(A[k] AND NOT X[i])

    if negative_d = 0 {
      //negative discrepancy is zero→we definitely take this row
      add A[k] to the selected rows subset
      goodsum = goodsum OR A[k]
    } else if goodbits > negative_d {
      //the row generates a negative discrepancy,
      //but it is lower than number of matching 1s,
      //so we are going to keep this row for further processing
      add A[k] to the candidate rows subset
    } else {
      //the row generates a negative discrepancy higher than
      //number of matching bits, so we can simply throw this row
      //away (we don't need to remember the row anymore)
    }
  }
}
```

For a given i in range $[1, n]$, we set goodbits to null vector and then repeat this piece of code for all k in range $[1, m]$. Figure 1 shows the same steps in UML diagram.

Important part of this code is function nzb, which computes number of nonzero bits in a binary vector. Unfortunately, this operation isn't naturally supported in CPUs. We need to compute nzb as fast as possible, so we prepare a precomputed table (an array) of 256 values for all 8-bit vectors. The nzb function then splits each incoming vector to 8-bit parts and uses the precomputed table. Final return value is the sum of nzb values of splitted 8-bit parts. (We can also use different lengths than 8-bit, it was just here just as an example.)

So, the above algorithm classifies (or separate) all the rows a_k to the three groups:

Selected: These ones are included in the resulting subset.
Discarded: These ones are not included in the resulting subset.
Candidate: These ones undergo further processing.

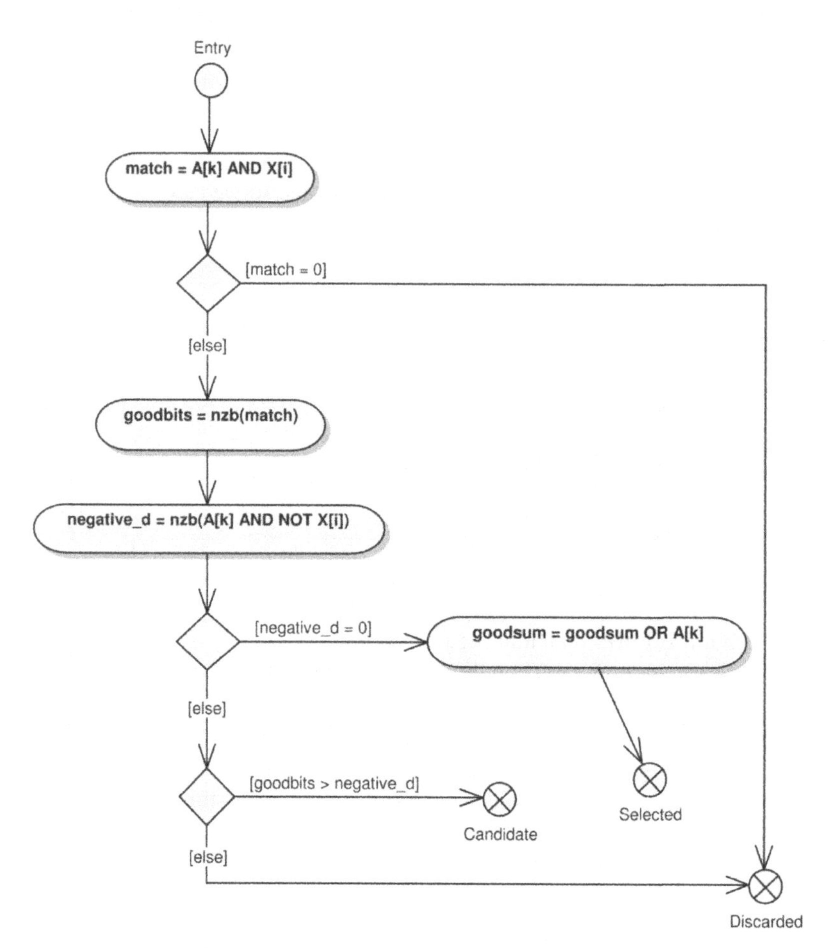

Fig. 1. Classification of candidates

At this point we have got a subset of rows of A generating no negative discrepancy, so their inclusion in the sought subset is safe (case 1 – the selected rows). Now we must subtract the superposition of these rows from x_i, and use the result in further searching. Let's name it rest.

rest = X[i] AND NOT goodsum

Now we can compute first approximation of f_i.

$$f_i = \bigvee (a_j : a_j \text{ is among the selected susbet rows}) \tag{4}$$

We can also rewrite formula 4 to the pseudo language:

```
F[i] = 0
for all selected_A[j] {
   F[i] = F[i] OR selected_A[j]
}
```

Now, if rest is null vector then the computation of f_i is finished, as there are no remaining bits in x_i, and we've got the exact representation of x_i, by the subset of rows of A.

If rest is a nonzero vector, we carry on with the next step. At first we restrict the rows of candidate subset (see case 3 above) to the bits not included in the goodsum.

```
for all candidate_A[j] {
  candidate_A[j] = candidate_A[j] AND rest
}
```

Now we take these restricted candidate rows, and find among them the subset with minimum discrepancy to the rest vector. This must be done by trying and checking all combinations of these vectors (including the empty and the whole set). The algorithm operates with the term "superposed subset", which refers to superposing of all vectors in a subset.

$$S = \bigvee_{j \in s} candidate\_A_j \, , s \in \text{selected subset}$$

```
best_vector = null vector
best_d = nzb(rest)
for all superposed subsets S of candidate_A {
  d = nzb(S XOR rest)
  if d < best_d {
    best_d = d
    best_vector = S
  }
} F[i] = F[i] OR best_vector
```

The second part of the algorithm is also shown in UML diagram at figure 2.

Speed of this last phase of the pseudo-division algorithm depends mainly of the number of candidate rows candidate_A. The experiments revealed that candidate_A is usually a very small subset of the rows, while the vast majority of them fall into groups of selected or discarded. (We will go into more detail later.)

The worst case time complexity of this algorithm is $O(2^m)$, which is identical to algorithm #2. Average case time complexity of this algorithm is much better, and we'll try to do more precise analysis in next section.

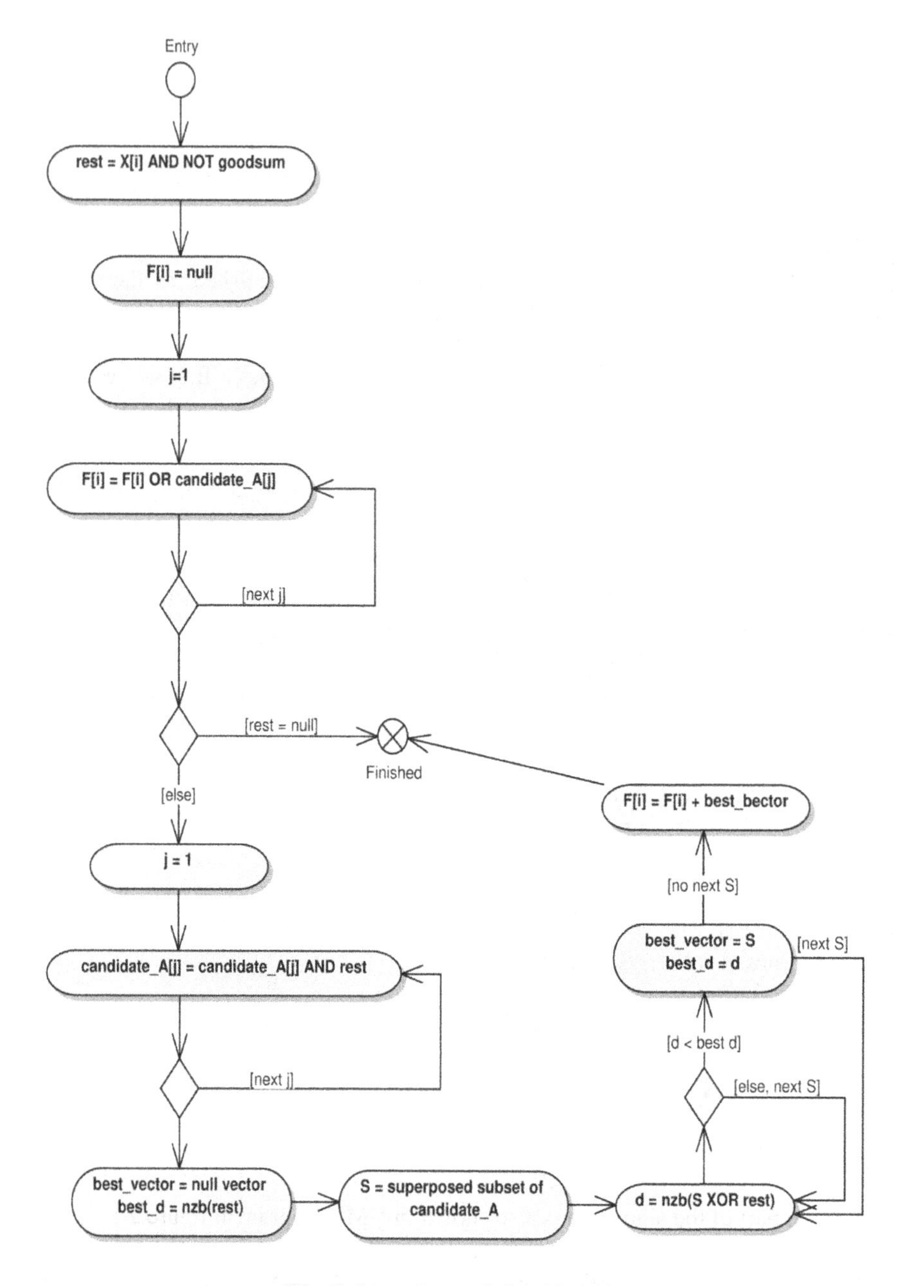

Fig. 2. Second part of algorithm #3

4 Complexity Analysis

Now we'll try to formally express time complexity of the three algorithms presented in the preceding section.

4.1 Algorithm #1: Simple Approach

Ordinary blind search of F is $O(2^{nm} \cdot nmp) \approx O(2^{nm})$, because we have to check out all bit combinations in $O(2^{nm})$, and compute discrepancy value based on the binary matrix product in $O(nmp)$.

4.2 Algorithm #2: Row-by-Row

The row-by-row finding of F is much faster; this algorithm is $O(n \cdot 2^m \cdot p) \approx O(2^m)$, because we check out all subsets of matrix A rows in $O(2^m)$ with comparison to the correspondent data row in $O(p)$, and we do it for all data rows in $O(n)$.

4.3 Algorithm #3: Classification of Candidates

However, the average case speed of the resulting algorithm is even higher. It depends mainly on the number of matrix A rows falling into candidate subset, as only these rows take part in the second part of the algorithm. The speed-up (compared to algorithm #2) is inversely proportional to the number of candidate rows. The effect of number of candidate rows to the resulting computation time and time complexity is strictly nonlinear as it is based on lowering the exponent N in the complexity class expression $O(2^N)$. This relation can hardly be expressed in general terms, so we need to analyze the behavior of the algorithm on real-world data. Table 2 shows the results of experiments revealing the proportion of selected, discarded and candidate rows among all matrix A rows participating in the computation.

The real-world experiments encompassed more than 2 milliards $2 \cdot 10^9$ matrix rows. The effort was to include equivalent amount of computation on three algorithms (genetic algorithm GABFA, concept lattice algorithm FCBFA, and Blind Search algorithm). First of all we can see the uneven number of actually performed divisions, based on particular algorithm. The vast majority of them were done by Blind Search. By contrast, less than 0.1% of divisions fall into FCBFA algorithm.

Candidate rows make 7.55% of the whole. We also need to take the practical importance of particular algorithms into account. Seeing that GABFA is the most often used BFA algorithm, we should take its 11.79% as more authoritative value. This number says that the algorithm #3 is $O(2^{0.1179m})$ in an average case. Theoretically, it's still $O(2^m)$ in the worst case, but there's a huge complexity gap between the algorithms #2 and #3[1], and the gap gets even bigger as the number of factors grows. For example, in the case of 35 factors the computation times change from days to minutes. (Here we compare algorithms #2 and #3.)

Also note that all computations can be made using bit-parallel Boolean instructions, which in praxis (on a classic x86 type CPU) means that 32 units of data are processed at once. This is also the reason why all algorithms are shown in vector-based pseudo code, as it is directly implementable in this form.

---

[1] Compare it to the well-known Quick Sort phenomenon.

Table 2. Pseudo-division experiments: The table shows the proportion of selected, discarded and candidate rows among all matrix A rows participating in the computation of more than $2 \cdot 10^9$ cases

	matrix A rows			
	selected	discarded	candidate	total
Genetic algorithm GABFA	8 224 812 9.71%	66 465 591 78.50%	9 978 397 **11.79%**	84 668 800
FCBFA:	331 068 14.43%	1 918 156 83.58%	45 726 **1.99%**	2 294 950
Blind Search BFA:	10 177 140 0.51%	1 839 111 783 92.12%	147 242 916 **7.37%**	1 996 531 839
all algorithms:	18 733 020 **0.90%**	1 907 495 530 **91.55%**	157 267 039 **7.55%**	2 083 495 589

5 Applications

The binary matrix pseudo-division algorithm was originally devised mainly for BFA computation, because it allows for several new interesting BFA algorithms to be used. While this work is aimed at BFA, the division algorithm itself is generally usable in Boolean binary algebra whenever needed. Let's just briefly summarize where it's used in BFA algorithms described in the recent chapters.

In this section we present a few application of binary matrix pseudo-division algorithm. All of them have one common factor: They are used for dimensionality reduction of binary spaces, extraction of features, and the search of clusters or factors in binary data. The common background of all these applications is due to the fact that the pseudo-division algorithm is a product of research in the field of binary factor analysis, while the presented algorithm is generic and fully usable anywhere where binary spaces and binary matrices are present.

5.1 Blind Search BFA

As mentioned above, the Blind Search BFA algorithm (see [2],[3]) is the most dependent on the division. It goes through all allowed bit combinations, and computes the pseudo-division X/A for each single candidate matrix A. The number of candidate matrices is very high, as seen in table 2.

An example: Many BFA-related papers use data set called p3 to test BFA computation. It's a data matrix of 100×100 bits, and it was formerly considered as a problematic one, as the computation of 5 factors took 7½ days. Now, with using Blind

Search and division algorithm #3, the computation takes only 8 seconds on the same computer.

5.2 Formal Concepts Based BFA

Another algorithm using binary matrix pseudo-division computes BFA using methods formal concept analysis (FCA, see [2],[3],[4]). It uses formal concepts to compile candidate matrices A, and then computes the pseudo-division to determine the quality of a candidate matrix. The experiments revealed (see table 2) that this algorithm needs just a very small number of divisions, compared to Blind Search, and only 1.99% of the rows are candidate ones. This leads to a nice result: Time complexity of pseudo-division is almost polynomial $O(nmp)$, instead of exponential $O(2^m)$. (We just check each particular row of A in $O(p)$, repeat it for all these rows in $O(m)$ and do that for all data rows in $O(n)$.)

5.3 Genetic Algorithm Based BFA

Genetic algorithm GABFA (see [3],[5]) is another example of using pseudo-division. Yet again, it solves the BFA problem and in contrast to above applications, genetic algorithms are randomized so that it is much harder to obtain a clear vision on how good it performs. The tests must be repeated many times to wipe out the effect of randomness. As you can see in table 2, both the share of candidate rows and the total number of divisions varies a lot in particular test runs. While still working with the same data, and still giving the same results, the number of pseudo-division operations was very diverse. The weighted average share of candidate rows was 11.79%, which means the approximate complexity is $O(2^{0.1179m})$.

5.4 Spurious Attractors of Neural Network

The research at Russian Academy of Sciences in last decade leads to the conclusion that, despite of the former assumptions, the dynamics similar to the function of Hopfield neural network are present in human brain (see [7]). One of the surprising results of this research is a modified Hopfield network which solves BFA problem. The main benefit of this algorithm is that more scalable and can process much larger data sets than other methods.

Mentioned neural network in its basic form doesn't use binary matrix pseudo-division. The drawback of the network is the presence of spurious attractors, which are natural element of Hopfield neural networks, and causes the algorithm to fail in some cases, when network's dynamic sometimes end up in a spurious attractor. This state is equivalent to an invalid row in matrix A. The pseudo-division algorithm isn't used by the network itself, but can be used to effectively detect these spurious attractors. If a row of A is invalid, the corresponding column of F is a null vector.

$$a_i \text{ is spurious} \longrightarrow \forall i: f_{ij} = 0$$

6 Summary

This paper presented a new algorithm for pseudo-division of binary matrices. It is a general purpose computational math algorithm, using native bit-parallel features of current microprocessors. Experimental complexity analysis revealed that the worst case time complexity is $O(2^m)$, which is the same as the old algorithm, but the average case time complexity is approximately $O(2^{0.1179m})$ which is a lot better. The paper also summarized the primary applications of this algorithm, in the first place the application in the solving of binary factorization problems, which showed a promising results, e.g. in factor analysis of a test case consisting of 35 factors was former computation time shortened from units of days to units of seconds.

Acknowledgement. This work has been supported by Project GAČR P403-12-1811: Unconventional Managerial Decision making Methods Development in Enterprise Economics and Public Economy.

References

[1] Húsek, D., Frolov, A.A., Řezanková, H., Snášel, V., Keprt, A.: O jednom neuronovém přístupu k redukci dimenze. In: Znalosti 2004, Brno, Czech Republic. VŠB – Technical University of Ostrava, Czech Republic (2004) ISBN 80-248-0456-5

[2] Keprt, A.: Using Blind Search and Formal Concepts for Binary Factor Analysis. In: Dateso 2004 – Proceedings of 4th Annual Workshop, Desná-Černá Říčka, Czech Republic, pp. 120–131. VŠB – Technical University of Ostrava, Czech Republic, CEUR WS – Deutsche Bibliothek, Aachen (2004) ISBN 80-248-0457-3, ISSN 1613-0073

[3] Keprt, A.: Algorithms for Binary Factor Analysis. Doctoral thesis, VŠB Technical University of Ostrava (2006)

[4] Keprt, A., Snášel, V.: Binary Factor Analysis with Help of Formal Concepts. In: Snášel, V., Bělohlávek, R. (eds.) CLA 2004 – Concept Lattices and their Applications, pp. 90–101. VŠB – Technical University of Ostrava, Czech Republic (2004) ISBN 80-248-0597-9

[5] Keprt, A., Snášel, V.: Binary Factor Analysis with Genetic Algorithms. In: Proceedings of 4th IEEE International Workshop on Soft Computing as Transdisciplinary Science and Technology – WSTST 2005, Muroran, Japan. Advances in Soft Computing, pp. 1259–1268. Springer, Heidelberg (2005) ISBN 3-540-25055-7, ISSN 1615-3871

[6] Mickey, M.R., Mundle, P., Engelman, L.: P8M – Boolean Factor Analysis. In: Dixon, W.J. (ed.) BMDP (Bio-Medical Data Processing) Manual, vol. 2. University of California Press, Berkeley (1990) (see also SPSS at `http://www.spss.com/`)

[7] Sirota, A.M., Frolov, A.A., Húsek, D.: Nonlinear Factorization in Sparsely Encoded Hopfield-like Neural Networks. In: ESANN 1999 Proceedings – European Symposium on Artificial Neural Networks, pp. 387–392. D-Factor Public., Bruges (1999) ISBN 2-600049-9-X

Nature Inspired Phenotype Analysis with 3D Model Representation Optimization

Lu Cao and Fons J. Verbeek

Section Imaging & BioInformatics, Leiden Institute of Advanced Computer Science, Leiden University, Leiden, The Netherlands
lucao@liacs.nl, fverbeek@liacs.nl

Abstract. In biology 3D models are made to correspond with true nature so as to use these models for a precise analysis. The visualization of these models helps in the further understanding and conveying of the research questions. Here we use 3D models in gaining understanding on branched structures. To that end we will make use of L-systems and will attempt to use the results of our analysis for the gaining of understanding of these L-systems. To perform our analysis we will have to optimize the 3D models. There are lots of different methods to produce such 3D model. For the study of micro-anatomy, however, the possibilities are limited. In planar sampling, the resolution in the sampling plane is higher than the planes perpendicular to the sampling plane. Consequently, 3D models are under sampled along, at least, one axis. In this paper we present a pipeline for reconstruction of a stack of images. We devised a method to convert the under sampled stack of contours into a uniformly distributed point cloud. The point cloud as a whole is integrated in construction of a surface that accurately represents the shape. In the pipeline the 3D dataset is processed and its quality gradually upgraded so that accurate features can be extracted from under sampled dataset.

The optimized 3D models are used in the analysis of phenotypical differences originating from experimental conditions by extracting related shape features from the model. We use two different sets of 3D models. We investigate the lactiferous duct of newborn mice to gain understanding of environmental directed branching. We consider that the lactiferous duct has an innate blue-print of its arborazation and assume this blueprint is kind of encoded in an innate L-system. We analyze the duct as it is exposed to different environmental conditions and reflect on the effect on the innate L-system. In order to make sure we can extract the branch structure in the right manner we analyze 3D models of the zebrafish embryo; these are simpler compared to the lactiferous duct and will ensure us that measuring features can result in the separation of different treatments on the basis of differences in the phenotype.

Our system can deal with the complex 3D models, the features separate the experimental conditions. The results provide a means to reflect on the manipulation of an L-system through external factors.

Keywords: biological model, 3D reconstruction, phenotype measurement, L-System.

A. Abraham et al. (eds.), *Innovations in Bio-inspired Computing and Applications*, Advances in Intelligent Systems and Computing 237,
DOI: 10.1007/978-3-319-01781-5_16,

1 Introduction

In recent studies [13, 15, 21] three-dimensional (3D) morphological information is used to find the various phenotypes in a sample population. Here the sample population consists of objects in the size range that requires microscope to study them; the differences in the population are found in the micro-anatomy. It is clear that 3D models derived from microscope images potentially offer new insights for analysis. Therefore, 3D images and 3D models are used more frequently over the past years. With image acquisition techniques in bright field microscopy and confocal microscopy it is possible to obtain 3D information from a stack of 2D slices. The sampling is realized by physical sections that are acquired to section images or by optical sections provided by the confocal microscope. The physical section technique is very suitable for modeling histological information whereas the confocal technique is more geared towards imaging specific structures.

From the 3D stack of images a model is derived by segmentation or manual delineation of structures of interest. Manual delineation is used when specific structural knowledge cannot directly be derived from the image; a specialist then selects the specific information through graphical annotation, aka delineation. A set of contours as extracted from the stack subsequently represents the 3D model. The general observation is that the output stack of 2D contours is (nearly) always under sampled perpendicular to the direction of sampling. In order to improve the model, an interpolation can be applied. The classical way of performing such interpolation between section images (slices) is an interpolation of the gray values in the slices so as to estimate the gray values in the missing slices [7]. A linear interpolation for estimating the missing slices, however, may lead to artifacts. A more advanced manner is a shape-based interpolation [7] which is applied directly to the contours of the model.

Techniques of 3D shape reconstruction from consecutive contours fall into two categories: contour stitching and volumetric methods. Contour stitching methods directly connect the vertices of adjacent contours and produce a mesh that passes through all contours such as the methods provided by Keppel [10] and Boissonnat [3]. The volumetric methods first interpolate intermediate gray-values and extract the iso-surfaces from a volumetric field. Representative methods are described by Levin [11], Barrett et al. [2] and Nilsson et al. [16]. Contour based reconstruction methods rely on the fact that the data is organized in parallel planes and performs best on contours that are closed. We tackle the problem of surface reconstruction in a more general way using a point cloud based surface reconstruction method. The merit of a point cloud based reconstruction method is that an abstraction from a specific case to a general problem sheds light on the critical aspects of the problem [8]. Our review paper provides an analytical evaluation of point cloud based reconstruction methods in order to find a method that can reconstruct the surface and suppress the noise with the information from three directions (xyz) [5]. The result of the evaluation shows that the Poisson reconstruction method fits the needs best. As a result, we develop a pipeline to firstly create a uniformly distributed point cloud from a set of plane-parallel contours through shape-based interpolation, then produce a precise and smooth surface using Poisson reconstruction method and finally extract important surface features to analyze the model that is derived from sampling under different conditions.

Once we have obtained an optimized model we can reduce it to a formalized system. To that end we investigate the connection from a arbor-like structure to a L-system [19, 20] and reason over the phenotypical effects as a result of different conditions to which the subjects are exposed.The remainder of this paper is structured as follows. In section 2 we introduce our methodology. In section 3 we first verify our method with a case study: the zebrafish embryo and subsequently continue with the analysis of the mouse lactiferous duct. In section 4 we present our conclusions and discuss the results.

2 Methodology for Model Optimization

Our pipeline is divided into several steps as schematically shown in Figure 1: i.e., (1) contour interpolation, (2) 3D surface reconstruction and (3) phenotype analysis. Subsequent to acquisition, the data are organized in a database and, structures apparent in the images are delineated by a contour. This is done either via segmentation or via a manual delineation for each slice using our dedicated annotation software (TDR); for manual delineation a WACOM digitizer tablet (WACOM, Cintiq LCD-tablet) is used. The contour stacks that are the basis of our models are well aligned by procedures prior to the modeling and in case of the confocal images the alignment is an intrinsic quality of the microscope. In our pipeline we use the stack of images with only the contours as input.

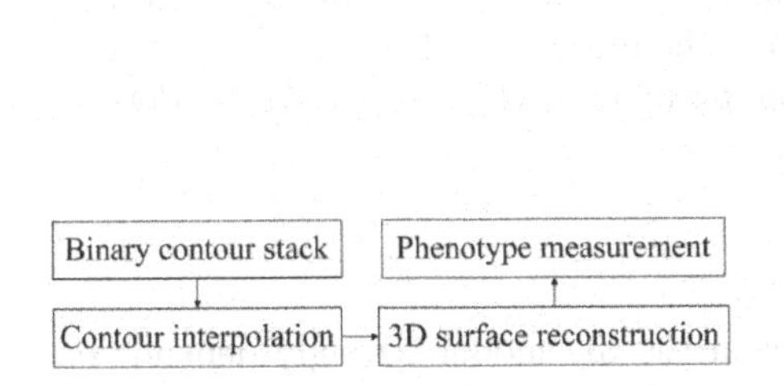

Fig. 1. Pipeline of our system

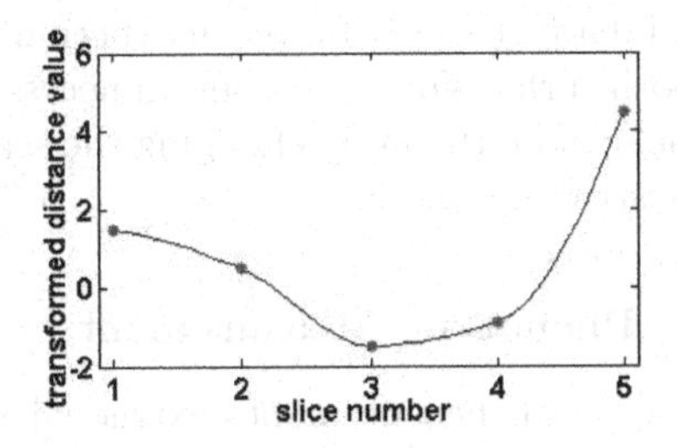

Fig. 2. Piecewise cubic Hermite interpolation

2.1 Contour Interpolation

For interpolation we make use of a shape-based contour interpolation method. To that end a distance transform is applied in the contour; each point in the resulting image represents the shortest distance to the contour. The points on the contour are zero. Now, the distances inside the contour are positive whereas the distances outside the contour are multiplied with -1 and thus these have a negative value. For each pixel-column in the z-direction, a 1D monotone piecewise cubic spline [6] is constructed to interpolate distance values in z-direction [4]. In Figure 2 an example for the construction of a 1D-interpolation in z-direction is shown. The reason to use the monotone piecewise cubic spline is that the output spline preserves the shape of the data and at the same time respects monotonicity. In this manner unwanted overshooting artifacts are eliminated by the method. Once the spline is constructed from each vertical column, the intensity of intermediate missing slices at the same column can be evaluated by providing different position values in the z-direction. Finally, the interpolated contour is extracted by setting the threshold of gray-value to zero for each slice. This approach results in an interpolated and equidistant sampled boundary for the model.

2.2 3D Surface Reconstruction

In order to be able to use the point cloud based reconstruction method, stack of binary contour images needs to be converted into point cloud in 3D space. Each pixel is converted to a point in 3D space taking the corresponding z-position of the slice into consideration. In this manner the point cloud is created without losing any details and at the same time is indisputably oversampled. From an evaluation of point cloud based reconstruction methods [5] that we conducted previously, we concluded that the Poisson reconstruction method [9] performs the best for both shape preservation and noise suppression. The Poisson reconstruction method, however, requires an oriented point cloud as an input. This means the method needs not only the location but also the normal of each point in the point cloud data. Therefore the normal for each point in the point cloud data is calculated using Hoppe's algorithm [8]. With the oriented point cloud data, the Poisson reconstruction method is applied to create a precise and smooth surface for the model. In necessary, the resolution of the resulting surface model can be tuned by changing the scale parameter which is part of the Poisson reconstruction method.

2.3 Phenotype Measurement

The type of measurements extracted from the 3D model strongly depends on the model at hand and the hypothesis posed to resolve phenotypical differences. We extract a range of different features, ones are derived directly from surface including global shape, i.e. volume and surface area, and local features such as surface curvature per point. Other features are extracted from a graph, i.e. skeleton and centerline. These features encode geometrical and topological shape properties in a faithful and intuitive manner [1]. The centerline is useful to describe the topology information of tubular structures such as blood vessels. We notice the L-systems are potentially embedded arbor-like lactiferous duct system. However, the morphological variation of the duct system is regulated by the hormones alteration. With these observations, we conduct the phenotype measurement by extracting the centerline from this arbor-like system. The centerline is a special case of the skeleton which have been studied extensively [12].

The skeleton is the basic representation of the L-system structure. Therefore, the features calculated from the centerline are representative for the arbor-like structure description. These features include number of branches, average branch-length, number of bifurcations, and so on. We would make use of these features for our phenotype analysis.

3 Result

The result of surface reconstruction method is used for the shape analysis. We will illustrate this with two different data sets; one from zebrafish and one from mouse development.

3.1 Case Study 1: Zebrafish Embryo, Measurement Verification

First we want to illustrate that the surface reconstruction process can be very well applied to a stack of images. As a test data set, confocal image stacks of zebrafish embryos are used. The images are part of an experiment that consists of two treatments and therefore two different groups; i.e., a control group that develops under normal oxygen levels and a treatment group that develops under a hypoxia condition. For each stack a 3D model of the embryo is constructed by extracting embryo contour per slice. In total the set consists of 21 embryo models (14 with normal treatment, 7 with hypoxia treatment). In Figure 3(a) and (b) the embryo models with different treatment are shown. For each model we calculated the surface area and the volume. The result from the point-cloud reconstruction is a triangulated surface and the surface area is computed by integration over all the triangle patches on the surface. The volume is calculated directly from the mesh using a known method [25]. For an objective assessment of the differences between two treatments, we use sphericity [24] as shape descriptor, described as follows:

$$\Psi = \frac{\pi^{\frac{1}{3}}(6V)^{\frac{2}{3}}}{A} \tag{1}$$

where V represents the volume of the object and A represents the surface area of the object. The sphericity is a normalized shape descriptor and for a sphere it equals 1. The higher the sphericity of the object the more it resembles a perfect sphere. Sphericity is computed for all our models and thus mean and standard deviation of the set are available. The results are listed in Figure 3(c). To test for differences, the Kolmogorov-Smirnov Test (KS-test) is used [14]. The test results indicate that sphericity of the control group is significantly smaller than that of hypoxia group. This is consistent with the biology; the hypothesis states that the embryo under normal level of oxygen is developing much faster than the one with hypoxia. As a result the embryo develops uneven in different directions and cannot resemble the spherical shape that it started from. The development of embryo with hypoxia, however, is restrained by the lack of oxygen and the shape remains spherical for a longer time.

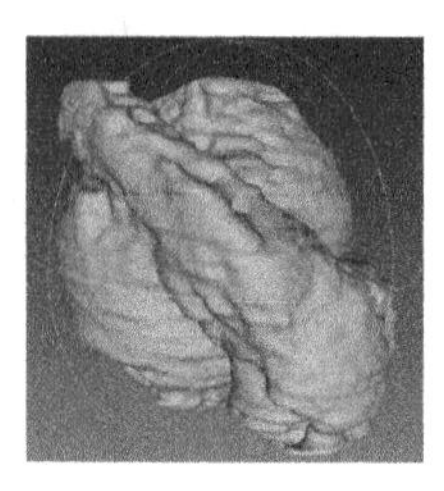
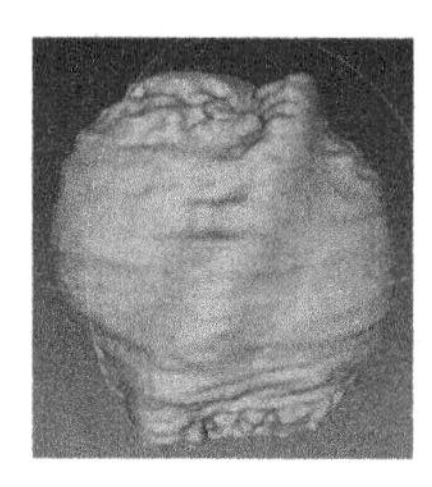
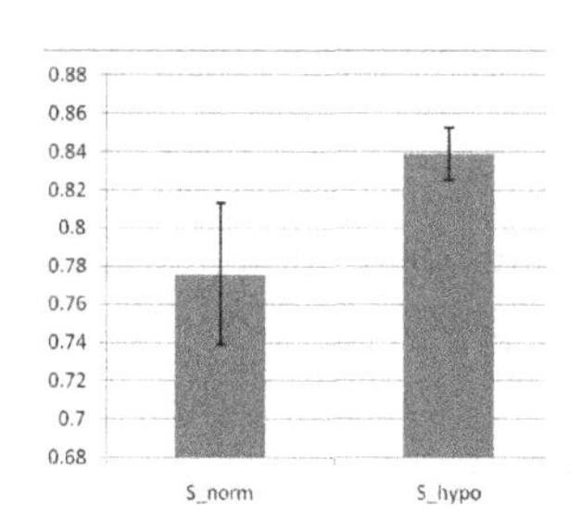

Fig. 3. Results of Zebrafish embryo (a) Normal condition (b) Hypoxia condition (c) Sphericity plot

3.2 Case Study 2: Mouse Lactiferous Duct, Phenotype Analysis

For our main focus, we shift to topological shape description. We have obtained experimental data from the development of the lactiferous duct; in early development a mouse embryo is exposed to different environmental factors. These factors are hormones or hormone-like substances, a.k.a. endocrine disruptors, and these factors will alter the structure of the duct. This means the innate coding of the L-system is altered and it will result in a different structure or phenotype. We measure the phenotype in order to show the effect of the environmental factor and at the same time we can reason from the result how the L-system is affected. Our study is based on 3D model of the lactiferous duct of newborn mice.

The 3D models are obtained from serial sections with a bright field microscope imaging setup. The images of the lactiferous duct are acquired and a stack of aligned images is used as input for the 3D model [22, 23]. For each stack the lactiferous duct structures are delineated by a specialist resulting in initial 3D models. In the data set used here we included 4 different conditions; meaning that the mother was exposed to these conditions and we would like to measure a maternal effect in the offspring. The control group is not exposed (WT) and consists of 7 models. A condition control group as shown in Figure 5(c),(d) is exposed to an inert component (OLIE) and consists of 12 models. One group is exposed to a specific concentration of diethylstilbestrol (DES) and consists of 8 models. And one group is exposed to a cocktail of estrogen and progesterone (EP) shown in Figure 5(a),(b) and consists of 8 models. From the reconstruction it shows that the mouse lactiferous duct has distinct tree-like structure as depicted in Figure 4. To assess the topological shape differences in the models we will derive the centerline of the lactiferous duct structure, subsequently the arborized structure is used in the analysis. To derive the topological structure a known centerline extraction method is used [17]. These centerlines are the basis of the L-system representation.

For all models interpolation (cf. 2.1) and surface reconstruction (cf. 2.2) is applied. Subsequently, the centerlines of all branches in the arbor-like structure are extracted. For each of the models from each global centerline we extract

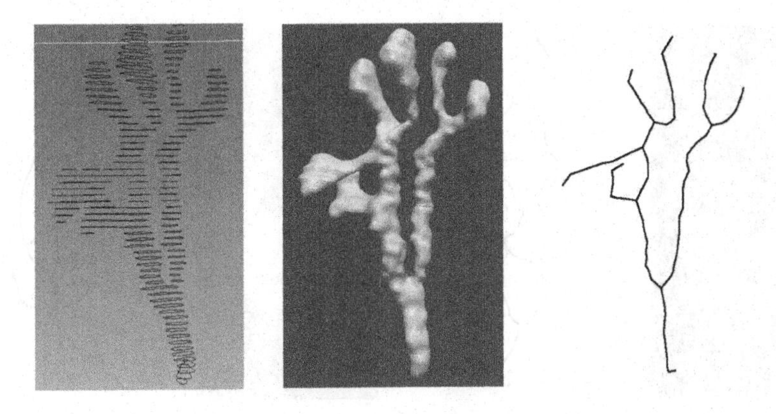

Fig. 4. (a) original contour sections; (b) surface model; (c) centerline

Table 1. Result of comparison

		length	curvature	torsion	tortuosity	min-radius	max-radius	mean-radius	median-radius
DES-EP	h	0	-1	0	0	0	0	0	0
	p	0.2027	0.0098	0.5588	0.8170	0.7951	0.6364	0.8275	0.9412
OLIE-WT	h	0	0	0	0	0	0	0	0
	p	0.2564	0.8537	0.7724	0.9246	0.0416	0.2951	0.1581	0.3943
DE-OW	h	-1	1	0	0	0	-1	-1	-1
	p	0.0043	0.0031	0.2714	0.5649	0.1007	4.99E-5	0.0026	0.0040

individual branches. For each branch we calculate phenotype measurements including branch length, curvature, torsion, tortuosity, minimal radius, maximal radius, mean of the radius and median of the radius. The curvature of a branch is defined as the average of curvatures of each of the branch-point. The curvature of a point is the inverse of the radius of the local osculating circle. The torsion is the degree by which the osculating plane rotates along the line [18]. The tortuosity is the ratio between the branch length and the distance of the branch endpoints. The radius is defined as the radius of maximum inscribed sphere for each point.

For analysis we construct a dataset by collecting all the branches with measurements from the same group and use the KS-test to check significant differences for every measurement. The hypothesis is that DES and EP will have an effect on the development of the lactiferous duct; this effect should not be seen in the OLIE and WT group in which development should not be affected at all. Consequently, the DES and EP group should be similar and so are the OLIE and WT group. At the same time, the DES and the EP should be significantly different from the OLIE and WT group. We therefore employ the kstest to compare three groups: one between the groups DES and EP (DES-EP); one between the groups OLIE and WT (OLIE-WT); the other one between the combined groups

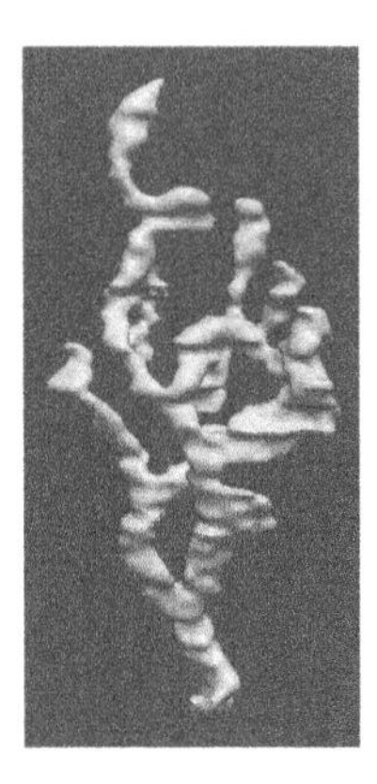
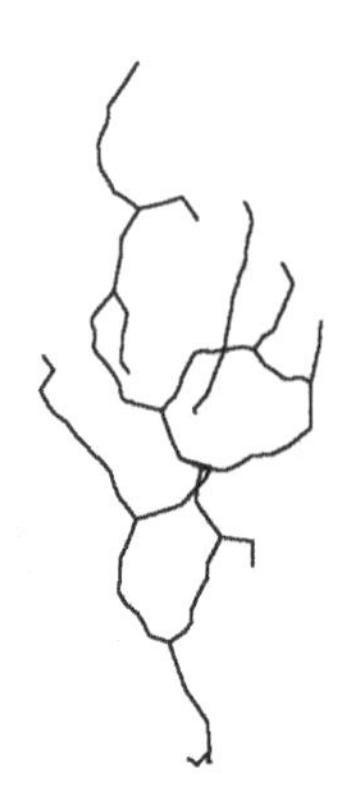

Fig. 5. Different branch structures of lactiferous duct (a) surface model from group EP; (b) centerline of the model from group EP; (c) surface model from group OLIE; (d) centerline of the model from group OLIE

DES with EP and OLIE with WT (DE-OW). In Table 1 the results are shown. The p-value refers to the significance level. Value h is 1 if the first dataset is significantly larger than the second. Value h is -1 if the first dataset is significantly smaller than the second dataset. Otherwise, value h is 0. The significance level was set to $p = 0.01$. From the results we can see group OLIE and WT are exactly the same. Group DES and group EP are almost the same except for curvature. Nevertheless, the combined group DES with EP and group OLIE with WT are significantly different in length, curvature, maximal radius, mean of the radius and median of the radius. These features seem to be prominent features over torsion, tortuosity and minimal radius. The results accept the hypothesis and also confirm the feasibility of our 3D analysis for such complex system.

4 Discussion and Conclusion

In this paper, we introduce a system for 3D representation and analysis from stack of images through a 3D model representation that is derived from that stack. The 3D model needs to be optimized and to that end a point cloud based reconstruction method is used. In this manner the stack of contours in the model is efficiently converted into a uniformly distributed point cloud in 3D space. As such it changes a specific contour based reconstruction problem to a much more generalized point cloud based reconstruction method. Since we take all the 3D points at once to construct the surface representation it makes full use of the boundary point cloud relationship in 3D space. In this way we overcome the restriction imposed by the stack of contours which does not efficiently use the relationship in the z-direction. From our former results we have learned that the Poisson reconstruction method performs well in both shape preserving and noise suppression. This is again confirmed in this current study.

Our purpose is representation and analysis of 3D biological models. The analysis pipeline presented here combines all best techniques and brings us close to realization. . The pipeline is applicable to different kinds of dataset that originate from different microscopes and sampling conditions. The complexity that one finds in micro-anatomy can be well covered by the pipeline. Henceforward, we will continue to develop this pipeline and include more techniques in order to be able to deal with the large variation of experimental settings and imaging typical for bio-medical research. The potential of our system in dealing with shape related studies in biology should be explored further.

The 3D models were used for different purposes. The zebrafish embryo models have been used to obtain information on the efficiency of our basic measurements. The lactiferous duct models were used to extract more complex features. In order to accommodate the branching structure that we analyze in a formal structure we invoke the L-system. This nature inspired formal system can help in the understanding of the branching; in fact, we consider an L-system to embody the blue-print of the normal development of the branching duct. Our experimental results illustrate the change of the L-system under environmental conditions and the features we have derived represent the change in phenotype. At the same time the features represent the mutations in the L-system. Thus, this analysis of induced phenotypes helps us in bringing new inspiration to the manipulation of the L-system in which the features provide a quantifiable twist to this system. Our research will be directed in this delicate interplay of nature inspired systems and nature driven models.

Acknowledgements. We would like to thank Dr F. Lemmen, J. Korving and W. Hage for their contribution in producing the initial 3D models. This work has been partially supported by the Chinese Scholarship Council and BSIK (Cyttron Project).

References

1. Akgül, C.B., Sankur, B., Yemez, Y., Schmitt, F.: 3D Model Retrieval Using Probability Density-Based Shape Descriptors. IEEE Trans. Pattern Anal. Mach. Intell. 31(6), 1117–1133 (2009)
2. Barrett, W., Mortensen, E., Taylor, D.: An Image Space Algorithm for Morphological Contour Interpolation. In: Proc. Graphics Interface, pp. 16–24 (1994)
3. Boissonnat, J.-D.: Shape reconstruction from planar cross sections. Comput. Vision Graph. Image Process. 44(1), 1–29 (1988)
4. Braude, I., Marker, J., Museth, K., Nissanov, J., Breen, D.: Contour-based surface reconstruction using implicit curve fitting, and distance field filtering and interpolation. In: Proc. International Workshop on Volume Graphics, pp. 95–102 (2006)
5. Cao, L., Verbeek, F.J.: Evaluation of algorithms for point cloud surface reconstruction through the analysis of shape parameters. In: Proceedings SPIE: 3D Image Processing (3DIP) and Applications 2012, pp. 82900G–82900G–10. SPIE Bellingham, WA (2012)
6. Fritsch, F.N., Carlson, R.E.: Monotone Piecewise Cubic Interpolation. SIAM Journal on Numerical Analysis 17(2), 238–246 (1980)

7. Herman, G.T., Zheng, J., Bucholtz, C.A.: Shape-Based Interpolation. IEEE Comput. Graph. Appl. 12(3), 69–79 (1992)
8. Hoppe, H., DeRose, T., Duchamp, T., McDonald, J., Stuetzle, W.: Surface reconstruction from unorganized points. SIGGRAPH Comput. Graph. 26(2), 71–78 (1992)
9. Kazhdan, M., Bolitho, M., Hoppe, H.: Poisson surface reconstruction. In: SGP 2006: Proceedings of the Fourth Eurographics Symposium on Geometry Processing, Aire-la-Ville, Switzerland, Switzerland, pp. 61–70. Eurographics Association (2006)
10. Keppel, E.: Approximating complex surfaces by triangulation of contour lines. IBM J. Res. Dev. 19(1), 2–11 (1975)
11. Levin, D.: Multidimen. In: Algorithms for Approximation, pp. 421–431. Clarendon Press, New York (1987)
12. Liu, J., Subramanian, K.: Accurate and robust centerline extraction from tubular structures in medical images 251, 139–162 (2009)
13. Long, F., Zhou, J., Peng, H.: Visualization and analysis of 3D microscopic images. PLoS Computational Biology 8(6), e1002519 (2012)
14. Massey, F.J.: The {K}olmogorov-{S}mirnov Test for Goodness of Fit. Journal of the American Statistical Association 46(253), 68–78 (1951)
15. Ng, A., Brock, K.K., Sharpe, M.B., Moseley, J.L., Craig, T., Hodgson, D.C.: Individualized 3D reconstruction of normal tissue dose for patients with long-term follow-up: a step toward understanding dose risk for late toxicity. International Journal of Radiation Oncology, Biology, Physics 84(4), e557–e563 (2012)
16. Nilsson, O., Breen, D.E., Museth, K.: Surface Reconstruction Via Contour Metamorphosis: An Eulerian Approach With Lagrangian Particle Tracking. In: 16th IEEE Visualization Conference (VIS 2005), October 23-28. IEEE Computer Society, Minneapolis (2005)
17. Piccinelli, M., Veneziani, A., Steinman, D.A., Remuzzi, A., Antiga, L.: A Framework for Geometric Analysis of Vascular Structures: Application to Cerebral Aneurysms. IEEE Trans. Med. Imaging 28(8), 1141–1155 (2009)
18. Pressley, A.: Elementary differential geometry. Springer, London (2010)
19. Rozenberg, G., Salomaa, A. (eds.): The book of L. Springer-Verlag New York, Inc., New York (1986)
20. Rozenberg, G., Salomaa, A.: Mathematical Theory of L Systems. Academic Press, Inc., Orlando (1980)
21. Rübel, O., Weber, G.H., Huang, M.-Y., Wes Bethel, E., Biggin, M.D., Fowlkes, C.C., Luengo Hendriks, C.L., Keränen, S.V.E., Eisen, M.B., Knowles, D.W., Malik, J., Hagen, H., Hamann, B.: Integrating data clustering and visualization for the analysis of 3D gene expression data. IEEE/ACM Transactions on Computational Biology and Bioinformatics/IEEE 7(1), 64–79 (2010)
22. Verbeek, F.J., Huijsmans, D.P.: A graphical database for 3D reconstruction supporting (4) different geometrical representations 465, 117–144 (1998)
23. Verbeek, F.J., Huijsmans, D.P., Baeten, R.J.A.M., Schoutsen, N.J.C., Lamers, W.H.: Design and implementation of a database and program for 3D reconstruction from serial sections: A data-driven approach. Microscopy Research and Technique 30(6), 496–512 (1995)
24. Wadell, H.: Volume, shape, and roundness of quartz particles. The Journal of Geology 43(3), 250–280 (1935)
25. Zhang, C., Chen, T.: Efficient Feature Extraction for 2D/3D Objects in Mesh Representation. In: Mesh Representation", ICIP 2001, pp. 935–938 (2001)

Multi-class SVM Based Classification Approach for Tomato Ripeness*

Esraa Elhariri[1], Nashwa El-Bendary[2], Mohamed Mostafa M. Fouad[2], Jan Platoš[3], Aboul Ella Hassanien[4], and Ahmed M.M. Hussein[5]

[1] Faculty of Computers and Information, Fayoum University, Fayoum - Egypt
Scientific Research Group in Egypt (SRGE)
eng.esraa.elhariri@gmail.com
http://www.egyptscience.net

[2] Arab Academy for Science,Technology, and Maritime Transport, Cairo - Egypt,
Scientific Research Group in Egypt (SRGE)
nashwa.elbendary@ieee.org, mohamed_mostafa@aast.edu
http://www.egyptscience.net

[3] Department of Computer Science, FEECS and IT4 Innovations
VSB-Technical University of Ostrava, Ostrava - Czech Republic
jan.platos@vsb.cz

[4] Faculty of Computers and Information, Cairo University, Cairo - Egypt,
Scientific Research Group in Egypt (SRGE)
aboitcairo@gmail.com
http://www.egyptscience.net

[5] Dept. of Genetics, Faculty of Agriculture, Minia University, Minya - Egypt
ahmed.mm.hussein@gmail.com

Abstract. This article presents a content-based image classification system to monitor the ripeness process of tomato via investigating and classifying the different maturity/ripeness stages. The proposed approach consists of three phases; namely pre-processing, feature extraction, and classification phases. Since tomato surface color is the most important characteristic to observe ripeness, this system uses colored histogram for classifying ripeness stage. It implements Principal Components Analysis (PCA) along with Support Vector Machine (SVM) algorithms for feature extraction and classification of ripeness stages, respectively. The datasets used for experiments were constructed based on real sample images for tomato at different stages, which were collected from a farm at Minia city. Datasets of 175 images and 55 images were used as training and testing datasets, respectively. Training dataset is divided into 5 classes representing the different stages of tomato ripeness. Experimental results showed that the proposed classification approach has obtained ripeness

* This work was partially supported by Grant of SGS No. SP2013/70, VSB - Technical University of Ostrava, Czech Republic., and was supported by the European Regional Development Fund in the IT4Innovations Centre of Excellence project (CZ.1.05/1.1.00/02.0070) and by the Bio-Inspired Methods: research, development and knowledge transfer project, reg. no. CZ.1.07/2.3.00/20.0073 funded by Operational Programme Education for Competitiveness, co-financed by ESF and state budget of the Czech Republic.

A. Abraham et al. (eds.), *Innovations in Bio-inspired Computing and Applications*, Advances in Intelligent Systems and Computing 237,
DOI: 10.1007/978-3-319-01781-5_17, © Springer International Publishing Switzerland 2014

classification accuracy of 92.72%, using SVM linear kernel function with 35 images per class for training.

Keywords: image classification, features extraction, ripeness, principal component analysis (PCA), support vector machine (SVM).

1 Introduction

Product quality is one of the prime factors in ensuring consistent marketing of crops. Determining ripeness stages is a very important issue in produce (fruits and vegetables) industry for getting high quality products. However, up to this day, optimal harvest dates and prediction of storage life are still mainly based on subjective interpretation and practical experience. Hence, automation of this process is a big gain at agriculture and industry fields. For agriculture, it may be used to develop automatic harvest systems and saving crops from damages caused by environmental changes. For industry, it is used to develop automatic sort system or checking the quality of fruits to increase customer satisfaction level [1].

So, an objective and accurate ripeness assessment of agricultural crops is important in ensuring optimum yield of high quality products. Also, identifying physiological and harvest maturity of a agricultural crops correctly, will ensure timely harvest to avoid cutting of either under- and over-ripe agricultural crops. Accordingly, monitoring and controlling ripeness of agricultural crops is a very important issue in produce industry, since the state of ripeness during harvest, storage, and market distribution determines the quality of the final product measured in terms of customer satisfaction [2].

This article presents a multi-class content-based image classification system to monitor the ripeness process of tomato via investigating and classifying the different maturity/ripeness stages based on the color features. The datasets used for experiments were constructed based on real sample images for tomato at different stages, which were collected from a farm at Minia city. Datasets of 175 images and 55 images were used as training and testing datasets, respectively. Training dataset is divided into 5 classes representing the different stages of tomato ripeness. The proposed approach consists of three phases; namely *pre-processing*, *feature extraction*, and *classification* phases. During pre-processing phase, the proposed approach resizes images to 250x250 pixels, in order to reduce their color index, and the background of each image will be removed using background subtraction technique. Also, each image is converted from RGB to HSV color space. For feature extraction phase, Principal Component Analysis (PCA) algorithm is applied in order to generate a feature vector for each image in the dataset. Finally, for classification phase, the proposed approach applied Support Vector Machine (SVM) algorithm classification of ripeness stages.

The rest of this article is organized as follows. Section 2 introduces some resent research work related to fruit ripeness monitoring and classification. Section 3 presents the core concepts of SVM and PCA algorithms. Section 4 describes the different phases of the proposed content-based classification system; namely pre-processing, feature extraction, and classification phases. Section 5 discusses the tested image dataset and presented the obtained experimental results. Finally, Section 6 presents conclusions.

2 Related Work

This section reviews current approaches tackling the problem of fruits/vegetables ripeness monitoring and classification.

In [3], Authors proposed an approach based on spectral images analysis to measure the ripeness of tomatoes for automatic sorting. The proposed approach compared hyper spectral images with standard RGB images for classification of tomatoes ripeness stages. That depends on individual pixels and includes gray reference in each image for obtaining automatic compensation of different light sources. The proposed approach in [3] applied the Linear discriminant analysis (LDA) as a classification technique depending on pixels values and proved that spectral images is better than standard RGB images for measuring ripeness stages of tomatoes via offering more discriminating power.

In [4], authors applied photogrammetric methodology in order to depict a relationship between the color of the palm oil fruits and their ripeness, then sorted them out physically. That approach was considered the first automation of palm oil grading system. Due to difficulty of using the average color digital number values at RGB color space for determining ripeness because of the fusion of the image of palm fruit with dirt and branches. The proposed approach applied the K-means clustering and segmented the fruit fresh bunches (FFB) colors in an automated fashion. Then, to differentiate ripe FFB from unripe fruits, the computed color value to R/G and R/B ratios of the digital number of the segmented images was used. On the other hand, in [2], authors proposed an assessment approach for ensuring optimum yield of high quality oil in order to overcome subjectivity and inconsistency of manual human grading techniques based on experience. Palm ripeness stages can be classified into under ripe, ripe and overripe depending on different color intensity. The developed approach is an automated ripeness assessment using RGB and fuzzy logic feature extraction and classification model to assess the ripeness of oil palm. It depends on color intensity. This approach achieved an efficiency of 88.74%.

Moreover, a neural network with image processing approach for color recognition is designed for identifying the ripe of banana fruit in [5]. The proposed system depends on RGB color components of the captured images of banana. It used four sets of bananas used with different type of sizes and ripeness. Each image of the banana is captured in four different positions and the images are captured daily until all bananas turn to be rotten. This research used supervised Neural Network model with utilizing the error back propagation model. It achieved an identification accuracy of 96%.

Also, an artificial neural network with image processing approach is designed in [6] to measure and determine the ripeness and quality of watermelon depending on its colors in YCbCr Color Space. It measures the ripeness by checking textured founded on the skin of watermelon, which was classified into ripe index from the segmented image depending on the amount of pixels at each region. This approach achieved an accuracy of 86.51%.

3 Preliminaries

3.1 Principal Component Analysis (PCA)

Principal component analysis is a statistical common technique, which is widely used in image recognition and compression for a dimensionality reduction, data representation and features extraction tool as it ensures better classification [7–10]. It basically reduces the dimensionality by avoiding redundant information, and reducing samples features space to features sub-space (smaller space which contains all independent variables which are needed to describe the data) by discarding all un-effective minor components. So, it's necessary to perform various pre-processing steps in order to utilize the PCA method for feature extraction. PCA algorithm can be performed by applying the following steps [8]:

- Step 1: Calculate the mean of all features.
- Step 2: Subtract the mean from each observation and calculate the covariance matrix C.
- Step 3: Calculate the eigenvectors and eigenvalues of the covariance matrix.
- Step 4: Rearrange the eigenvectors and eigenvalues and select a subset as a basis vectors.
- Step 5: Project the data.

3.2 Color Features

A widely used feature in image retrieval and image classification problems is the color, which is as well an important feature for image representation [11]. In this research two color descriptors will be used; namely *colormoment* and *colorhistogram*. Moreover, the first three color moments, which are mean, standard deviation, and skewness [11, 12], have been proved to be efficient and effective way for representing color distribution in any image.

Mean, standard deviation, and skewness for a colored image of size $N \times M$ pixels are defined by the following equations.

$$\bar{x}_i = \frac{\sum_{j=1}^{M \cdot N} x_{ij}}{M \cdot N} \tag{1}$$

$$\partial_i = \sqrt{\frac{1}{M \cdot N} \sum_{j=1}^{M.N} (x_{ij} - \bar{x}_i)^2} \tag{2}$$

$$S_i = \sqrt[3]{\frac{1}{M \cdot N} \sum_{j=1}^{M.N} (x_{ij} - \bar{x}_i)^2} \tag{3}$$

where x_{ij} is the value of image pixel j of color channel i (e.g RGB, HSV and etc..), $\bar{x}_i$ is the mean for each channel i=(H,S and V), ∂_i is the standard deviation, and S_i is the skewness for each channel [11, 12]. On the other hand, colored histogram is a color descriptor that shows representation of the distribution of colors in an image. It represents the number of pixels that have colors in each range of colors [11].

3.3 Support Vector Machine (SVM)

The Support Vector Machine (SVM) is a Machine Learning (ML) algorithm that is used for classification and regression of high dimensional datasets with great results [13–15]. SVM solves the classification problem via trying to find an optimal separating hyperplane between classes. it depends on the training cases which are placed on the edge of class descriptor this is called support vectors, any other cases are discarded [16–18].

SVM algorithm seeks to maximize the margin around a hyperplane that separates a positive class from a negative class [13–15]. Given a training dataset with n samples $(x_1, y_1), (x_2, y_2), \ldots, (x_n, y_n)$, where x_i is a feature vector in a v-dimensional feature space and with labels $y_i \in -1, 1$ belonging to either of two linearly separable classes C_1 and C_2. Geometrically, the SVM modeling algorithm finds an optimal hyperplane with the maximal margin to separate two classes, which requires to solve the optimization problem, as shown in equations (4) and (5).

$$maximize \sum_{i=1}^{n} \alpha_i - \frac{1}{2} \sum_{i,j=1}^{n} \alpha_i \alpha_j y_i y_j . K(x_i, x_j) \quad (4)$$

$$Subject - to : \sum_{i=1}^{n} \alpha_i y_i, 0 \leq \alpha_i \leq C \quad (5)$$

where, α_i is the weight assigned to the training sample x_i. If $\alpha_i > 0$, x_i is called a support vector. C is a regulation parameter used to trade-off the training accuracy and the model complexity so that a superior generalization capability can be achieved. K is a kernel function, which is used to measure the similarity between two samples. Different choices of kernel functions have been proposed and extensively used in the past and the most popular are the gaussian radial basis function (RBF), polynomial of a given degree, linear, and multi layer perceptron MLP. These kernels are in general used, independently of the problem, for both discrete and continuous data.

4 The Proposed Classification System

The proposed ripeness classification system consists of three phases; namely *preprocessing*, *feature extraction*, and *classification* phases. Figure 1 describes the general structure of the proposed approach.

The datasets used for experiments were constructed based on real sample images for tomato at different ripeness stages, which were collected from a farm at Minia city. Collected datasets contains colored JPEG images of resolution $3664x2748$ pixels that were captured using Kodak C1013 digital camera of 10.3 megapixels resolution. Datasets of 175 images and 55 images were used as training and testing datasets, respectively. Training dataset is divided into 5 classes representing the different stages of tomato ripeness.

4.1 Pre-processing Phase

During pre-processing phase, the proposed approach resizes images to 250x250 pixels, in order to reduce their color index, and the background of each image will be removed using background subtraction technique. Also, each image is converted from RGB to HSV color space as it is widely used in the field of color vision and close to the categories of human color perception [19].

4.2 Feature Extraction Phase

As previously stated, since tomato surface color is the most important characteristic to asset the ripeness, this system uses HSV colored histogram and color moments for ripeness stages classification. For this phase, Principal Component Analysis (PCA) algorithm is applied as features extraction technique in order to generate a feature vector for each image in the dataset.

It transforms the input space into sub-spaces for dimensionality reduction, after completing the previous 1D 16x4x4 HSV histogram, 16 level for hue and 4 level for each of saturation and value. In addition, nine color moments, three for each channel (H, S and V channels) (mean, standard deviation, and skewness), will be computed. Then, a feature vector will be formed as a combination of HSV 1D histogram and the nine color moments.

4.3 Classification Phase

Finally, for classification phase, the proposed approach applied SVM algorithm for classification of ripeness stages. The inputs are training dataset feature vectors and their corresponding classes, whereas the outputs are the ripeness stage of each image in the testing dataset.

Since SVM is a binary class classification method and our problem is an N-class classification problem, so in this research the SVM algorithm is applied to Multi-class problem [20]. We used one-against-all approach to do that. That approach worked as follows:

1. Construct N binary SVM.
2. Each SVM separates one class from the rest classes.
3. Train the $i^t h$ SVM with all training samples of the $i^t h$ class with positive labels, and training samples of other classes with negative labels.

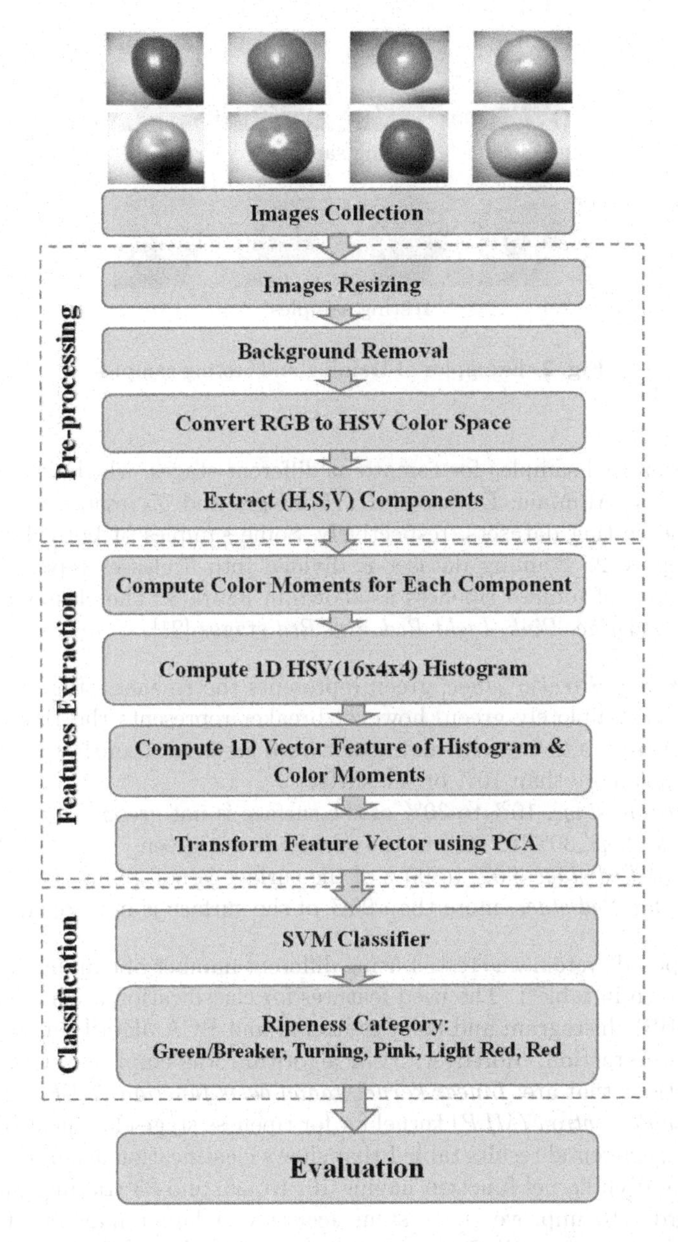

Fig. 1. Architecture of the proposed ripeness classification approach

5 Experimental Results

Simulation experiments in this article are done on a PC with Intel corei7 Q720 @ 1.60 GHZ CPU and 6GB memory. Our approach is designed with Matlab on the operation system Windows 7. The datasets used at this research were

Fig. 2. Examples of training and testing samples

prepared from real samples for tomato at different stages, which were collected from a farm at Al-minia. Datasets of 175 images and 55 images were used as training and testing datasets, respectively. Some samples of both datasets are shown in figure 2. Training dataset is divided into 5 classes representing the different stages of tomato ripeness as shown in figure 3. The classes are *Green & Breaker*, *Turning*, *Pink*, *Light Red*, and *Red* stages [21].

- For *Green & Breaker stage*, green represents the ripeness stage where fruit surface is completely green, however breaker represents the ripeness stage where there is a definite break in color from green to tannish-yellow, pink or red on not more than 10% of the surface.
- For *Turning stage*, 10% to 30% of the surface is not green.
- For *Pink stage*, 30% to 60% of the surface is not green.
- For *Light Red stage*, 60% to 90% of the surface is not green.
- Finally, for *Red stage*, more than 90% of the surface is not green.

The proposed system was tested using different number of training images per class, as shown in table 1. The used features for classification are a combination of colored HSV histogram and color moments and PCA algorithm was applied for features extraction. Moreover, SVM algorithm was employed with different kernel functions that are: *Linear kernel*, *radial basis function (RBF)* kernel, and *Multi-Layer Perceptron (MLP)* kernel for for ripeness stage classification. Figure 4 depicts experimental results table I that shows classification accuracy obtained via applying each kernel function having 10, 15, 25, and 35 training images per class. In order to improve the system accuracy, a larger number of training images per class were used. The results of the proposed PCA-SVM classification system are evaluated against human expert assessment for measuring obtained accuracy. As shown in figure 4, with the linear kernel function being used for SVM algorithm and the number of training images per class is 35, the proposed classification system achieved 92.72 % accuracy for all ripeness stages.

$$Accuracy = \frac{Numberofcorrectlyclassifiedimages}{Totalnumberoftestingimages} \tag{6}$$

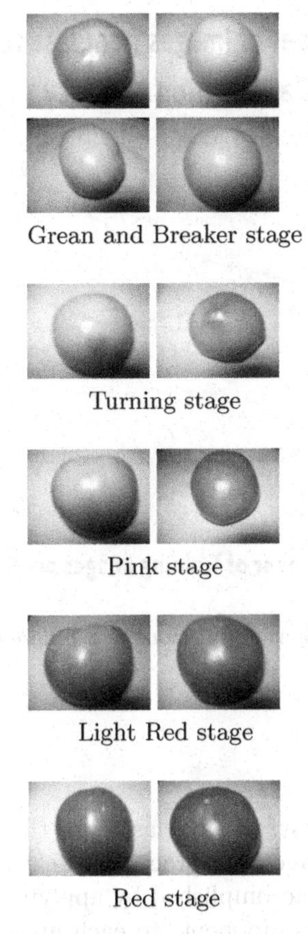

Fig. 3. Examples of tomato ripeness stages

Table 1. Classification accuracy for different kernel functions

Number of Training Images per Class	Kernel Function	Accuracy
10	**Linear**	56.36%
	RBF	45.45%
	MLP	41.18%
15	**Linear**	63.64%
	RBF	47.27%
	MLP	41.18%
25	**Linear**	72.72%
	RBF	41.18%
	MLP	41.18%
35	**Linear**	92.72%
	RBF	52.72%
	MLP	41.18%

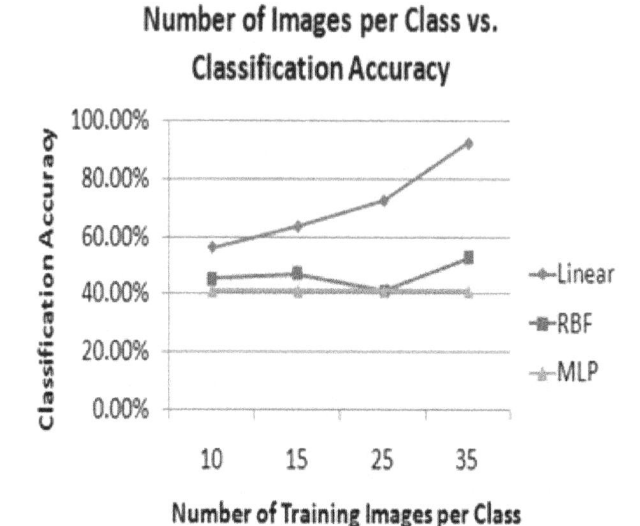

Fig. 4. Results for different kernel functions and different sizes for training class

6 Conclusions

In this article, we developed a system for classifying the ripeness stages of tomato. The system has three main stages; pre-processing, feature extraction and ripeness classification. The work was accomplished by applying resizing, background removal, and extracting color components to each image then feature extraction is applied to each pre-processed image, HSV histogram and color moments are obtained as a feature vector and used as a PCA inputs for transformation. finally SVM model is developed for ripeness stage classification. Based on the obtained the results, the ripeness classification accuracy is 92.72% for using SVM linear kernel function and 35 images per class for training.

References

1. Brezmes, J., Llobet, E., Vilanova, X., Saiz, G., Correig, X.: Fruit ripeness monitoring using an electronic nose. Sensors and Actuators B-Chem. Journal 69(3), 223–229 (2000)
2. May, Z., Amaran, M.H.: Automated ripeness assessment of oil palm fruit using RGB and fuzzy logic technique. In: Demiralp, M., Bojkovic, Z., Repanovici, A. (eds.) Proc. the 13th WSEAS International Conference on Mathematical and Computational Methods in Science and Engineering (MACMESE 2011), Wisconsin, USA, pp. 52–59 (2011)

3. Polder, G., van der Heijden, G.W.A.M., Young, I.T.: Spectral Image Analysis for Measuring Ripeness of Tomatoes. Transactions-American Society of Agricultural Engineers International Journal 45(4), 1155–1162 (2002)
4. Jaffar, A., Jaafar, R., Jamil, N., Low, C.Y., Abdullah, B.: Photogrammetric Grading of Oil Palm Fresh Fruit Bunches. International Journal of Mechanical & Mechatronics Engineering (IJMME) 9(10), 18–24 (2009)
5. Paulraj, M.P., Hema, C.R., Krishnan, R.P., Radzi, S.S.M.: Color Recognition Algorithm using a Neural Network Model in Determining the Ripeness of a Banana. In: Proc. the International Conference on Man-Machine Systems (ICoMMS), Penang, Malaysia, pp. 2B7-1–2B7-4 (2009)
6. Rizam, S., YAsmin, A.R.F., Ihsan, M.Y.A., Shazana, K.: Non-destructive Watermelon Ripeness Determination Using Image Processing and Artificial Neural Network (ANN). International Journal of Intelligent Technology 4(2), 130–134 (2009)
7. Suganthy, M., Ramamoorthy, P.: Principal Component Analysis Based Feature Extraction, Morphological Edge Detection and Localization for Fast Iris Recognition. Journal of Computer Science 8(9), 1428–1433 (2012)
8. Ada, RajneetKaur: Feature Extraction and Principal Component Analysis for Lung Cancer Detection in CT scan Images. International Journal of Advanced Research in Computer Science and Software Engineering 3(3) (2013)
9. El-Bendary, N., Zawbaa, H.M., Hassanien, A.E., Snasel, V.: PCA-based Home Videos Annotation System. The International Journal of Reasoning-based Intelligent Systems (IJRIS) 3(2), 71–79 (2011)
10. Xiao, B.: Principal component analysis for feature extraction of image sequence. In: Proc. International Conference on Computer and Communication Technologies in Agriculture Engineering (CCTAE), Chengdu, China, vol. 1, pp. 250–253 (2010)
11. Shahbahrami, A., Borodin, D., Juurlink, B.: Comparison between color and texture features for image retrieval. In: Proc. 19th Annual Workshop on Circuits, Systems and Signal Processing (ProRisc 2008), Veldhoven, The Netherlands (2008)
12. Soman, S., Ghorpade, M., Sonone, V., Chavan, S.: Content Based Image Retrieval using Advanced Color and Texture Features. In: Proc. International Conference in Computational Intelligence (ICCIA 2012), New York, USA (2012)
13. Wu, Q., Zhou, D.-X.: Analysis of support vector machine classification. J. Comput. Anal. Appl. 8, 99–119 (2006)
14. Zawbaa, H.M., El-Bendary, N., Hassanien, A.E., Abraham, A.: SVM-based Soccer Video Summarization System. In: Proc. the Third IEEE World Congress on Nature and Biologically Inspired Computing (NaBIC 2011), Salamanca, Spain, pp. 7–11 (2011)
15. Zawbaa, H.M., El-Bendary, N., Hassanien, A.E., Kim, T.-H.: Machine learning-based soccer video summarization system. In: Kim, T.-H., Gelogo, Y. (eds.) MulGraB 2011, Part II. CCIS, vol. 263, pp. 19–28. Springer, Heidelberg (2011)
16. Tzotsos, A., Argialas, D.: A support vector machine approach for object based image analysis. In: Proc. International Conference on Object-based Image Analysis (OBIA 2006), Salzburg, Austria (2006)
17. Zhang, Y., Xie, X., Cheng, T.: Application of PSO and SVM in image classification. In: Proc. 3rd IEEE International Conference on Computer Science and Information Technology (ICCSIT), Chengdu, China, vol. 6, pp. 629–631 (2010)

18. Suralkar, S.R., Karode, A.H., Pawade, P.W.: Texture Image Classification Using Support Vector Machine. International Journal of Computer Applications in Technology 3(1), 71–75 (2012)
19. Yu, H., Li, M., Zhang, H.-J., Feng, J.: Color texture moments for content-based image retrieval. In: Proc. International Conference on Image Processing, New York, USA, vol. 3, pp. 929–932 (2002)
20. Liu, Y., Zheng, Y.F.: One-against-all multi-class SVM classification using reliability measures. In: Proc. IEEE International Joint Conference on Neural Networks (IJCNN 2005), Montreal, Quebec, Canada, vol. 2, pp. 849–854 (2005)
21. U.S.D.A. United States Standards for Grades of Fresh Tomatoes, U.S. Dept. Agric./AMS, Washington, DC (1991), http://www.ams.usda.gov/standards/vegfm.htm (accessed: March, 2013); Nose, Sensors and Actuators B 2000 69, 223–229; Signals and fruit quality indicators on shelf-life measurements with pinklady apples. Sensors and Actuators B 2001 80, 41–50

Solving Stochastic Vehicle Routing Problem with Real Simultaneous Pickup and Delivery Using Differential Evolution

Eshetie Berhan[1], Pavel Krömer[2,3],
Daniel Kitaw[1], Ajith Abraham[3], and Václav Snášel[2,3]

[1] Addis Ababa University
Addis Ababa Institute of Technology
School of Mechanical and Industrial Engineering
Addis Ababa, Ethiopia
{eshetie_ethio,danielkitaw}@yahoo.com
[2] Faculty of Electrical Engineering and Computer Science
VŠB Technical University of Ostrava
Ostrava, Czech Republic
[3] IT4 Innovations
VŠB Technical University of Ostrava
Ostrava, Czech Republic
{pavel.kromer,vaclav.snasel}@vsb.cz

Abstract. In this study, Stochastic VRP with Real Simultaneous Pickup and Delivery (SVRPSPD) is attempted the first time and fitted to a public transportation system in Anbessa City Bus Service Enterprise (ACBSE), Addis Ababa, Ethiopia. It is modeled and fitted with real data obtained from the enterprise. Due to its complexity, large instances of VRP and/or SVRPSPD are hard to solve using exact methods. Instead, various heuristic and metaheuristic algorithms are used to find feasible VRP solutions. In this work the Differential Evolution (DE) is used to optimize bus routes of ACBSE. The findings of the study shows that, DE algorithm is stable and able to reduce the estimated number of vehicles significantly. As compared to the traditional and exact algorithms it has exhibited better used fitness function.

Keywords: vehicle routing problem, pickup and delivery, machine learning, differential evolution, real-world application.

1 Introduction

The problem of designing a minimum cost set of routes to serve a collection of customers with a fleet of vehicles is a fundamental challenge in the field of logistics, distribution and transportation [1]. This is because transportation and distribution contribute approximately 20% to the total costs of a product [2]. The task of designing delivery or pickup routes to service customers in the transport and supply chain is known in the literature as a Vehicle Routing Problem [1]. It

A. Abraham et al. (eds.), *Innovations in Bio-inspired Computing and Applications*, Advances in Intelligent Systems and Computing 237,
DOI: 10.1007/978-3-319-01781-5_18,

was the first time proposed by [3] under the title "Truck dispatching problem" with the objective to design optimum routing of a fleet of gasoline delivery trucks between a bulk terminal and a large number of service stations supplied by the terminal. Often the context is that of delivering goods located at a central depot to customers who have placed orders for such goods, but the area of application of VRP is also versatile and are used in many areas in real world life.

This paper is trying to develop a VRP model that addresses the stochastic nature of passengers' pickup and delivery services in Anbessa City Bus Service Enterprise (ACBSE), Addis Ababa, Ethiopia. ACBSE is a urban public bus transport enterprise which provides a public transport service in the city of Addis Ababa, Ethiopia. The model developed is called Stochastic VRP with Real Simultaneous Pickup and Delivery (SVRPSPD). It is the first of its kind in the literature of VRP due to the fact that it considered stochastic pickup and delivery of commuters', and both the pickup and delivery services are also performed simultaneously during the transportation services. The SVRPSPD model is simulated and solved using Differential Evolution with real data collected from ACBSE.

2 Literature Review

In general, VRPs are represented in a graph theory. The general model is defined as: let $G = (V, A)$ be a directed or asymmetric graph where $V = \{0, ...n\}$ is a set of vertices representing *cities* with *depot* located at vertex 0, and A is the set of arcs. With every arc $(i, j), i \neq j$ is associated a non-negative distance matrix $C = (c_{ij})$. In some context, c_{ij} can be interpreted as a *travel cost* or as a *travel time* [4]. When C is symmetrical, it is often convenient to replace A by a set E of undirected edges, the graph being called symmetric or undirected graph. In many practical cases, the cost or the distance matrix satisfies the triangular inequality such that $c_{ik} + c_{kj} \geq c_{ij}, \forall_{i,j,k} \in V$ [5]. The general or classical VRP consists of designing a set of at most K delivery or collection routes such that each route starts and ends at the depot, each customer is visited exactly once by exactly one vehicle, the total demand of each route does not exceed the vehicle capacity and the total routing cost is minimized [5].

Moreover, because of the complexity and the practical relevance of VRPs, vast literature is devoted to the Bus Scheduling Problem (BSP) and many optimization models have been proposed [6]. The models tried to achieve several near optimal solutions with a reasonable amount of computational effort [1]. Various extensions for the Vehicle Schedule Problem (VSP) or VRP with different additional requirements were also covered in the literature over the last fifty years [5]. Among others the existence of one depot [7] or more than one depot [8], a heterogeneous fleet with multiple vehicle types [7] the permission of variable departure times of trips, VRP with Stochastic Demand (VRPSD) [1,9,10] is also one of the major variants of VRP in the literature and the VRP with incapacitated vehicle [11,12,10] are the very few examples in the VRP literature.

The basic version of VRP, Stochastic VRP (SVRP) and/or VRP with stochastic Customer Demand (VRPSD), which are stated above are either a pure pickup or a pure delivery problems [13,14]. The pure pickup or pure delivery types of VRP has been extensively studied in the literature with many application areas [15].

The VRP with delivery and pickup (VRPDP) is also a well studied in the literature but with the assumption of deterministic demand or with modifications of as classical VRP models separately for the pickup and for the delivery [16]. As it is evident in [17] and the surveying of [18], the problem can be divided into two independent CVRPs [16]; one for the delivery (linehaul) customers and one for the pickup (backhaul) customers, such that some vehicles would be designated to linehaul customers and others to backhaul customers.

A similar models and approach was also studied VRPSDP by [19] with consideration of first delivery and then followed by pickup service but named as simultaneous delivery and pickup VRP problem [20,21,22]. This assumption is more clearly illustrated by [18] and [19] with a symbol representation and mathematical model. The model illustrated first with symbols ▼ as delivery and second with symbols ▲ as pickup along each route [19]. Where as in the work of [18,20,21,23] it considered P as a set of backhauls or pickup vertices, $P = \{1, \ldots, n\}$ and D as a set of linehauls or delivery vertices, $D = \{n+1, \ldots, n+\tilde{n}\}$. As it can be seen from this consideration, the assumption is that, in VRPSPD all delivered goods must be originated from the depot and served to n nodes ($\{1, \ldots, n\}$) and all pickup goods must be transported back to the depot $\{n+1, \ldots, n+\tilde{n}\}$ or even viceversa.

A real simultaneous delivery and pick up with deterministic demand was noted on the work of [24,25,26] bust with deterministic demand. In addition to this drawback, previous works on VRPSDP has also limitation on the consideration of demand. In other work of the same authors [27], studied simultaneous pickup and delivery service but deliveries are supplied from a single depot at the beginning of the service followed by pickup loads to be taken to the same depot at the conclusion of the service.

2.1 Differential Evolution

The DE is a versatile and easy to use stochastic evolutionary optimization algorithm [28]. It is a population-based optimizer that evolves a population of real encoded vectors representing the solutions to given problem. The DE was introduced by Storn and Price in 1995 [29,30] and it quickly became a popular alternative to the more traditional types of evolutionary algorithms. It evolves a population of candidate solutions by iterative modification of candidate solutions by the application of the differential mutation and crossover [28]. In each iteration, so called trial vectors are created from current population by the differential mutation and further modified by various types of crossover operator. At the end, the trial vectors compete with existing candidate solutions for survival in the population.

The DE Algorithm. The DE starts with an initial population of N real-valued vectors. The vectors are initialized with real values either randomly or so, that they are evenly spread over the problem space. The latter initialization leads to better results of the optimization [28].

During the optimization, the DE generates new vectors that are scaled perturbations of existing population vectors. The algorithm perturbs selected base vectors with the scaled difference of two (or more) other population vectors in order to produce the trial vectors. The trial vectors compete with members of the current population with the same index called the target vectors. If a trial vector represents a better solution than the corresponding target vector, it takes its place in the population [28].

There are two most significant parameters of the DE [28]. The scaling factor $F \in [0, \infty]$ controls the rate at which the population evolves and the crossover probability $C \in [0, 1]$ determines the ratio of bits that are transferred to the trial vector from its opponent. The size of the population and the choice of operators are another important parameters of the optimization process.

The basic operations of the classic DE can be summarized using the following formulas [28]: the random initialization of the ith vector with N parameters is defined by

$$x_i[j] = rand(b_j^L, b_j^U), \quad j \in \{0, \dots, N-1\} \tag{1}$$

where b_j^L is the lower bound of jth parameter, b_j^U is the upper bound of jth parameter and $rand(a, b)$ is a function generating a random number from the range $[a, b]$. A simple form of the differential mutation is given by

$$v_i^t = v_{r1} + F(v_{r2} - v_{r3}) \tag{2}$$

where F is the scaling factor and v_{r1}, v_{r2} and v_{r3} are three random vectors from the population. The vector v_{r1} is the base vector, v_{r2} and v_{r3} are the difference vectors, and the ith vector in the population is the target vector. It is required that $i \neq r1 \neq r2 \neq r3$. The uniform crossover that combines the target vector with the trial vector is given by

$$l = rand(0, N-1) \tag{3}$$

$$v_i^t[m] = \begin{cases} v_i^t[m] & \text{if } (rand(0,1) < C) \text{ or } m = l \\ x_i[m] \end{cases} \tag{4}$$

for each $m \in \{1, \dots, N\}$. The uniform crossover replaces with probability $1 - C$ the parameters in v_i^t by the parameters from the target vector x_i.

There are also many other modifications to the classic DE. Mostly, they differ in the implementation of particular DE steps such as the initialization strategy, the vector selection, the type of differential mutation, the recombination operator, and control parameter selection and usage [28].

Recent Applications of DE to the Vehicle Routing Problem. Large VRP instances are due to the NP-hardness of the problem hard to solve using exact methods. Instead, various heuristic and metaheuristic algorithms are employed to find approximate VRP solutions in reasonable time [31,32]. A categorized bibliography of different metaheuristic methods applied to VRP variants can be found in [32].

The DE has proved to be an excellent method for both, continuous and discrete optimization problems. This section provides a short overview of recent applications of DE to different variants of the VRP published in 2012. Hou et al. introduced in [33] a new discrete differential evolution algorithm for stochastic vehicle routing problems with simultaneous pickups and deliveries. The proposed algorithm used natural (integer) encoding with the symbol 0 as sub-route separator and fitness function incorporating routing objective and constraints. Besides the traditional DE operators, new bitwise mutation was proposed. The algorithm also utilized an additional *revise* operator to eliminate illegal chromosomes that might have been created during the evolution. The experiments conducted by the authors have shown that the proposed algorithm delivers better solutions and converges faster than other DE-based and GA-based VRP solvers.

Liu et al. [34] used a memetic differential evolution algorithm to solve vehicle routing problem with time windows. The algorithm used a real-valued source space and discrete solution space. A source vector was translated into an solution vector by modifications of the source vector (e.g. insertion of sub-route separator '0' in feasible locations) and optimized by three local search algorithms. The fitness of the best routing found by the local searches was called generalized fitness of the source vector. The experiments performed by the authors have shown that the proposed modifications improve the quality of solutions found by the DE and that the new algorithm is especially suitable for solving VRP instances with clustered locations.

Xu and Wen [35] used differential evolution for unidirectional logistics distribution vehicle routing problem with no time windows. The authors approached the task as an multi-objective optimization problem (although none of the traditional multi-objective DE variants was used) and established an encoding scheme that mapped the real-valued candidate vector to a routing of K vehicles.

3 Model Formulation

The presented model is the first of its kind considering VRP with real simultaneous pickup and delivery at each bus stop where the pickup and delivery demand at each bus stop is treated as stochastic and random. The presented model assumes:

1. The number of passengers expected to be picked up and dropped is uncertain, but follows a poisson probability distribution.
2. The cumulative number of passengers picked up along the route will not exceed the vehicle capacity. Moreover, split delivery is not allowed.
3. The fleet consists of homogeneous vehicles with limited capacity operating from a single depot.
4. Each vehicle can be used repeatedly within the planning horizon.

The mathematical model for SVRP which considered a simultaneous pickup and delivery techniques is formulated as a mixed integer LP problem. Consider a fleet of K vehicles with $k = \{1, 2, \ldots, K\}$ with identical vehicle capacity of Q serving a set of passengers with demand (to be picked or dropped) in passenger's location $V \backslash \{0\}$; $v_0 =$ depot, each passenger must be completely served by a single vehicle. Each vehicle starts from the depot v_0 and picks and/or drops passengers on each visited node v_i except the depot. Moreover, the first node v_1 is treated as pickup-only node and the last node before the depot v_n is treated as drop-only node.

Suppose a vehicle starts from the depot v_0 and travels along a certain path until it reaches node v_n. Along the path $(v_1, v_2, \ldots, v_n)$, the vehicle will pick and/or drop passengers up to the last node v_n. The cumulative number of passengers picked by vehicle k denoted as C_p and the cumulative number of passengers dropped along the path v_n $(v_1, v_2, \ldots, v_n)$ denoted as C_d are given by:

$$C_p(V_n) = \sum_{i \in V(0, V_n)} p_i \tag{5}$$

$$C_d(V_n) = \sum_{i \in V(0, V_n)} d_i \tag{6}$$

where p_i is the number of passengers picked up at node i and d_i is the number of passengers dropped at the same node i. At the depot, $C_p = C_d = 0$ and the vehicle capacity equals to Q. The path becomes infeasible if the cumulative load exceeds the vehicle capacity Q, that is when: $C_d \geq Q$ and $C_p \geq Q$.

Each feasible route will be formed when: $C_d(v_n) \leq Q$ and $C_d(v_{n+1}) \geq Q$ and $C_p(v_n) \leq Q$ and $C_p(v_{n+1}) \geq Q$. The other consideration checks whether the net load of a bus for any consecutive nodes will not exceed the bus capacity after the bus visiting node v_n. Let the net load picked is $L_p(v_n)$ it is computed as $L_p(v_n) = C_p(v_n) + L_p(v_{n-1}) - C_d(v_n)$.

The solution will be feasible if a vehicle served all the demand (for pick up or drop off) at each node along the path or route without exceeding its capacity. That is the net load in transit between any consecutive nodes or vertices should not exceed the vehicle capacity $(L_p(v_n) \leq Q)$. The objective is to determine a route for each vehicles that serve a set of nodes (v_i) so that the total distance traveled is minimized. To formulate the model of SVRP with simultaneous pickup and delivery as Mixed Integer LP problem, the following notations and definitions are used:

V is set of nodes or vertices $V = \{v_0, v_1, v_2, \ldots, v_n\}$ treated as bus stops; $v_0 =$ depot
A is set of arcs $(i, j) \in A$
K is number of vehicles $k = \{1, 2, \ldots, K\}$
c_{ij} is the distance traversing from node i to node j
p_i is the number of passengers (demand) to be picked at node i, which is a random non-negative integer
d_i is the number of passengers (serviced) to be dropped at node i which is a random non-negative integer

Q is the vehicle capacity
n is total number of nodes or vertices or bus stops included in the model

Decision Variables

y_i^k is the cumulative number of passengers picked by vehicle k when leaving from node i.
z_i^k is the number of passengers remaining in vehicle k when leaving from node i.
$x_{ij}^k = 1$ if vehicle k travels from node i to node j; 0 otherwise

Then the general model is represented as follows:

$$\text{Minimize} \sum_{k=1}^{K}\sum_{i=0}^{n}\sum_{j=0}^{n} c_{ij}x_{ij}^k \tag{7}$$

Subject to

$$\sum_{j=1}^{n} x_{0j}^k \leq 1; \tag{8}$$

$$\sum_{i=0}^{n} x_{ij}^k = 1; \tag{9}$$

$$\sum_{i=0}^{n} x_{ij}^k - \sum_{i=0}^{n} x_{ji}^k = 0; \tag{10}$$

$$z_i^k + y_i^k \leq Q; \tag{11}$$

$$(z_i^k - d_j - z_j^k)x_{ij}^k = 0; \tag{12}$$

$$(y_i^k + p_j - y_j^k)x_{ij}^k = 0; \tag{13}$$

$$z_0^k = y_0^k = 0; \tag{14}$$

$$z_i^k \geq 0; \tag{15}$$

$$y_i^k \geq 0; \tag{16}$$

$$x_{ij}^k \in \{0, 1\}; \tag{17}$$

$$\forall_k = 1, \ldots, K \quad and \quad i, j = 1, \ldots, n \tag{18}$$

The decision variables are given above and equation 7 is the objective function to be minimized. Equation 8 ensures that each vehicle is used at most once, equation 9 indicates that each node has to be visited exactly by one vehicle, equation 10 shows that the same vehicle arrives and departs from each node it serves, 11 ensures that the load on vehicle k when departing from node i is always less than the vehicle capacity. Equation (12) and eq. (13) are the transit load constraint, which indicate that when arc (i, j) is traversed by vehicle k, the number of passengers to be dropped by the vehicle has to be decreased by d_j while the number of passengers picked-up has to be increased by p_j. 14 ensures that the remaining and the cumulative number of passengers when a vehicle k departs from the depot is always zero; indicates that the vehicle is empty and available with full capacity. Constraints 15 and 16 are a non-integer and non-negative sign restriction and the last equation 17 is an restriction on the non-negative integer value.

4 Model Input Parameters

To run and evaluate the model, different input parameters that have to be substituted to the model are required. These inputs are either collected or generated/computed. The from-to-distance, the demand realization probability, the demand distributions are computed or generated whereas the longitude and latitude value for each location point v_i is collected from the Google Earth. Each of them are briefly explained and presented in section 4.1 and 4.2

4.1 From-to-Distance

Origin-destination of 59 locational points that depart and end from Merkato Terminal (lat. 9°1'50"N and long. 38°44'15"E), are used in the model validation. The origin-destination points are modified in such a way that they can fit to the model but without losing its information. The from-to-distance computational input parameters of each location i is computed by taking the longitude and latitude location of each point using Great Circle distance formula that considers the circular nature of earth.

Table 1. From-to-distance matrix (C_{ij})

v_{ij}	1	2	3	4	5	6	7	8	.	.	.	57	58	59
1	0	8	17	7	7	8	1	1	.	.	.	17	20	19
2	8	0	11	7	2	8	9	7	.	.	.	22	14	23
3	17	11	0	11	13	10	18	15	.	.	.	33	20	21
.	.	.	.	.	.	.	.	.	.	.	.	.	.	.
.	.	.	.	.	.	.	.	.	.	.	.	.	.	.
.	.	.	.	.	.	.	.	.	.	.	.	.	.	.
58	20	14	20	21	14	22	21	20	.	.	.	27	0	37
59	19	23	21	15	24	15	20	18	.	.	.	34	37	0

Each c_{ij} is defined as the distance from i to j, which can be directly considered as the cost associate to transport passenger demand including depot 1. Further, it assumes that the distance is symmetric, $c_{ij} = c_{ji}$, and $c_{ii} = 0$. The sample output is shown in Table 1.

4.2 Stochastic Passengers Demand

The other input parameter is the number of passengers picked up and dropped off at each location point called passengers demand. The passengers' demand collected in 10 routes of ACBSE are used to fit the demand behavior of passengers in the remaining routes. The snap shot of the demand distribution of passengers picked and dropped along the 10 selected routes were used to fit the demand distribution of the remaining location points.

The volumes of passengers picked up at each vertex V are integer-valued random variables with known probability distributions denoted by vector $\widetilde{p}(i) \in Z$[36]. In the demand based modeling individual demands are estimated by using parameter $\gamma = \{0, 1\}$, which determines the risk preferences [14].

Therefore, the demand used for computing the solutions is evaluated using $\hat{p} = \gamma p_i^{min} + (1 - \gamma)p_i^{max}$. Finally, for convenience of notation, it can be defined that $\underline{p}(0) = \overline{p}(0) = 0$. Similarly, $\hat{d}$ is also computed using the same formula. $\gamma = 0$ clearly implies the risk averse (risk free) case where failure can never occur, while $\gamma = 1$ is the other extreme, which is risk seeking. Moreover, $\gamma = 0.5$ corresponds to computing solutions with the expected demand. According to the simulation run, the demand distribution of the number passengers picked up and delivered for some nodal points are given below in Table 2.

Table 2. Sample demand distribution and location data

v_i	Location(decimal)		Expected Passengers	
	Longi.	Lati.	$\hat{p_i}$	$\hat{d_i}$
1	38.8	9.16	na	na
2	38.9	9.1	8	4
3	38.9	8.94	9	3
.	.	.	.	.
.	.	.	.	.
58	39.0	9.13	9	4
59	38.8	9.04	10	3

5 Differential Evolution for SVRPSPD

The version of VRP considered in this work can be seen as a combinatorial optimization problem. The goal of the optimization is to find a set of routes connecting selected locations (bus stops) so that each location is visited by a vehicle exactly once, each route starts and terminates in a special location (depot), considered constraints are satisfied, and selected objective function is minimized. In this work we represent a set of routes as a permutation of considered locations (without the depot) and separate each sub-route by a special sub-route separator similarly as e.g. in [31].

The DE proposed in this work uses permutation-based VRP representation, automatically selects the number of vehicles when an upper bound is given, and avoids the creation of illegal candidate solutions.

Encoding. There is a variety of possible encoding schemes for modelling permutations for populational metaheuristic algorithms [37]. The DE uses real-encoded candidate solutions so a modified version of the random key (RK) encoding [38] was chosen. An RK encoded permutation is represented as a string of real numbers (random keys), whose position changes after sorting correspond to the permutation gene. The advantage of RK encoding is that it is at a large extent prone to creation of illegal solutions in course of the artificial evolution (e.g. by the crossover operator in Genetic Algorithms). The drawbacks of the RK encoding include computational complexity as it is necessary to perform a sorting of the random keys every time the candidate solution is decoded. Moreover, RK encoding translates a discrete combinatorial optimization problem into a real-valued optimization problem with a larger continuous search space.

The routing of a maximum of k buses for n locations (without the depot) is encoded as $\mathbf{x} = (x_1, x_2, \ldots x_{k+n-1}), x_i \in \mathbb{R}$. Routing $\mathcal{R}$ is from the encoded vector $\mathbf{x}$ created according to Algorithm 1. During the decoding process, $k-1$ largest values of $\mathbf{x}$ are interpreted as route separators. The remaining values are used as random keys and translated into permutation of n locations π. The values of π are split into k routes, each of which starts in the depot and terminates in the depot. Empty routes can be created when the vector $\mathbf{x}$ contains two or more route separators next to each other. The set of non-empty routes defines the routing $\mathcal{R}$.

```
1  Sort candidate vector: x^s = sort(x);
2  Use the k − 1 largest values of x as route separators: x_sep = x^s_{n−1};
3  Create key vector k and vector with route sizes s:
4  route_size = 0;
5  for i ∈ {0, . . . , k + n − 1} do
6      if x_i > x_sep then
7          Append route_size to s;
8          route_size = 0;
9      else
10         Append x_i to k;
11         route_size = route_size + 1;
12     end
13 end
14 Translate key vector k to permutation π;
15 index = 0;
16 for i ∈ {0, . . . , k} do
17     if s_i > 0 then
18         for j ∈ {0, . . . , s_i} do
19             Append location π_index to route i;
20             index = index + 1;
21         end
22         Add i to routing R;
23     end
24 end
```

Algorithm 1. Decoding of routing $\mathcal{R}$

Fitness Function. Fitness function used in this work is based on covered distance, number of routes, and penalty for bus capacity violation.

$$fit(\mathcal{R}) = \frac{\sum_{r \in \mathcal{R}} dist(r)}{|\mathcal{R}|} \tag{19}$$

where $dist(r)$ is distance of route r. A penalty is applied (route distance is artificially decreased) when the capacity of the bus is depleted. $|\mathcal{R}|$ represents the number of routes in $\mathcal{R}$.

5.1 Experiments

The proposed algorithm was implemented in C++ and used to optimize routings of the ACBSE company. The DE was executed with population size 100, maximum number of vehicles 20, bus capacity 70, number of generations 1000, and parameters $F = 0.9$ and $C = 0.4$ respectively. The parameter values were set on the basis of initial experiments and algorithm tuning. The optimization was repeated 30 times due to the stochastic nature of the algorithm.

The results of the optimization were: average number of routes 5.433, minimum number of routes 5 and maximum number of routes 7. The fitness values in each generation of the 30 independent runs of the algorithm are shown in fig. 1. The results show that the algorithm is stable and able to reduce the estimated number of vehicles significantly. Moreover, the solutions found in the DE algorithm are better (in terms of used fitness function) than a traditional savings algorithm for the VRP [15].

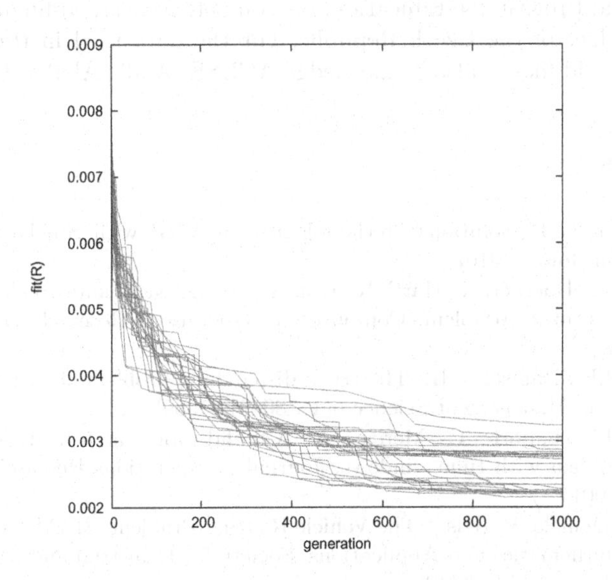

Fig. 1. The evolution of fitness function

6 Conclusions

This work introduced a new variety of VRP, Stochastic VRP with real Simultaneous Pickup and Delivery and used it to describe the situation of a real-world transportation company operating in Addis Ababa, Ethiopia. A new metaheuristic model based on the Differential Evolution was proposed to optimize vehicle routing in ACBSE network. The DE encoded set of tours as a permutation and automatically minimized the number of buses from an initial upper estimate.

In contrast to a number of previous metaheuristic algorithms, all solutions generated by the proposed algorithm are valid routings through the set of bus stops and computational resources were not wasted on processing of invalid solutions. However, solutions that violate constraints such as bus capacity can be still obtained in course of the evolution.

The routing found by the algorithm was compared to a VRP solution obtained by a traditional (savings) VRP algorithm and it was found better with regard to the used fitness function. The results presented in this study are promising and metaheuristic solvers with various permutation-based representations of candidate solutions [37] will be investigated in the future.

Acknowledgement. This work was supported by the European Regional Development Fund in the IT4Innovations Centre of Excellence project (CZ.1.05/1.1.00/02.0070) and by the Bio-Inspired Methods: research, development and knowledge transfer project, reg. no. CZ.1.07/2.3.00/20.0073 funded by Operational Programme Education for Competitiveness, co-financed by ESF and state budget of the Czech Republic. For the data used in this work, the researchers would like to also acknowledge ACBSE, Addis Ababa, Ethiopia.

References

1. Christopher, G.J.: Solutions Methodologies for VRP with Stochastic Demand. Dessirtation, Iowa (2010)
2. Reimann, M., Doerner, K., Hartl, R.: D-ants: Savings based ants divide and conquer the vehicle routing problem. Computers & Operations Research 31(4), 563–591 (2003)
3. Dantzig, G.B., Ramser, J.H.: The truck dispatching problem. Journal of Management Science, Management Science 6(1), 80–91 (1959)
4. Cordeau, J.F., Laporte, G., Mercier, A.: A unifid tabu search heuristic for vehicle routing problem with time windows. Journal of Operations Research Society 53, 928–936 (2001)
5. Paolo, T., Daniele, V. (eds.): The Vehicle Routing Problem. SIAM Monographs on Discrete Mathematics and Applications. Society for Industrial and Applied Mathematics, Philadelphia (2002)
6. Dror, M., Trudeau, P.: Savings by split delivery routing. Transportation Science 23, 141–145 (1989)
7. Calvete, H.I., Carmen, G., María, J.O., Belén, S.-V.: Vehicle routing problems with soft time windows: an optimization based approach. Journal of Monografías del Seminario Matemático García de Galdeano 31, 295–304 (2004)
8. Bertsimas, D.: A vehicle routing problem with stochastic demand. Journal of Operations Research 40(3), 554–585 (1991)
9. Secomandi, N.: A rollout policy for the vehicle routing problem with stochastic demands. Operations Research 49, 796–802 (2001)
10. Laporte, G., Louveaux, F., van Hamme, L.: An integer l-shaped algorithm for the capacitated vehicle routing problem with stochastic demands. Operations Research 50, 415–423 (2002)

11. Gendreau, M., Laporte, G., Seguin, R.: Stochastic vehicle routing. European Journal of Operational Research 88, 3–12 (1996a)
12. Kenyon, A.S., Morton, D.P.: Stochastic vehicle routing with random travel times. Journal of Transportation Science 37(1), 69–82 (2003)
13. Bertsimas, D.: A vehicle routing problem with stochastic demand. Operations Research 40, 574–585 (1992)
14. Reimann, M.: Analyzing a vehicle routing problem with stochastic demand using ant colony optimization. In: EURO Working Group on Transportation (2005)
15. Clarke, G., Wright, J.: Scheduling of vehicles from a central depot to a number of delivery points. Operations Research 12, 568–581 (1964)
16. Ropke, S., Pisinger, D.: An adaptive large neighborhood search heuristic for the pickup and delivery problem with time windows. Transportation Science 40(4), 455–472 (2006)
17. Linong, C.Y., Wan, R.I., Khairuddin, O., Zirour, M.: Vehicle routing problem: Models and solutions. Journal of Quality Measurement and Anlysis 4(1), 205–218 (2008)
18. Parragh, S., Doerner, K., Hartl, R.: A survey on pickup and delivery problems. Journal für Betriebswirtschaft 58, 81–117 (2008), 10.1007/s11301-008-0036-4
19. Kanthavel, K., Prasad, P.S.S., Vignesh, K.P.: Optimization of vehicle routing problem with simultaneous delivery and pickup using nested particle swarm optimization. European Journal of Scientific Research 73(3), 331–337 (2012)
20. Goksal, F.P., Karaoglan, I., Altiparmak, F.: A hybrid discrete particle swarm optimization for vehicle routing problem with simultaneous pickup and delivery. Computers & Industrial Engineering 65(1), 39–53 (2013)
21. Liu, R., Xie, X., Augusto, V., Rodriguez, C.: Heuristic algorithms for a vehicle routing problem with simultaneous delivery and pickup and time windows in home health care. European Journal of Operational Research (2013)
22. Wang, H.F., Chen, Y.Y.: A genetic algorithm for the simultaneous delivery and pickup problems with time window. Computers & Industrial Engineering 62(1), 84–95 (2012)
23. Karaoglan, I., Altiparmak, F., Kara, I., Dengiz, B.: The location-routing problem with simultaneous pickup and delivery: Formulations and a heuristic approach. Omega 40(4), 465–477 (2012)
24. Zhang, T., Chaovalitwongse, W.A., Zhang, Y.: Scatter search for the stochastic travel-time vehicle routing problem with simultaneous pick-ups and deliveries. Computers & Operations Research 39(10), 2277–2290 (2012)
25. Cruz, R., Silva, T., Souza, M., Coelho, V., Mine, M., Martins, A.: Genvns-ts-cl-pr: A heuristic approach for solving the vehicle routing problem with simultaneous pickup and delivery. Electronic Notes in Discrete Mathematics 39, 217–224 (2012)
26. Fermin, A.T., Roberto, D.G.: Vehicle routing problem with simultaneous pick-up and delivery service. Operational Research Society of India (OPSEARCH) 39(1), 19–34 (2002)
27. Fermin, A.T., Roberto, D.G.: A tabu search algorithm for the vehicle routing problem with simultaneous pick-up and delivery service. Operational Research Society of India (OPSEARCH) 33(1), 595–619 (2006)
28. Price, K.V., Storn, R.M., Lampinen, J.A.: Differential Evolution A Practical Approach to Global Optimization. Natural Computing Series. Springer, Berlin (2005)
29. Storn, R., Price, K.: Differential Evolution- A Simple and Efficient Adaptive Scheme for Global Optimization over Continuous Spaces. Technical report (1995)
30. Storn, R.: Differential evolution design of an IIR-filter. In: Proceeding of the IEEE Conference on Evolutionary Computation, ICEC, pp. 268–273. IEEE Press (1996)

31. Alba, E., Dorronsoro, B.: Solving the vehicle routing problem by using cellular genetic algorithms. In: Gottlieb, J., Raidl, G.R. (eds.) EvoCOP 2004. LNCS, vol. 3004, pp. 11–20. Springer, Heidelberg (2004)
32. Gendreau, M., Potvin, J.Y., Bräysy, O., Hasle, G., Løkketangen, A.: Metaheuristics for the vehicle routing problem and its extensions: A categorized bibliography. In: Golden, B., Raghavan, S., Wasil, E. (eds.) The Vehicle Routing Problem: Latest Advances and New Challenges. Operations Research/Computer Science Interfaces, vol. 43, pp. 143–169. Springer, US (2008)
33. Hou, L., Hou, Z., Zhou, H.: Application of a novel discrete differential evolution algorithm to svrp. In: 2012 Fifth International Joint Conference on Computational Sciences and Optimization (CSO), pp. 141–145 (2012)
34. Liu, W., Wang, X., Li, X.: Memetic differential evolution for vehicle routing problem with time windows. In: Tan, Y., Shi, Y., Ji, Z. (eds.) ICSI 2012, Part I. LNCS, vol. 7331, pp. 358–365. Springer, Heidelberg (2012)
35. Xu, H., Wen, J.: Differential evolution algorithm for the optimization of the vehicle routing problem in logistics. In: 2012 Eighth International Conference on Computational Intelligence and Security (CIS), pp. 48–51 (2012)
36. Anbuudayasankar, S., Ganesh, K.: Mixed-integer linear programming for vehicle routing problem with simulatneous delivery and pick-up with maximum route-length. The International Journal of Applied Management and Technology 6(1), 31–52 (2008)
37. Krömer, P., Platoš, J., Snášel, V.: Modeling permutations for genetic algorithms. In: Proceedings of the International Conference of Soft Computing and Pattern Recognition (SoCPaR 2009), pp. 100–105. IEEE Computer Society (2009)
38. Snyder, L.V., Daskin, M.S.: A random-key genetic algorithm for the generalized traveling salesman problem. European Journal of Operational Research 174(1), 38–53 (2006)

An Intelligent Multi-agent Recommender System*

Mahmood A. Mahmood[1], Nashwa El-Bendary[2], Jan Platoš[3],
Aboul Ella Hassanien[4], and Hesham A. Hefny[1]

[1] ISSR, Computer Sciences and Information Dept., Cairo University, Cairo - Egypt,
Scientific Research Group in Egypt (SRGE)
mahmood.moneim@egyptscience.net,
h.hefny@ieee.org
http://www.egyptscience.net

[2] Arab Academy for Science, Technology, and Maritime Transport, Cairo - Egypt,
Scientific Research Group in Egypt (SRGE)
nashwa.elbendary@ieee.org
http://www.egyptscience.net

[3] VSB-Technical University of Ostrava, Department of Computer Science,
Ostrava - Czech Republic
jan.platos@vsb.cz

[4] Information Technology Dept.,
Faculty of Computers and Information, Cairo University,
Cairo - Egypt,
Scientific Research Group in Egypt (SRGE)
aboitcairo@gmail.com
http://www.egyptscience.net

Abstract. This article presents a Multi-Agent approach for handling the problem of recommendation. The proposed system works via two main agents; namely, the matching agent and the recommendation agent. Experimental results showed that the proposed rough mereology based Multi-agent system for solving the recommendation problem is scalable and has possibilities for future modification and adaptability to other problem domains. Moreover, it succeeded in reducing the information overload while recommending relevant decisions to users. The system achieved high accuracy in ranking using users profile and information system profiles. The resulted value of the Mean Absolute Error (MAE) is acceptable compared to other recommender systems applied other computational intelligence approaches.

Keywords: rough mereology, multi-agent, recommender system.

* This work was partially supported by Grant of SGS No. SP2013/70, VSB - Technical University of Ostrava, Czech Republic., and was supported by the European Regional Development Fund in the IT4 Innovations Centre of Excellence project (CZ.1.05/1.1.00/02.0070) and by the Bio-Inspired Methods: research, development and knowledge transfer project, reg. no. CZ.1.07/2.3.00/20.0073 funded by Operational Programme Education for Competitiveness, co-financed by ESF and state budget of the Czech Republic.

A. Abraham et al. (eds.), *Innovations in Bio-inspired Computing and Applications*,
Advances in Intelligent Systems and Computing 237,
DOI: 10.1007/978-3-319-01781-5_19,

1 Introduction

During the last decade, the amount of information available online increased exponentially. Recommender systems addressed the problem of filtering information that is likely of interest to individual users. Building such applications implies to consider two main issues: collecting information from different sources on the Internet, and solving the problem itself. Typically, users profiles are employed to predict ratings for items that have not been considered [1]. Depending on the application domain, items can be web pages, movies, or any other products found on a web store.

Various researches have been proposed for solving complex problems addressed by recommender systems such as the approach presented in [2] for handling the problem of arranging meetings and scheduling travels. In [2], authors have implemented a Multi-Agent recommender system that plans meetings using agenda's information and transportation schedules. Agents communicated with each other using the Constraint Choice Language (CCL), FIPA compliant content language is utilized for modeling problems using constraint satisfaction formalism, and the Java Constraint Library (JCL) was used for solving complex problems in Multi-Agent systems. Also, authors in [3] presented a recommender system using agents with two different algorithms (associative rules and collaborative filtering). Both algorithms are incremental and work with binary data. Results showed that the proposed Multi-Agent approach combining different algorithms is capable of improving user's satisfaction. Moreover, authors in [4] presented a Multi-agent system for solving the training course recommendation problem for reducing the information overload while recommending relevant courses to users. The proposed system achieved high accuracy in ranking using user information and course information. The final system is scalable and has possibilities for future modification and adaptability to other problem domains. Furthermore, authors in [5] presented a TV recommender system, named AVATAR, conceived as a Multimedia Home Platform (MHP) application. For this system, authors have described an open Multi-Agent architecture that easily allows including modules with additional functionalities to enhance the recommendations. The proposed approach improved the previous TV recommendation tools by incorporating reasoning capabilities about the content semantics. For that purpose, the TV-Anytime initiative has been used to describe the TV contents and a TV ontology to share and reuse the knowledge efficiently. To discover semantic associations among different TV contents, the system employed a query language that infers data from the AVATAR knowledge base.

The intelligent multi-agent recommender system proposed in this article works via two main agents; namely, the matching agent and the recommendation agent. Users interact with the system, through the user interface, in order to access information of attributes to the system. The matching agent utilizes rough mereology to search the dataset for similar users who have at least one common attribute, while the recommendation agent proposes a recommendation value for the decision attribute. The proposed Multi-Agent approach aims at providing scalable and adaptable recommender system that can be utilized in various problem domains. Moreover, the proposed recommender system worked on reducing the information overload while recommending relevant decisions to users. The rest of this article is organized as follows. Section 2 presents an overview of the rough mereology approach and recommendation algorithms. Section 3 describes the proposed Multi-Agent recommender system along with its different agents and modules. Section 4 discusses evaluation of the proposed recommender system. Finally, section 5 summarizes conclusions and discusses future work.

2 Background

2.1 Rough Mereology

Rough mereology can be classified according to the measurement of similarity. The similarity functions [6] that satisfy certain similarity properties, namely Monotonic (MON), Identity (ID), Extreme, or proportionality (EXT), can be defined as follows:

- (MON) if similarity (x, y, 1) then for each z, where $d(z,x) > d(z,y)$, from similarity (z, x, r) it follows that similarity (z, y, r).
- (ID) similarity (x, x, 1) for each x.
- (EXT) if similarity (x, y, r) and $s \leq r$ then similarity (x, y, s).

Rough mereology proposed by Lesniewski in [7] as the theory of concept, where the relation of mereology is a part of relation, e.g. x mereology y means x is a part of y, according to Polkowski [8], the mereology relation described in equation (1), where $\pi(u,w)$ is a partial relation (proper part) and $ing(u,w)$ is ingredient relation means an improper part.

$$ing(u,w) \Leftrightarrow \pi(u,w) \ or \ u = w \tag{1}$$

$\mu(x,y,r)$ means rough mereology relation x is part of y at least degree r, also described as shown in equation (2).

$$\mu(x,y,r) = sim_{\delta}(x,y,r) \Leftrightarrow \rho(x,y) \leq (1 - -r) \tag{2}$$

Computing the indiscernibility relation to get the object can be achieved by using rough inclusion, which is of less complexity time than the indiscernibility relation computed by rough set technique. Rough inclusion from metric, according to Polkowski in [8], can be computed by the Euclidean metric space or Manhattan space, where

$$\mu_h(x,y,r) \Leftrightarrow \rho(x,y) \leq 1 - -r$$

Rough inclusion technique satisfies the similarity properties. Datasets are formalized as decision systems of the form of triple (U, A, d) or information system of the form (U, A), where U is a finite set of objects, A is a finite set of attributes, each attribute $a \in A$ described as mapping $a : U \rightarrow V_a$ of objects in U into the value set of a, and $d \notin A$ is the decision. The indiscernibility relation Ind can be computed as shown in equation (3).

$$Ind(x,y) = \frac{|IND(x,y)|}{|A|} \tag{3}$$

Then equation (2) becomes:

$$\mu_h(x,y,r) \Leftrightarrow Ind(x,y) \geq r \tag{4}$$

and

$$IND(x,y) = a \in A : a(x) = a(y) \tag{5}$$

where a is an attribute(s) in an information system A, $a(x)$ is the value of tuple x in attribute a, $a(y)$ is the value of tuple y in attribute a, and $|A|$ is the cardinality of a set A.

2.2 Recommendation Algorithms

Collaborative Filtering. Within the last decade, Collaborative Filtering (CF) has been improved continually and finally became one of the most prominent personalization techniques in the field of recommendation systems. It is the process of evaluating information using the opinion of other people [9]. Originally CF was inspired by the nature of human being, and the fact that individuals share opinions with each other [9]. Typically, predictions about user interests are made by collecting taste information from many other similar users, assuming that having those individuals agreed in the past tend to agree again in the future. CF systems usually need to process huge amounts of information, including large-scale datasets, such as in electronic commerce and web applications. Nowadays, computers and the Internet allow us to consider the opinions of large interconnected communities with thousands of members [9]. Individuals can profit from a community, in that they get access to the knowledge of other users and their experience about diverse items. Furthermore, this information can help individuals to develop their own personalized view or to make a decision regarding the items that where rated. The most popular non-probabilistic CF approach is the Nearest Neighborhood (NNH) algorithm. In general there are two basic NNH techniques [9]:

- *User-based nearest neighborhood*: that check who shared the same rating pattern with the active user, then use the ratings of the like-minded users to calculate the predictions.

- *Item-based nearest neighborhood*: that are the transpose of user-based algorithms, which generate predictions based on similarities between items.

For user-based nearest neighborhood algorithms, in order to generate more accurate predictions, rating values of neighbor are assigned with weights according their similarity to the target user. Generally, user similarity is calculated by means of the *Pearson Correlation Coefficient*, which compares ratings for items rated by both target user and neighbor. On the other hand, for item-based nearest neighborhood algorithms, given a user-item pair (u,i), the prediction is composed of a weighted sum of the user us ratings for the item most similar to i. The general item-based prediction algorithm has been formalized according to equation (6):

$$pred(u,i) = \frac{\sum_{j \in ratedItems(u)} itemSim(i,j).r_{ui}}{\sum_{j \in ratedItems(u)} |itemSim(i,j)|} \tag{6}$$

Several variations existed for calculating the similarity between two items. The Person Correlation Coefficient alias Adjusted-Cosine Similarity [9,10] is considered the most accurate similarity metric, as shown in equation (7):

$$itemSim(i,j) = \frac{\sum_{u \in ratedBoth(i,j)} (r_{ui} - \overline{r}_i)(r_{uj} - \overline{r}_j)}{\sigma_i . \sigma_j} \tag{7}$$

where,

$$\sigma_i = \sqrt{\sum_{u \subset ratedBoth(i,j)} (r_{ui} - \overline{r}_i)^2}$$

and

$$\sigma_j = \sqrt{\sum_{u \subset ratedBoth(i,j)} (r_{uj} - \overline{r}_j)^2}$$

The major drawback of the CF approach is the high memory usage for the generated item similarity matrix ($|items|^2$). However, the size of the model can be substantially reduced via considering the correlations with more than a minimum number of co-ratings, or by only retaining the top n correlations for each item [9].

Fig. 1 presents the interaction of a user with an online collaborative recommender system through a web interface. In order to suggest products to a user, web server and recommender system need to communicate with each other. Usually, the web server application forwards user feedback to the recommender system, and receives personalized recommendations in return. User ratings and item correlations are both stored on the recommender platform to ensure real-time results.

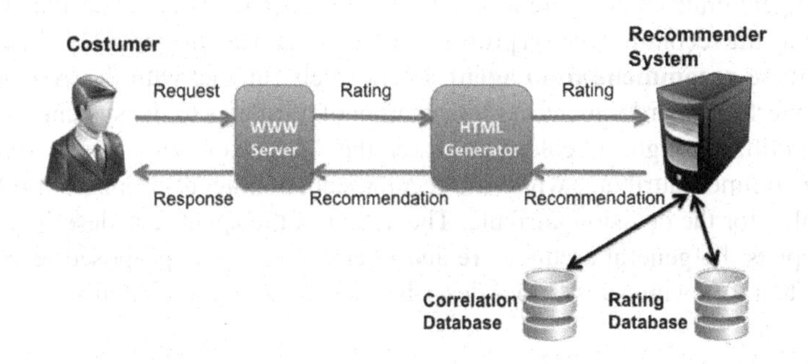

Fig. 1. Collaborative Recommender System Architecture [11]

Content-Based Filtering. Content-based filtering (CBF) techniques make suggestions upon item features and user interest profiles [12]. Typically, personalized profiles are created automatically through user feedback, and describe the type of items a person likes. In order to determine what items to recommend, collected user information is compared against content features of the items to examine. Usually, items are stored in database tables, whereas each row describes one item and each column represents a specific item feature. Depending on the domain, features can be represented either by boolean values or by a set of restricted values ($i.e.x \in N$). As previously mentioned, most CBF recommender systems store user interests in personalized profiles that contain descriptions about the items an individual likes most or information about user interactions with the recommendation system. Interactions may include saving user queries as well as capturing user activities related to particular items [1]. The history of user interactions cannot only facilitate the user returning to viewed items, but also can assist the recommender system to filter out item that were already considered by the user. In addition, user history may serve as training data for machine learning algorithms, which employ the gathered information to generate a more sophisticated user model [12]. User information might be provided explicitly by the individuals (e.g. common feedback techniques) or gathered implicitly by a software agent with no additional intervention by the user during the process of constructing profiles (e.g. automatically updates as the user interacts with the system) [1].

The challenge with CBF is to extract those item features that are most predictive. Once a user profile has been constructed form the items a user expressed interest in, then new item can be compared to the profile to make recommendations. In contrast, CF assumes that people with similar tastes will rate things similarly [13]. Hence, CF needs ratings for an item in order to predict for it, however CBF can predict relevance for items without rating. The main drawback of content-based filtering approaches is that they tend to overspecialize, because solely items that match the content features in the user profile are recommended.

3 The Proposed Multi-agent Recommender System

The intelligent multi-agent system proposed in this article, which is designed and built for handling the recommendation problem, has two main agents ; namely, the **matching agent** and the **recommendation agent**. Users solely interact with the system through the user interface in order to access information of attributes to the system. The matching agent utilizes rough mereology to search the dataset for similar users who have at least one common attribute, while the recommendation agent proposes a recommendation value for the decision attribute. The action of the agents are described in fig. 2, which depicts the general architecture and interactions of the proposed recommender system. Then following subsections describe each agent in more details.

3.1 Matching Agent

The matching agent uses the dataset information and preferences in users profile to produce a list of matching information. This is attained by searching the dataset using user profile and collecting the matching information into a new set. The matching agent uses three modules to achieve its goal; namely *user module*, *data information module*, and *rough mereology module*.

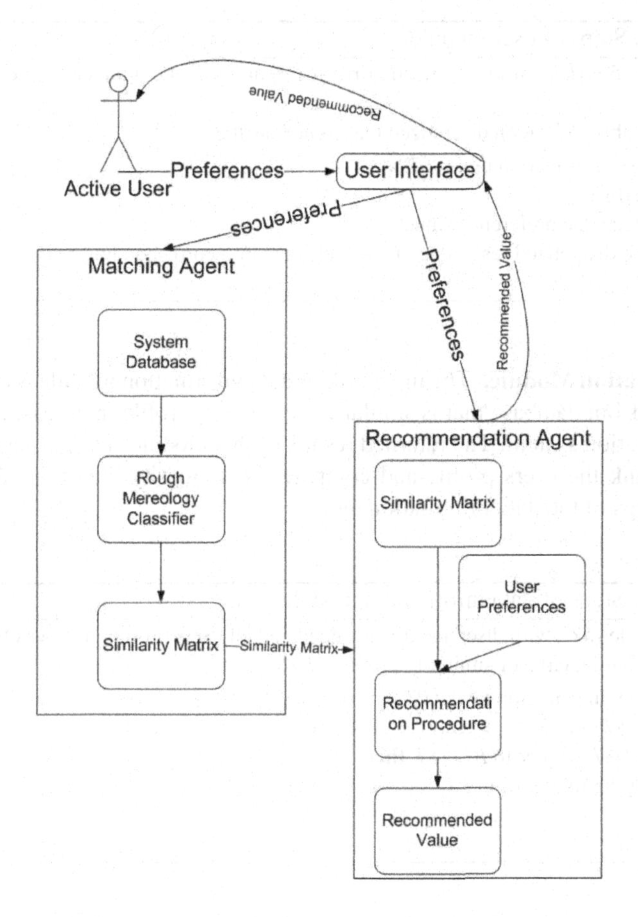

Fig. 2. Architecture of the proposed recommender system

User Module. User module is responsible to collect user information. The users of the system have to be modeled so as to use them in determining what they would consider as good information about the system.The ones chosen for the recommendation system are (interests, abilities), where interests refers to the professional and personal preferences, while abilities refers to the experience in ranking a field according to previous information about this field. These attributes then make up the user profile that will be created to each user. The returned results are then sent to data information module in order to get list of matching information set to user preferences profile. Algorithm 1 shows the steps of the user module.

Algorithm 1. Steps of user module

Input: User profile (U) with real valued attributes A_i and n is the number of intervals for each attribute.
Output: User table (ST) with discretized real valued attribute

1: Create a new information table (ST)
2: **for** $A_i \in U$ **do**
3: User selects the preference value
4: Insert the discretized real value of each attribute into corresponding (ST)
5: **end for**

Data Information Module. The main task of data information module is to collect the users profile from database that is similar to active user profile in at least one attribute of the information system. The returned results of this module are then sent to the next module to rank the users profile and compute the similarity measures. Algorithm 2 shows the steps of the data information module.

Algorithm 2. Steps of data information module

Input: User table (ST) with discretized real valued and all users profile in dataset table (DST)
Output: Information table of similarity profiles (PT)

1: Create a new information table (PT)
2: **for** $A_{ij} \in DST$ **do**
3: **if** $A_{ij} \in DST$ similar to $B_j \in ST$ **then**
4: Insert the row of A_i in PT
5: **end if**
6: **end for**

Rough Mereology Module. The goal of the rough mereology module is ranking the dataset, using the equations of rough mereology previously described in subsection 2.1, in order to obtain a useful recommendation to active user. The returned result of this module is a similarity items matrix that is corresponding to each users profiles without active user profile. Accordingly, this items similarity designed in symmetric matrix like a cross table. Algorithm 3 shows the steps of the rough mereology module.

Algorithm 3. Steps of rough mereology module

Input: Information table of similarity profiles (PT) **Output:** Similarity symmetric table (SM)

1: Create a new similarity table (SM)
2: **for** Each column $A_j \in DST$ **do**
3: Compute indiscernibility value according to equation (3) for each A_j and access to table (PT)
/* That represents the similarity value between each attribute and other attributes */
4: **end for**

3.2 Recommendation Agent

The main task of recommendation agent is to recommend the decision value to active user. The recommendation agent receives the similarity symmetric table from the matching agent and the active user profile, then it computes the recommended value for decision attribute, as shown in equation (8), where $sim(i, j)$ is a similarity value computed by the matching agent of attribute i to another attribute j, and r_{ui} is a rating value of active user u to an attribute i.

$$pred(u,i) = \frac{\sum_{j \in ratedItems(u)} (sim(i,j).r_{ui})}{\sum_{j \in ratedItems(u)} (sim(i,j)} \quad (8)$$

3.3 Integration of Agents and User Interface

The agents and their modules are integrated so that the ranked value and user interface can be accessed. The proposed framework utilized the Java Virtual Machine (JVM) paradigm along with Rapid Application Development (RAD) for building a software application. The JVM paradigm applied an object oriented and modular approach for putting the final recommendation system together with a well built user interface. The model represents the logical architecture of the application, whereas the view is the interface to the application and the controller connects both of them together. This allows developers to focus on system logic regardless the interface's view and vice-versa. Thus, the resultant complete system was integrated with features such as data security and validation added in without much effort. The parallel building of the interface was designed according to the JVM paradigm. THere is no need for programmers to know about other modules or actions to build their part of application. Also, the usage of object relational database was encouraged by the framework. Moreover, all the databases used in the recommender system are relational databases, which allows speed and data integrity through using keys.

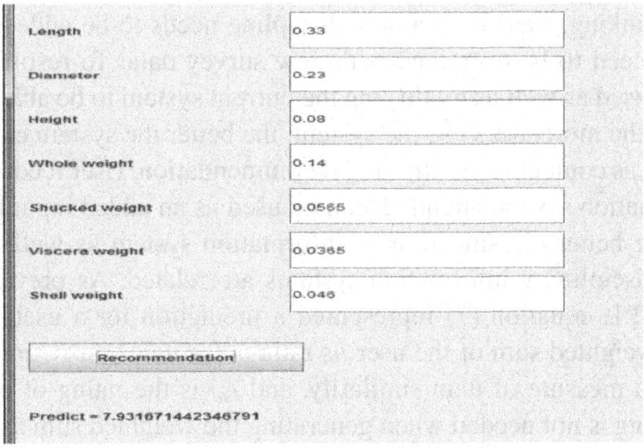

Fig. 3. Sample of user interface

User interface for the proposed system was designed to be a user-friendly interface. Ease of use in such recommender systems is paramount so that the user need only to focus on the task at hand and not on learning/train how to use the system. When the user has registered and logged on to the system, his/her profile information is viewed, as well as recommended decision value and the search. This is illustrated in fig. 3 showing a screen-shot of the user interface for the proposed recommender system. Users can simply edit their profiles, which will automatically then refresh the recommended decision value. If there are errors in data entry then the system will display a warning message to the user. Moreover, users information in not accessible to other users.

4 Evaluation and Discussion

Evaluating the recommendation quality of the proposed recommender system is mainly based on *statistical precision* and *decision supporting precision* measurement methods [14]. Statistical precision measurement method adopts the Mean Absolute Error (MAE) in order to measure the recommendation quality [15]. MAE is a commonly used recommendation quality measure that calculates the irrelevance between the recommendation value predicted by the recommender system and the actual evaluation value. Each pair of interest predicted rank values is represented as $< p_i, q_i >$, where p_i is the system predicted value and q_i is the user evaluation value. Based on the entire set of $< p_i, q_i >$ pairs, MAE calculates the absolute error value $|p_i - q_i|$ and the sum of all the absolute error value. Then, the average value is calculated subsequently. Small MAE values represent a good recommendation quality indicators. The predicted values of rating sets can be represented as $p_1, p_2, ..., p_N$ and its corresponding actual testing rating set can be represented as $q_1, q_2, ..., q_N$. The MAE can be defined as shown in equation (9) [16]:

$$MAE = \frac{\sum_{i=1}^{N} |p_i - q_i|}{N} \tag{9}$$

One limitation of the proposed Multi-Agent recommender system is the use of rough mereology for ranking. That is, if a new discipline needs to be added then the rough mereology will need to be re-trained with new survey data. To resolve this problem, methods can be used as well as modifying the current system to be able to provide user feedback. Thus, the more users use the system, the better the system can perform. This would be termed as content based filtering recommendation. User feedback, in terms of ratings of information system attended, can be used as an added input into the system. This would offer better classification of information system as well as offer insight into how multidisciplinary information systems are related. As previously described in subsection 2.2.1, equation (7) represented a prediction for a user u and item i is composed of a weighted sum of the user us ratings for items most similar to i, where, $itemSim(i, j)$ is a measure of item similarity, and r_{ui} is the rating of user u to item i. Average correcting is not needed when generating the weighted sum as the component ratings are all from the same target user.

One of the most frequently suggested methods for the calculation of predicted ratings is the weighted sum method [17], calculated using equation (10), where $pred(u, i)$

represents the predicted vote for item i calculated for user u, $\overline{r_u}$ refers to the average rating of the active user.

$$pred(u,i) = \overline{r_u} + \frac{\sum_{j \in ratedItems(u)} itemSim(i,j).(r_{ui} - \overline{r_j})}{\sum_{j \in ratedItems(u)} (itemSim(i,j))} \tag{10}$$

While this method generates good results, our experiments indicated that, in some situations, it fails to generate a good range of possible suggestions. The most obvious case is when a user put only very high or very low ratings into the system. His/her average rating is then either very high or very low, accordingly. In either case, the problem is the sum of weights from the nearest neighbors. The system relies on this sum to indicate whether the item will be appropriate or not. The problem with these two situations is that this sum will be much smaller than the user's average rating. This means that the final predicted rating will have almost the same value as the user's average rating. In the case of a high average rating, this means that all the items recommended by neighbors will be automatically recommended and in the case of a low average rating no recommendations will be made because no object will be deemed appropriate. Accordingly, equation (11) normalizes the prediction through dividing by the sum of the neighbors similarities, where $\overline{r_j}$ is the mean of each user rated item.

$$pred(u,i) = \frac{\sum_{j \in ratedItems(u)} itemSim(i,j).(r_{ui} - \overline{r_j})}{\sum_{j \in ratedItems(u)} (itemSim(i,j))} \tag{11}$$

In fact, equation (11) is just a modified version of equation (10). The logic behind this is that in the above-mentioned situations the value of predicted vote is less important than finding out whether the content will be appropriate or not. Therefore, we concentrated on the sum of weights obtained from the nearest neighbors. Whereas, if a neighbor is liked, the content of his/her weights will be positive (they rated the content with a rating that is higher than their average rating). On the other hand, weights will be negative if they disliked it. If the sum of all the weights is positive this means that most neighbors liked the content and it should therefore be recommended. If the sum is negative, this means that they either did not like it at all or did not consider it to be very interesting. Therefore, it can be discarded.

To evaluate the proposed rough mereology based recommender system, we considered in this article the Abalone dataset in UCI [18] that consists of 4177 objects for our experiments to predict the age of abalone from physical measurements based on the number of rings.

Table 1, clarifies a sample part of the Abalone dataset, which consists of 8 attributes: *Length* (longest shell measurement), *Diameter* (perpendicular to length), *Height* (with meat in shell), *Whole weight* (whole abalone), *Shucked weight* (weight of meat), *Viscera weight* (gut weight, after bleeding), *Shell weight* (after being dried), and *Rings* (the number of rings is the value to predict the age in years). Our experiments showed that this measure is well suited for situations where the average rating of the user is very high or low, while in other cases the results produced are slightly worse than normal as shown in table 1. The MAE of the system is about 3.63 by using equation (8), 3.59 by using equation (10), and 3.43 by using equation 1(11), which is the mean of 1000 tests that represents the active users and 3177 training data representing the users profile.

Table 1. Comparative analysis for prediction equations

Length	Diameter	Height	Whole	Shucked Weight	Viscera Weight	Shell Weight
0.51	0.41	0.14	0.66	0.29	0.13	0.20
0.38	0.3	0.1	0.25	0.11	0.05	0.08
0.27	0.20	0.07	0.10	0.05	0.01	0.03
0.37	0.30	0.1	0.27	0.12	0.06	0.08
0.5	0.39	0.14	0.55	0.23	0.07	0.21
0.65	0.51	0.17	1.31	0.43	0.26	0.52
0.57	0.44	0.12	0.92	0.40	0.18	0.29
0.67	0.55	0.18	1.71	0.70	0.39	0.58
0.59	0.42	0.15	0.88	0.37	0.23	0.24
0.47	0.36	0.11	0.50	0.24	0.13	0.13
Actual Ring	**Predicted Ring Eq. (8)**	**Predicted Ring Eq. (10)**	**Predicted Ring Eq. (11)**	**MAE Eq. (8)**	**MAE Eq. (10)**	**MAE Eq. (11)**
15	11.49	8.84	8.57	3.51	6.16	6.43
8	7.66	7.52	7.35	0.35	0.48	0.65
8	12.01	11.86	11.53	4.01	3.86	3.53
7	10.58	10.44	10.16	3.58	3.44	3.16
11	7.34	7.19	7.04	3.66	3.81	3.96
10	8.80	8.66	8.45	1.20	1.34	1.55
11	12.85	12.69	12.32	1.85	1.69	1.32
13	11.88	11.73	11.37	1.12	1.27	1.63
11	10.45	10.31	10.03	0.55	0.69	0.97
6	12.64	12.48	12.08	6.64	6.48	6.08

5 Conclusions and Future Work

In this article, a Multi-Agent approach for solving the problem of recommendation has been presented. The proposed system works via two main agents; namely, the matching agent and the recommendation agent. Experimental results showed that the proposed rough mereology based Multi-agent system for solving the recommendation problem is scalable and has possibilities for future modification and adaptability to other problem domains. Moreover, it succeeded in reducing the information overload while recommending relevant decisions to users. The system achieved high accuracy in ranking using users profile and information system profiles. The resulted value of the Mean Absolute Error (MAE) is acceptable compared to other recommender systems applied other computational intelligence approaches. Based on its performance, the proposed recommendation system can be considered as a good platform for future research into the use of computational intelligence in recommender systems via applying further improvements to it.

References

1. Gauch, S., Speretta, M., Chandramouli, A., Micarelli, A.: User Profiles For Personalized Information Access. In: Brusilovsky, P., Kobsa, A., Nejdl, W. (eds.) Adaptive Web 2007. LNCS, vol. 4321, pp. 54–89. Springer, Heidelberg (2007)
2. Macho, S., Torrens, M., Faltings, B.: A Multi-Agent Recommender System For Planning Meetings. In: Proc. of the 4th International Conference on Autonomous Agents, Workshop on Agent-based Recommender Systems, WARS 2000 (2000)

3. Morais, A.J., Oliveira, E., Jorge, A.M.: A multi-agent recommender system. In: Omatu, S., Paz Santana, J.F., González, S.R., Molina, J.M., Bernardos, A.M., Rodríguez, J.M.C. (eds.) Distributed Computing and Artificial Intelligence. AISC, vol. 151, pp. 281–288. Springer, Heidelberg (2012)
4. Marivate, V.N., Ssali, G., Marwala, T.: An Intelligent Multi-Agent Recommender System For Human Capacity Building. In: Proc. of the 14th IEEE Mediterranean Electrotechnical Conference, pp. 909–915 (2008)
5. Blanco-Fernández, Y., Pazos-Arias, J.J., Gil-Solla, A., Ramos-Cabrer, M., Barragáns-Martínez, B., López-Nores, M., García-Duque, J., Fernández-Vilas, A., Díaz-Redondo, R.P.: AVATAR: An Advanced Multi-agent Recommender System of Personalized TV Contents by Semantic Reasoning. In: Zhou, X., Su, S., Papazoglou, M.P., Orlowska, M.E., Jeffery, K. (eds.) WISE 2004. LNCS, vol. 3306, pp. 415–421. Springer, Heidelberg (2004)
6. Veltkamp, R.C., Hagedoorn, M.: Shape Similarity Measures, Properties and Constructions. In: Laurini, R. (ed.) VISUAL 2000. LNCS, vol. 1929, pp. 467–476. Springer, Heidelberg (2000)
7. Lesniewski, S.: On the foundations of set theory. Topoi 2, 7–52 (1982)
8. Polkowski, L., Artiemjew, P.: Granular Computing in the Frame of Rough Mereology. A Case Study: Classification of Data into Decision Categories by Means of Granular Reflections of Data. International Journal of Intelligent Systems 26(6), 555–571 (2011)
9. Schafer, J.B., Frankowski, D., Herlocker, J., Sen, S.: Collaborative Filtering Recommender Systems. In: Brusilovsky, P., Kobsa, A., Nejdl, W. (eds.) Adaptive Web 2007. LNCS, vol. 4321, pp. 291–324. Springer, Heidelberg (2007)
10. Linden, G., Smith, B., York, J.: Amazon.Com Recommendations: Item-To-Item Collaborative Filtering. IEEE Internet Computing 7(1) (2003)
11. Sarwar, B.M., Karypis, G., Konstan, J.A., Riedl, J.T.: Application of Dimensionality Reductio in Recommender System - A Case Study. In: Acm Webkdd Workshop (2000)
12. Pazzani, M.J., Billsus, D.: Content-Based Recommendation Systems. In: Brusilovsky, P., Kobsa, A., Nejdl, W. (eds.) Adaptive Web 2007. LNCS, vol. 4321, pp. 325–341. Springer, Heidelberg (2007)
13. Koren, Y.: Tutorial on Recent Progress in Collaborative Filtering. In: Pu, P., Bridge, D.G., Mobasher, B., Ricci, F. (eds.) Proceedings of the 2008 Acm Conference on Recommender Systems, Recsys 2008, Lausanne, Switzerland, October 23-25, pp. 333–334 (2008)
14. Ettouney, R.S., Mjalli, F.S., Zaki, J.G., El-Rifai, M.A., Ettouney, H.M.: Forecasting Ozone Pollution using Artificial Neural Networks. Mgmt. Environ. Quality 20, 668–683 (2009)
15. Abdul-Wahab, S., Bouhamra, W., Ettouney, H., Sowerby, B., Crittenden, B.D.: Predicting Ozone Levels: A Statistical Model for Predicting Ozone Levels. Environ. Sci. Pollut. Res. 3, 195–204 (1996)
16. Pawlak, Z., Grzymala-Busse, J., Slowinski, R., Ziarko, W.: Rough Sets. Communications of the ACM 38(11), 88–95 (1995)
17. Breese, J.S., Heckerman, D., Kadie, C.M.: Empirical Analysis of Predictive Algorithms for Collaborative Filtering. In: UAI, Technical report MSR-TR-98-12, pp. 43–52 (1998)
18. UCI ML Repository Datasets, http://www.ics.uci.edu/~mlearn/databases/

Geodata Scale Restriction Using Genetic Algorithm

Jiří Dvorský[2], Vít Pászto[1], and Lenka Skanderová[2]

[1] Department of Geoinformatics, Palacký University,
Třída Svobody 26, 771 46, Olomouc, Czech Republic
vit.paszto@gmail.com

[2] Department of Computer Science, VŠB – Technical university of Ostrava,
17. listopadu 15, 708 33 Ostrava – Poruba, Czech Republic
{jiri.dvorsky,lenka.skanderova}@vsb.cz

Abstract. With recent advances in computer sciences (including geosciences) it is possible to combine various methods for geodata processing. There are many methods established for geodata scale restriction, but none of these take into account the concept of information entropy. Our research focused on using genetic algorithm that calculates information entropy in order to set an optimal number of intervals from original non-restricted geodata. We used fitness function by minimizing information entropy loss and we compared the results with commonly used classification method in geosciences. We propose an experimental method that provides promising approach for geodata scale restriction and consequent proper visualization, which is very important for geographical phenomena interpretation.

1 Introduction

Every single research starts with the measurement of a phenomenon. Especially, in quantitative geosciences, there are plenty of various data that comes from direct measurements of a phenomenon (e.g. rainfall, air temperature, soil humidity). This primary measured data are in many cases either too big or precise (in terms of decimal numbers), which is sometimes not necessary for consequent geographical analysis and visualization; and slow-down the data processing and further understanding of a phenomenon nature. It is then appropriate to reduce the amount of data using multivariate statistics or just simply restrict measurement scale. The second approach is commonly used in geography and especially in cartography by transforming primary data into interval data. This transformation enables the map reader to easily understand underlying analysis displayed in a map. Nevertheless, during every transformation process there is a certain information loss. Thus, the question might often arise – how much information do we lose by primary measurement scale restriction?

Shortly after the World War II, Claude E. Shannon defined the concept of entropy as a measure of information [12]. The *information entropy* quantifies the amount of information in a transmitted message. This concept has been

A. Abraham et al. (eds.), *Innovations in Bio-inspired Computing and Applications*, Advances in Intelligent Systems and Computing 237,
DOI: 10.1007/978-3-319-01781-5_20,

widely used in computer science as well as in *geographical information science* (GIScience). The geovisualization via maps could be understood as a message carrying the information. There are rules in cartography that defines how to construct a map. But these rules do not preserve the optimal amount of information that could be passed onto the map reader. Therefore it is important to properly set up intervals of displayed phenomenon by maximizing the information derived from entropy values.

The aim of the paper is to use some kind of bioinspired computation, genetic algorithm in this case, to find out optimal distribution of measured value into several intervals. The optimal division means maximal preservation of information with respect to the original data. The distribution of air temperatures in the Czech Republic is used as the model example. Consequently, cartographic visualizations were made in order to evaluate results of the experiment and for comparison with existing commonly used method in cartography.

The paper is organized as follows. In Sect. 2 scale restrictions in geosciences and genetic algorithms are described. A Sect. 3 deals with scale restriction using genetic algorithms, problem formalization. In Sect. 4 experimental data, the results of the experiments and comparison of genetic algorithm results with exhaustive search are mentioned and Sect. 5 contains the conclusion.

2 State of the Art

2.1 Scale Restrictions in Geosciences

In geosciences, it is common to restrict geodata by transformation into interval data. There are several ways, how to set up intervals for measured dataset [10,4,13]. First of all, it is obvious that newly created intervals must cover whole measurement scale. Secondly, the proper classification method has to be chosen. Most of the *Geographical Information Systems* (GIS) provide several basic classification schemes – e.g. equal interval, quantile interval, geometrical interval, standard deviation interval (σ,$\frac{1}{2}\sigma$, $\frac{1}{3}\sigma$ and $\frac{1}{4}\sigma$, where σ is a standard deviation of input data) and natural breaks – Jenks classification [3]. Obviously, it is also possible to classify geodata into intervals manually.

Equal intervals method divides the geodata scale range into equal-size intervals and user specifies only the number of desired intervals. Quantile interval classification scheme divides the geodata scale range into intervals with an equal number of features according to frequency distribution. Geometrical interval method creates intervals from former scale range by minimizing the square sum of element per class. Standard deviation classification scheme divides the geodata into intervals showing measured values deviation from the mean. And finally, natural breaks method developed by an American geographer and cartographer G. F. Jenks defines intervals by minimizing each interval (values in given interval) average deviation from the interval mean, while maximizing each interval deviation from the means of the other groups. This approach reflects natural groupings inherent in the geodata and is one of the mostly used in geosciences.

It is very purpose-dependent, which method to use. For example, quantile classification is suitable to apply on linearly distributed geodata, while geometrical interval method is applicable for continuous geodata and natural breaks (Jenks method) is suitable for geodata that are specific in its structure. Nevertheless, any of these methods do not take information loss (the entropy rate) into consideration.

2.2 Genetic Algorithms

Genetic algorithms (GA) belong to the family of evolution algorithms. Their principles are based on the nature principles – natural selection, crossing and mutation. At the beginning of the genetic algorithm, new population is created. It contains individuals. During the evolution individuals are crossed and mutated. Better individuals survive while worse die. Each individual consists of parameters, each parameter has its own lower and upper bound, and each individual has its fitness, the value of the cost function. Fitness says how good this individual within population is [14]. In [6] authors combine genetic algorithm with data envelopment analysis to determine optimal resource levels in surgical services. In [2] genetic algorithm is used to solve decentralized Partially Observable Markov Decision Process problems and in [9] Sand Pile modelling. Genetic algorithms are often used in connection with neural networks, see [1,8,11] and complex networks, see [7,5].

In this paper the genetic algorithm is used to find the optimal distribution of the measured temperatures to the intervals so that information, i.e. entropy, preservation is the biggest.

3 Scale Restriction Using Genetic Algorithms

As it was mentioned above, genetic algorithms are based on three basic principles – natural selection, crossing according to the Darwin's theory and mutation according to the Mendel's theory. These principles are applied on population consisting of group of individuals. A representation of possible solution of given problem as particular individual should be determined at first. This is the main problem in application of the evolutionary algorithms – how to represent the individual and how to compute fitness value. The second step is the definition, how the parent choosing, crossing and mutation will be going on. When the first population is generated, reproduction cycle can begin. The parents are chosen from the current population of individuals and the new individuals are created.

In this paper, individual is denoted as I_j and it is represented by its parameters and fitness as it is usual, but the parameters of individuals are not real or integer numbers, the parameters are created by N intervals, $I_j = (V_i^j, \ldots, V_N^j)$. Each interval has its own lower and upper bound and lower and upper bound can be the same value in one interval. Intervals are sorted in ascending order and each follows the next one, e.g the second interval follows the first one, the third follows the second etc. Each value can be used as a bound at most once. It is not possible to have value, which represents upper bound of the one interval and lower bound of the following one.

At the beginning new population is generated. Each individual has its own randomly chosen parameters – intervals. From this parameters, the fitness value is computed according to the Eq. (15). The evolution cycle follows these steps:

- For each individual random neighbour is chosen.
- **Crossing:** the actual individual is crossed with the random neighbour by this way: First parameter (interval) is taken from the random neighbour and next parameters (intervals) are taken from the actual one.
- **Mutation:** It is clear that if we cross these individuals, their intervals will probably not follow each other, because the upper bound of the first parameter can be greater than lower bound of the second parameter. In this case, the first parameter in the individual stays and next parameters are generated randomly.
- The fitness value of the new individual is computed.
- If the fitness value of the new individual is smaller than actual individual's fitness, new individual will be added to the new generation. Otherwise the actual individual is added to the new generation.

3.1 Number of Possible Scale Restrictions – An Estimation

Genetic algorithms are usually used to solve problems, where analytical solution is not known or it is intractable in reasonable time. The problem of scale restriction is the second kind of problem. Let's suppose that we have n different unique input values and the restriction is done to N intervals, see Sect. 3.2. It can be easy show that number of all possible scale restrictions $R(n, N)$ can be expressed recursively as:

$$R(a,b) = \begin{cases} a-1 & \text{for } b = 2 \\ \sum\limits_{i=1}^{a-b+1} R(a-i, b-1) & \text{for } b > 2 \end{cases} \tag{1}$$

To illustrate number of possible scale restrictions some values of function R are provided in Table 1.

3.2 Problem Formalization

Formally, all possible values of input data can be represented as a finite set $V = \{v_1, v_2, \ldots, v_n\}$, where $v_i \in \mathbb{R}$. It is also supposed that $v_i < v_{i+1}$ for $i = 1, \ldots, n-1$. For each possible input data value v_i we define its frequency f_i. At this point we need to distinguish between unique input value v_i, e.g. air temperature, and number of occurrence of given input value f_i, e.g. number of points at selected region or area with given air temperature.

Performing scale restriction on set of input values V, the range of n unique input values is restricted into a sequence of N non-overlapping intervals $V_1, \ldots, V_N$.

Table 1. Number of possible scale restrictions

n	N	$R(n, N)$
10	5	126
20	10	92,378
50	10	$\approx 2 \times 10^9$
100	10	$\approx 1.7 \times 10^{12}$
100	50	$\approx 5 \times 10^{28}$
200	10	$\approx 1 \times 10^{15}$
200	100	$\approx 5 \times 10^{58}$

Let's define two vectors $\boldsymbol{l} \in \mathbb{N}^N$ and $\boldsymbol{u} \in \mathbb{N}^N$, where $\boldsymbol{l}_i$ represents index of lower bound of interval V_i and $\boldsymbol{u}_i$ represents index of its upper bound, formally:

$$V_i = [v_{\boldsymbol{l}_i}, v_{\boldsymbol{u}_i}], \tag{2}$$

In order to fully cover the set V using the intervals $V_1, V_2, \ldots, V_N$ the following properties must be satisfied:

$$1 \leq \boldsymbol{l}_i \leq n \tag{3}$$
$$1 \leq \boldsymbol{u}_i \leq n \tag{4}$$
$$\boldsymbol{l}_1 = 1 \tag{5}$$
$$\boldsymbol{u}_N = n \tag{6}$$
$$\boldsymbol{l}_{j+1} = \boldsymbol{u}_j + 1 \tag{7}$$
$$\boldsymbol{l}_i \leq \boldsymbol{u}_i \tag{8}$$
$$V = \bigcup_{i=1}^{N} V_i, \tag{9}$$

for all $i = 1, \ldots, N$ and for all $j = 1, \ldots, N-1$. If $\boldsymbol{l}_i = \boldsymbol{u}_i$ then interval V_i contains only one element $v_{\boldsymbol{l}_i}$.

Each possible sequence of intervals $V_1, V_2, \ldots, V_N$ determines one possible scale restriction on original set of input data V and also defines the individual in proposed usage of genetic algorithm. Formally, population consists of M individuals $I_1, I_2, \ldots, I_M$. Each individual is characterized by particular sequence of intervals, so

$$I_j = (V_1^j, \ldots, V_N^j) \tag{10}$$

for all $j = 1, \ldots, M$.

The proposed approach for scale restriction is based on entropy preserving, so entropy of input data set and entropy of given individual from current population have to be defined.

Let $p_i = \frac{f_i}{F}$ be a probability of value v_i, where $F = \sum_{i=1}^{n} f_i$. Then mean entropy H_0 of the input data set V is given as:

$$H_0 = -\sum_{i=1}^{n} p_i \log_2 p_i \tag{11}$$

The definition of entropy of individual $I_j = (V_1^j, \ldots, V_N^j)$ is based on entropies of intervals V_i^j and subsequently interval's entropy is based on probability of occurrences of all values from given interval. The probability of the interval is sum of probabilities of particular input values from the interval:

$$f_{V_i^j} = \sum_{k=\boldsymbol{l}_i}^{\boldsymbol{u}_i} f_k \tag{12}$$

$$p_{V_i^j} = \frac{f_{V_i^j}}{F} \tag{13}$$

The mean entropy of individual $I_j = (V_1^j, \ldots, V_N^j)$ is

$$H_j = -\sum_{i=1}^{N} p_{V_i^j} \log_2 p_{V_i^j} \tag{14}$$

Fitness function $f(I_j)$ of individual I_j is given as

$$f(I_j) = H_0 - H_j \tag{15}$$

4 Experimental Results

4.1 Experimental Data

As it is experimental study, geodata used for this paper are integer values of long-term mean annual air temperatures from 1961 to 2000 in the Czech Republic. The geodata scale ranges from 1 to 10°C. The geodata is in raster format (ESRI grid) with 100 meters cell size and 4,865 columns and 2,780 rows. This dataset was used due to its explicit interpretation in order to evaluate and visualize experimental results.

Table 2. Experimental input data set V

v_i [°C]	f_i
1	1968
2	12680
3	50204
4	160566
5	468285
6	1564203
7	3128512
8	2135303
9	382978
10	950

In Table 2 the experiments values are mentioned. The first column denotes the measured air temperature values and in the second one the number of map's pixels are presented. Each air temperature value has its own color in the map. The question is how the information will be distort when the measured air temperatures will be categorized to the intervals. The experiments have been done for 2 – 9 intervals and we have observed how much the value of entrophy decreased.

For $N = 10$ intervals, where each interval represents only one air temperature, the entrophy is $H_0 = 2.135$ bits. The distortions, which are the consequences of the categorization of the air temperatures to the intervals are mentioned in Table 3. From the view of evolutionary algorithms the dimension has been changing according to the intervals number, number of individuals M has been set to 2000 and number of generation cycles G to 1500. For smaller number of individuals in the population evolution in cases with small number of intervals reached the worse results.

4.2 Comparison of GA Results with Jenks Algorithm

Our method using genetic was visually and numerically compared with the natural breaks (Jenks) method. Since the geodata scale range is from 1 to 10°C, there is 8 reasonable numbers of intervals that can be set. Maximal number of possible interval distributions was found for 5 (126 possible options). From geoscientific point of view, the best results of presented fitness approach were achieved just for this number of intervals (Fig. 1), where uncertainty of possible interval settings is the highest. The resulting interval distribution is also different when 2, 6, 7 and 8 intervals are desired, but with lower significance. Generally, as the number of intervals grows the compared methods resulted into very similar interval distributions, thus presented approach did not bring added value in this case.

Table 3. The interval distribution. The smallest number of intervals was 2, the biggest one is then 9. The origin entropy value H_0 denotes the entropy of the input data set V, N is the number of interval, I_{Best} denotes the best individual in the end of algorithm and $f(I_{Best})$ is the fitness of I_{Best}.

N	I_{Best}	$f(I_{Best})$ [bits]	H_0 [bits]
2	$[1,2,3,4,5,6,7]$ $[8,9,10]$	1.232	2.135
3	$[1,2,3,4,5,6]$ $[7]$ $[8,9,10]$	0.564	2.135
4	$[1,2,3,4,5]$ $[6]$ $[7]$ $[8,9,10]$	0.309	2.135
5	$[1,2,3,4,5]$ $[6]$ $[7]$ $[8]$ $[9,10]$	0.113	2.135
6	$[1,2,3,4]$ $[5]$ $[6]$ $[7]$ $[8]$ $[9,10]$	0.033	2.135
7	$[1,2,3]$ $[4]$ $[5]$ $[6]$ $[7]$ $[8]$ $[9,10]$	0.009	2.135
8	$[1,2]$ $[3]$ $[4]$ $[5]$ $[6]$ $[7]$ $[8]$ $[9,10]$	0.002	2.135
9	$[1]$ $[2]$ $[3]$ $[4]$ $[5]$ $[6]$ $[7]$ $[8]$ $[9,10]$	0.001	2.135

In Fig. 1, the resulting maps using two distinct classification methods are quite similar, but individual values of air temperature are divided differently. The best result of genetic algorithm (H_{Best}) preserves the amount of optimal information and splits the geodata into intervals in a way that is suitable for cartographic visualization and consequent understanding of the phenomena. Visualization using our method carries more information than using Jenks method, better displays lower air temperature values and makes the map richer for consequent user interpretation.

On the other hand, the results depicted in Fig. 1 appears to be very similar, but it is caused by experimental data resolution. As the number of input values

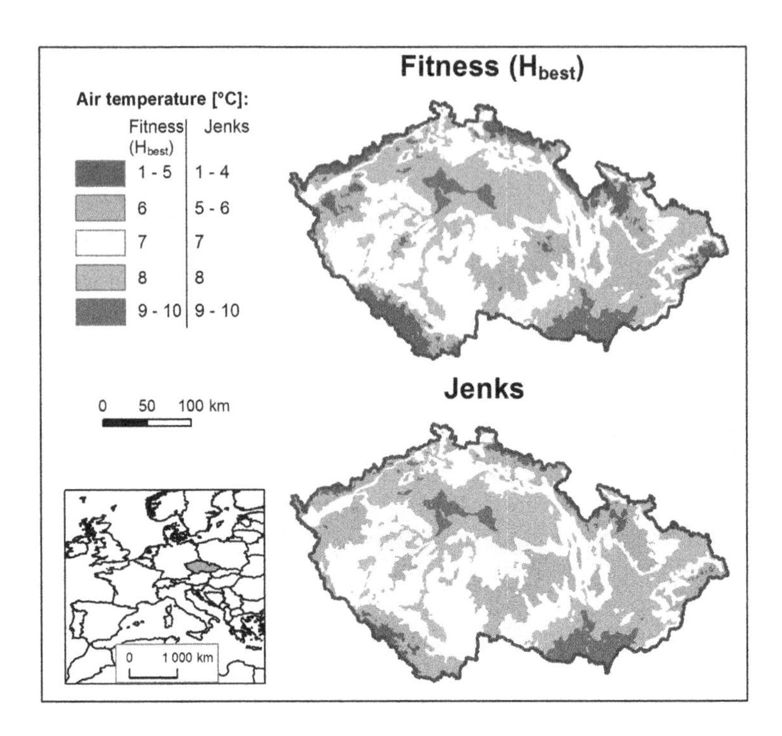

Fig. 1. Comparison of interval distributions into 5 intervals for the best fitness (H_{best}) and natural breaks (Jenks)

increases (e.g. air temperature values with decimal numbers), the number of possible scale restriction also increases, see Table 1 and traditional approach using natural breaks (Jenks) method could not be applicable. Therefore our method provides a very promising way how to handle large (geo)datasets.

5 Conclusion

Our method for setting the interval distribution using genetic algorithm was tested on experimental integer data of air temperatures. For future work, a larger geodata (in terms of magnitude order) is going to be used. Uncertainty of proper interval distribution settings is significantly greater when examining larger geo-data, thus our GA based method could contribute to reduce this uncertainty of cartographic visualization process. Another aspect of GA based method usage is the sensitivity for extremes detection by maximizing fitness function Eq. (15), which is opposite to presented approach. This is going to be studied more in detail in the future.

Acknowledgement. This work was supported by the European Regional Development Fund in the IT4 Innovations Centre of Excellence project (CZ.1.05/1.1.00/02.0070) and by the Development of human resources in research and development of latest soft computing methods and their application in practice project, reg. no. CZ.1.07/2.3.00/20.0072 funded by Operational Programme Education for Competitiveness, co-financed by ESF and state budget of the Czech Republic.

References

1. Chen, H.Q., Zeng, Z.G.: Deformation Prediction of Landslide Based on Improved Back-propagation Neural Network. Cognitive Computation 5, 56–62 (2013)
2. Eker, B., Akin, H.L.: Solving decentralized POMDP problems using genetic algorithms. Autonomous Agents and Multi-Agent Systems 27, 161–196 (2013)
3. Jenks, G.F.: The Data Model Concept in Statistical Mapping. International Yearbook of Cartography 7, 186–190 (1967)
4. Kraak, M.J., Ormeling, F.: Cartography: Visualization of Spatial Data, 3rd edn., 248 p. Prentice Hall (2009)
5. Li, Y.F., et al.: Non-dominated sorting binary differential evolution for the multi-objective optimization of cascading failures protection in complex networks. Reliability Engineering & System Safety 111, 195–205 (2013)
6. Lin, R.C., et al.: Multi-objective simulation optimization using data envelopment analysis and genetic algorithm: Specific application to determining optimal resource levels in surgical services. Omega-International Journal of Management Science 41, 881–892 (2013)
7. Liu, D.Y., et al.: Genetic Algorithm with a Local Search Strategy for Discovering Communities in Complex Networks. International Journal of Computational Intelligence Systems 6, 354–369 (2013)
8. Mehmood, H., Tripathi, N.K.: Optimizing artificial neural network-based indoor positioning system using genetic algorithm. International Journal of Digital Earth 6, 158–184 (2013)
9. Pina, J.M., et al.: Sand Pile Modeling of Multiseeded HTS Bulk Superconductors: Current Densities Identification by Genetic Algorithms. EEE Transactions on Applied Superconductivity 23(Pt. 3) (2013)
10. Robinson, A.H., et al.: Elements of Cartography, 6th edn., 688 p. John Wiley & Sons (1995)
11. Senygit, E., et al.: Heuristic-based neural networks for stochastic dynamic lot sizing problem. Applied Soft Computing 13, 1332–1339 (2013)
12. Shannon, C.E.: A Mathematical Theory of Communication. Bell System Technical Journal 27, 379–423, 623–656 (1948)
13. Slocum, T.A., et al.: Thematic Cartography and Geographic Visualization, 2nd edn., 528 p. Prentice Hall (2004)
14. Whitley, D.: A genetic algorithm tutorial. Statistics and Computing 4, 65–85 (1994)

Principal Component Analysis Neural Network Hybrid Classification Approach for Galaxies Images*

Mohamed Abd. Elfattah[1], Nashwa El-Bendary[2], Mohamed A. Abou Elsoud[1], Jan Platoš[3], and Aboul Ella Hassanien[4]

[1] Faculty of Computers and Information, Mansoura University, Mansoura - Egypt, Scientific Research Group in Egypt (SRGE)
abdelfatah@egyptscience.net, moh_soud@mans.edu.eg
http://www.egyptscience.net

[2] Arab Academy for Science, Technology, and Maritime Transport, Cairo - Egypt, Scientific Research Group in Egypt (SRGE)
nashwa.elbendary@ieee.org
http://www.egyptscience.net

[3] Department of Computer Science, FEECS and IT4 Innovations VSB-Technical University of Ostrava, Ostrava - Czech Republic
jan.platos@vsb.cz

[4] Faculty of Computers and Information, Cairo University, Cairo - Egypt, Scientific Research Group in Egypt (SRGE)
aboitcairo@gmail.com
http://www.egyptscience.net

Abstract. This article presents an automatic hybrid approach for galaxies images classification based on principal component analysis (PCA) neural network and moment-based features extraction algorithms. The proposed approach is consisted of four phases; namely image denoising, feature extraction, reduct generation, and classification phases. For the denoising phase, noise pixels are removed from input images, then input galaxy image is normalized to a uniform scale and Hu seven invariant moment algorithm is applied to reduce the dimensionality of the feature space during the feature extraction phase. Subsequently, for reduct generation phase, attributes in the information system table that is more important to the knowledge is generated as a subset of attributes. Rough set is used as feature reduction approach. The subset of attributed, which is called a reduct, is fully characterizing the knowledge in the database. Finally, during the classification phase, principal component analysis neural network algorithm is utilized for classifying the input galaxies

* This work was partially supported by Grant of SGS No. SP2013/70, VSB - Technical University of Ostrava, Czech Republic., and was supported by the European Regional Development Fund in the IT4Innovations Centre of Excellence project (CZ.1.05/1.1.00/02.0070) and by the Bio-Inspired Methods: research, development and knowledge transfer project, reg. no. CZ.1.07/2.3.00/20.0073 funded by Operational Programme Education for Competitiveness, co-financed by ESF and state budget of the Czech Republic.

A. Abraham et al. (eds.), *Innovations in Bio-inspired Computing and Applications*, Advances in Intelligent Systems and Computing 237,
DOI: 10.1007/978-3-319-01781-5_21, © Springer International Publishing Switzerland 2014

images into one of four obtained source catalogue types. Experimental results showed that combining PCA and rough set as feature reduction techniques along with invariant moments for feature extraction provided better classification results than having no rough set feature reduction technique applied. It is also concluded that a small set of features is sufficient to classify galaxy images and provide a fast classification.

Keywords: PCA neural network, rough set, moment invariant, fisher score, galaxy images.

1 Introduction

Galaxies can be considered as massive system of stars, gas, dust, and other forms of matter bound together gravitationally as a single physical unit [5]. The morphology of galaxies is generally an important issue in the large scale study of the Universe [1]. Galaxy classification is considered as a staring point for astronomers towards a greater understanding of the origin and formation process of galaxies, better understanding of galaxy image properties, and accordingly the evolution processes of the Universe. Generally, galaxy classification is important for two main reasons. Firstly, for producing large catalogues for statistical and observational programs, and secondly for discovering underlying physics [2].

Morphological galaxy classification is a system used by astronomers to classify galaxies based on their structure and appearance [3]. It is becoming an important issue because astrophysicists frequently make use of large database of information either to test existing theories, or to form new conclusions to explain the physical processes governing galaxies, star-formation, and the nature of the universe. Galaxies have a wide variety of appearances. Some are smooth and some are lumpy. Moreover, some have a well-ordered symmetrical spiral pattern.

This article presents an automatic hybrid approach for galaxies images classification based on principal component analysis (PCA) neural network and moment-based features extraction algorithms. The proposed approach is consisted of four phases; namely image denoising, feature extraction, reduct generation, and classification phases. For the denoising phase, noise pixels are removed from input images, then input galaxy image is normalized to a uniform scale and Hu seven invariant moment algorithm is applied to reduce the dimensionality of the feature space during the feature extraction phase. Subsequently, for reduct generation phase, attributes in the information system table that is more important to the knowledge is generated as a subset of attributes. Rough set is used as feature reduction approach. The subset of attributed, which is called a reduct, is fully characterizing the knowledge in the database. Finally, during the classification phase, principal component analysis (PCA) neural network algorithm is utilized for classifying the input galaxies images into to one of four obtained source catalogue type. The rest of this paper is structured as follows. Section 2 presents presents the basic concepts of rough sets and PCA neural network algorithms. Section 3 presents the proposed approach for galaxies images classification and describes its phases. Section 4 discusses experimental result. Finally, section 5 addresses conclusions and presents future work.

2 Preliminaries

2.1 Rough Sets

Rough set theory [12–14, 11] is a fairly new intelligent technique for managing uncertainty that has can be used for the discovery of data dependencies, evaluation of the importance of attributes, discovery of patterns in data, reduction of attributes, and the extraction of rules from databases. Such rules have the potential to reveal new patterns in the data and can also collectively function as a classifier for unseen data sets. Unlike other computational intelligence techniques, rough set analysis requires no external parameters and uses only the information present in the given data. One of the interesting features of rough sets theory is that it can tell whether the data is complete or not based on the data itself. If the data is incomplete, it suggests more information about the objects to be collected in order to build a good classification model. On the other hand, if the data is complete, rough sets can determine the minimum data needed for classification. This property of rough sets is important for applications where domain knowledge is limited or data collection is very expensive/laborious because it makes sure the data collected is just good enough to build a good classification model without sacrificing the accuracy of the classification model or wasting time and effort to gather extra information about the objects [12–14, 11].

In rough sets theory, the data is collected in a table, called a decision table (DT). Rows of the decision table correspond to objects, and columns correspond to attributes. In the data set, we assume that the a set of examples with a class label to indicate the class to which each example belongs are given. We call the class label the decision attributes, and the rest of the attributes the condition attributes. Rough sets theory defines three regions based on the equivalent classes induced by the attribute values: *lower approximation*, *upper approximation* and *boundary*. Lower approximation contains all the objects, which are classified surely based on the data collected, and upper approximation contains all the objects which can be classified probably, while the boundary is the difference between the upper approximation and the lower approximation. So, we can define a rough set as any set defined through its lower and upper approximations. On the other hand, indiscernibility notion is fundamental to rough set theory. Informally, two objects in a decision table are indiscernible if one cannot distinguish between them on the basis of a given set of attributes. Hence, indiscernibility is a function of the set of attributes under consideration. For each set of attributes we can thus define a binary indiscernibility relation, which is a collection of pairs of objects that are indiscernible to each other. An indiscernibility relation partitions the set of cases or objects into a number of equivalence classes. An equivalence class of a particular object is simply the collection of objects that are indiscernible to the object in question.

2.2 Principal Component Analysis (PCA) Neural Network

The basic problem in automated classification system is to found data features that are important for the classification (feature extraction). One wishes to transform the input samples into a new space (the feature vector) where the information about the samples is retained, but the dimensionality is reduced. This will make the classification is high accuracy, as each galaxy image is presented by 7 feature (moment invariant) [10].

Principal component analysis neural networks approach combines unsupervised and supervised learning in the same topology. Principal component analysis is an unsupervised linear procedure that finds a set of uncorrelated features, principal components, from the input. On the other hand, a multilayer perceptron (MLP) neural network is a supervised procedure to perform the nonlinear classification from these components [9]

Principal component analysis (PCA) also called Karhunen-Loeve transform of Singular Value Decomposition (SVD) is such a technique. PCA finds an orthogonal set of directions in the input space and provides a way of finding the projections into these directions in an ordered fashion. The first principal component is the one that has the largest projection. The orthogonal directions are called the eigenvectors of the correlation matrix of the input vector, and the projections the corresponding eigenvalues. Since PCA orders the projections, we can reduce the dimensionality by truncating the projections to a given order. The reconstruction error is equal to the sum of the projections (eigenvalues) left out. The features in the projection space become the eigenvalues. Note that this projection space is linear. PCA is normally done by analytically solving an eigenvalue problem of the input correlation function. However, Sanger and Oja demonstrated that PCA can be accomplished by a single layer linear neural network trained with a modified Hebbian learning rule. For a network having p inputs/components and m¡p linear output PEs. The output is given by the following equation (1) [10].

$$y_i(n) = \sum_{i=0}^{p-1} W_{ij}(n)X_i(n). \tag{1}$$

where j= 0.1,..., m-1.

To train the weights, we will use the following modified Hebbian rule2.

$$\triangle W_{ij}(n) = \eta[y_i(n)X_i(n) - y_i(n)\sum_{k=0}^{i} W_{kj}(n)y_k(n)] \tag{2}$$

Where h is the step size.

The importance of PCA analysis is that the number of inputs for the MLP classifier can be reduced a lot, which positively impacts the number of required training patterns, and the training times of the classifier.

3 The Proposed Approach for Galaxies Images Classification

The proposed system is generally composed of four phases, (1) **image denoising**, which is utilized to remove noise pixels, (2) **feature extraction**, where input galaxies images are normalized to a uniform scale, then the Hu seven invariant moment algorith is applied in order to reduce the dimensionality of the feature space, and (3) **reduct generation**, where attributes in the information system table that is more important to the knowledge is generated as a subset of attributes. The subset of attributed, which is called a reduct, is fully characterizing the knowledge in the database. Finally (4) **classification phase**, where principal component analysis (PCA) neural network algorithm is utilized for classifying the input galaxies images into to one of four obtained source catalogue type. These four phases are described in details in the following section along with the steps involved and the characteristics feature for each phase. General architecture of the introduced approach is described in figure 1.

3.1 Image Denoising Phase

In order to calculate the invariant moment features the image must be denoised so that it contains only the desired galaxy. The type of noise that dominates the Galaxy image is spike noise. Simple edge detection is the only denoising process that is required as the spikes have sharp edges that distinguish them from local intensity variation in the image.

3.2 Feature Extraction Phase

Create a 7x1 feature vector through the process of feature extraction that is performed by computing the Hu seven invariant moments for all of the training and test galaxy images. Then Principal Component Analysis (PCA) neural network as artificial neural network based classifier to classify the galaxy image.

Moment Invariant. Invariant moment is a feature extraction from galaxies images, that algorithm has been proven its best techniques because gives feature invariant to scale, position and orientation of galaxy image. Traditionally, Invariant moments are computed based on the information provided by both the boundary and interior of the shape [7].

Traditionally, moment invariants are computed based on the information provided by both the shape boundary and its interior region. Given a continuous function $f(x, y)$ defined on a region D in the x, y plane, these regular moments are defined as in equation (3):

$$m_{pq} = \iint_D x^p y^q f(x, y) dx dy \tag{3}$$

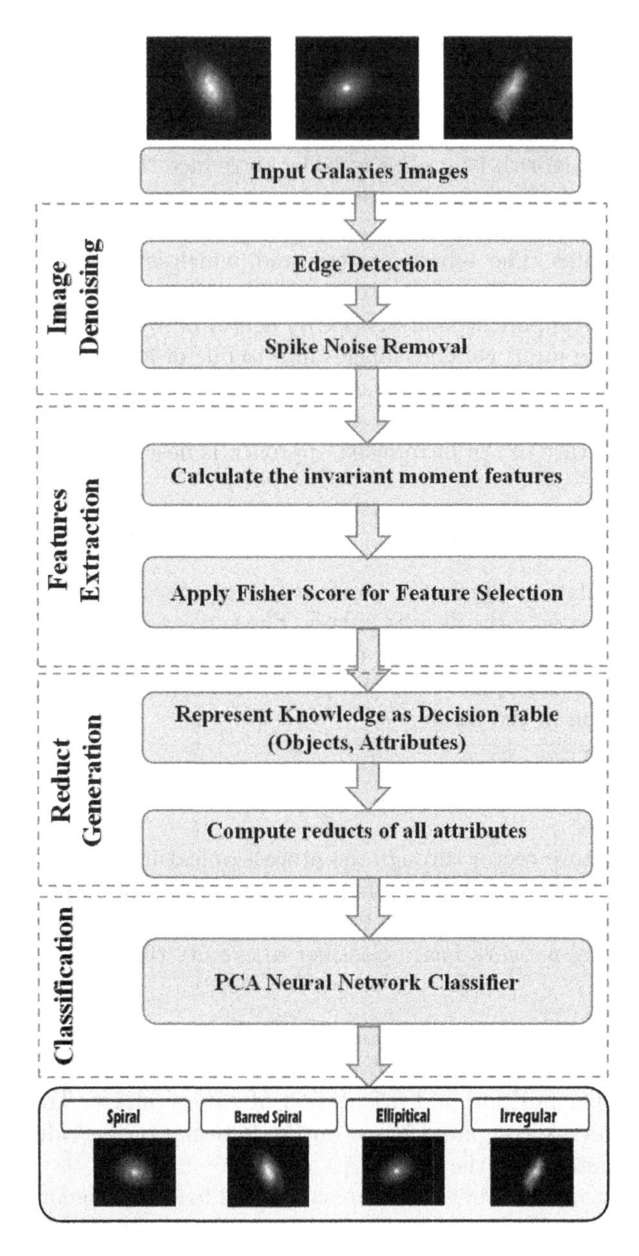

Fig. 1. Architecture of the proposed ripeness classification approach

Although the moments used to construct the moment invariants are defined for continuous space two dimensional functions f(x, y), however practical images are defined on discrete two dimensional space in the form of pixels and hence they are computed in discrete form as in equation (4):

$$m_{pq} = \sum_{j=0}^{N-1} \sum_{i=0}^{M-1} (i \triangle x)^p (j \triangle y)^q f(i \triangle x, j \triangle y) \tag{4}$$

Where are pixel indices in the x and y directions respectively, are the corresponding space sampling intervals and are the number of pixels in the x and y directions respectively. Strictly speaking, the seven moments defined by Hu are computed by the normalized central moments defined as the following equations (5) and (6):

$$\eta_{pq} = \frac{\mu_{pq}}{\mu_{00}^{\gamma}} \tag{5}$$

$$\mu_{pq} = \sum_{j=0}^{N-1} \sum_{i=0}^{M-1} (i \triangle x - \bar{x})^p (j \triangle y - \bar{y})^q f(i \triangle x, j \triangle y) \tag{6}$$

Where $\bar{x}$ and $\bar{y}$ are the image centroids defined as the equations (7) and (8), μ_{00}^{γ} is the normalization constant defined as the equation (9):

$$\bar{x} = \frac{m_{10}}{m_{00}} \tag{7}$$

$$\bar{y} = \frac{m_{01}}{m_{00}} \tag{8}$$

$$\gamma = \frac{p+q}{2} + 1 \tag{9}$$

In terms of the central moments, the seven moments are given as the following equations (10), (11), (12), (13), and (14):

$$M_1 = \eta_{20} + \eta_{02} \tag{10}$$

$$M_2 = (\eta_{20} - \eta_{02})^2 + 4\eta_{11}^2 \tag{11}$$

$$M_3 = (\eta_{30} - 3\eta_{12})^2 + (3\eta_{21} - \eta_{03})^2 \tag{12}$$

$$M_4 = (\eta_{30} - \eta_{12})^2 + (3\eta_{21} - \eta_{03})^2 \tag{13}$$

$$\begin{aligned} M_7 = {} & (3\eta_{21} - \eta_{03})(\eta_{30} + \eta_{12})[(\eta_{30} + \eta_{12})^2 \\ & -(3\eta_{21} - \eta_{03})^2] - (\eta_{30} - 3\eta_{12})(\eta_{21} \\ & -\eta_{03})[3(\eta_{30} - 3\eta_{12}) - (\eta_{21} - \eta_{03})]^2 \end{aligned} \tag{14}$$

The most interesting property of this type of features is that all of the above seven moments are invariant to galaxy scale, position, and orientation. The feature vector of the input image to the procedure is thus defined as equation (15):

$$X = [\mathrm{M}_1\ \mathrm{M}_2......\mathrm{M}_7]^{\mathrm{T}} \tag{15}$$

Where $(.)^{\mathrm{T}}$ denotes the matrix transpose operation.

Fisher Score. In this section, we briefly review the Fisher score for feature selection and discuss its shortcomings. The key idea of the Fisher score [17] is to find a subset of features such that, in the data space spanned by the selected features, the distances between data points in different classes are as large as possible, while the distances between data points in the same class are as small as possible. For a given set of selected features, the Fisher score
the fisher score computed as the following equations (16), (17), and (18):

$$F(X_j) = \frac{\sum_{k=1}^{c} n_k(\mu(j,k) - \mu(j))^2}{\sigma_j^2} \tag{16}$$

$$\sigma_j^2 = \sum_{k=1}^{C} n_k(\sigma(j,k))^2 \tag{17}$$

$$\mu(j) = \sum_{k=1}^{C} n_k(\mu(j,k))^2 \tag{18}$$

Where C is the number of classes, n_k is the size of the kth class respectively in the data space, i.e., $k = 1, 2, ...n_k$, $\mu(j,k)$ is the mean vector, and $\mu(j)$ is the overall mean vector of the reduced data. The fisher score of the feature is easily compute [8].

3.3 Reduct Generation Phase

The issue of knowledge representation is of primary importance in current research in Artificial Intelligence (AI), for computational reason we need syntactic representation of knowledge which suitable for computer processing. Table 1 considered knowledge representation system (Decision Table) columns of which are labelled by attributes, rows are labelled by objects,and each row represent a piece of information about the corresponding object. In this phase, an interesting question is whether there are attributes in the information system table is more important to the knowledge represented in the equivalence class structure than other attributes. Often, we wonder whether there is a subset of attributes which can, by itself, fully characterize the knowledge in the database; such an attribute set is called a reduct.

Computation of reducts of all attributes which equivalent to elimination of some column from the decision table .

Table 1. Information System Table

Object	Attributes							Decision
	M1	M2	M3	M4	M5	M6	M7	
1	0.7445	0.0993	0.0053	0.0098	0	-0.0012	0	1
2	0.699	0.342	0.0002	0	0	0	0	1
69	1.1701	0.3443	0.7452	0.2793	0.0564	-0.0828	-0.0233	2
93	0.6135	0.0005	0.0001	0.0033	0	-0.0001	0	3
108	0.6888	0.0051	0.0029	0.0203	0.0002	0.0013	0.0001	4

3.4 Classification Phase

An institutive goal of classification is to discriminate between different galaxies types or to measure structural properties of these galaxies. Classification phase based on a set of selected features or compute. The proposed system for galaxies images classification based on the invariant moment features using Principal Component Analysis Neural Network. The details about PCA and the performance measures, which calculations are presented in the next subsection. The PCA classifier was evaluated and assessed based on a set of selected feature as described in experimental result section.

4 Experimental Results and Discussion

4.1 Data Collection

The images used in this paper were obtained from the ZsoltFrei Catalogue provided by the department of Astrophysical Sciences at Princeton University [4]. This catalogue contains approximately 113 different galaxy images and is often used as a benchmark for astronomical study as the images are carefully calibrated. Images taken in several passbands and a color-composite image are included for each galaxy. Images of 31 galaxies were taken with the 1.5 meter telescope of the Palomar observatory in 1991; images of the other 82 galaxies were taken with the 1.1 meter telescope of the Lowell observatory in 1989. At Palomar the camera had an 800x800 TI CCD; at Lowell the camera had an RCA 512x320 CCD. Palomar images are available in three passbands of the Thuan-Gunn system: g, r, and i. At Lowell images are in two passbands (J and R) of the filter system developed by Gullixon et al. [15]

The selected data set of Zsolt frei catalogue has the following properties (i) **high resolution** (ii)**good quality** (iii) **careful calibration**. Which doesnt require preprocessing phase, such as gamma correction or histogram equalization.

The proposed algorithm is highly sensitive to the noise as the number of tested features is very small, only seven invariant moments. Hence, an image denoising technique is required to remove noisy pixels that dont belong to the galaxy pixels. The advantage of the used algorithm is the dependence on a very small dataset for each class. Only six training images were used for the training phase and a small number of features was tested without the need to select the most top ranked features from the Fisher score matrix. Although color images were used, no color features were used in this study as the images are transformed to the gray scale. The four different measured show the proposed algorithm is fast and more accuracy than other attempts used before to solving the classification problem of galaxy images.

4.2 Performance Measures

Four different measures have been used to evaluate the performance of the artificial neural network (ANN) based classification techniques as the following:

- **Mean Squared Error (MSE):** The mean squared error is calculating the difference between values implied by an estimator and the true values of the quantity being estimated. MSE is a risk function If is a vector of n predictions, and is the vector of the true values, then the MSE of the predictor as in the equation (19):

$$MSE = \frac{\sum_{J=0}^{n}(P_{ij} - T_j)^2}{n} \tag{19}$$

Where P_{ij} is the value predicted by the individual case i for fitness case j (out of n fitness cases or sample cases), and T_j is the target value for fitness case j.

- **Normalized Mean Squared Error (NMSE):** The normalized mean square error is an estimator of the overall deviations between predicted and measured values as defined in the equation (20):

$$NMSE = \sum_{J=0}^{n}(P_{ij} - T_j)^2(n \times P \times T) \tag{20}$$

Where $P = \sum_{i=1}^{n} P_{ij}/n$ and $P = \sum_{j=1}^{n} T_j/n$

- **Correlation Coefficient (R)**
The correlation coefficient (R) is a quantity that gives the quality of a least squares fitting to the original image. For two data sets x, y, the auto correlation is given as in equation (21):

$$R = \frac{cov(x.y)}{\sigma x \times \sigma y} \tag{21}$$

Where x and y are the standard deviation of image x and y.

Fig. 2. Image Denoising Result

Table 2. Performance Measure

ANN name	PCA Based Classifier	PCA Based Classifier Rough set
MSE	0.001215621822	0.000461733078
NMSE	0.003999405670	0.001519105578
r	0.998153586134	0.999373416338
%error	1.9110336092777	1.131480628624

- **Error Percentage (E):** The error percentage is calculated as the percentage difference between the measured value and the accepted value [6].

The proposed approach was tested using a dataset of 108 galaxy images (spiral, barred spiral, elliptical and irregular) representing the most common types of galaxies. The images were first classified manually by the author. The author classication is assumed to be 100 perfect as the human recognition system is supposed to be an absolute reference for classication. The galaxies in the dataset were divided into two groups, a training set consisting of 24 images (six images for each galaxy type) and 68 test images. All images were selected randomly by the MATLAB simulation program from the full dataset.

An advantage of this study is that both monochromatic and color images can be used, as the image is rstly transformed to a monochromatic black and white image. It is also independent of the galaxy position, orientation, and size inside the image which con. This contrasts with the Galaxy Zoo monochrome images [16]. Another advantage of the pro- posed algorithm is that it relies on a very small dataset for each class, i.e. only six training images were used for the training phase and a small number of features were tested without the need to select the top-ranked features from the Fisher score matrix. However, because only a small number of features was tested, i.e., only seven invariant moments. Figure 2 shows an original galaxy image and the resulted denoised image. As can be seen in figure 2, edge detection has successfully identified almost all the spikes in the sample image. Fortunately, the area of the image that contain the galaxy is characterized by slow variation in intensity such that there is no sharp boundary between the area of the galaxy and it neighborhood. This helped a lot in the process of separating the desired content from undesired object such as distant stars, comets or any other astronautical objects that is observed in the same image.

The proposed algorithm is highly sensitive to noise. This noise appears in the form of spikes, scattered at different positions in the image as a large-scale salt-and-pepper noise. Hence, an image denoising technique is required to remove noise pixels. The results of the simulation over the whole available data set show that about 98. of the galaxy images were classified correctly to the galaxy classes. The sher score of all seven invariant moments used for classification are arranged in a descending order. The top-ranked feature is the first central moment and there is a clear distinction between the value of the Fisher score for this feature and the remaining six features. This implies that the MSE classification may depend only on one feature and ignoring the remaining features will not strongly affect the outcome of the class prediction with more number of image greater than 2000 images. Table 2 shows comparison between classification accuracy using combination of PCA and rough set as feature reduction techniques along with invariant moments for feature extraction and classification accuracy having no rough set feature reduction technique applied. It shows using invariant moments for feature extraction in combination with PCA and rough set provided better classification rate compared to utilizing PCA neural network algorithm only.

5 Conclusions and Future Work

We have presented the invariant moment as promising and good feature for galaxy classification. The aim of this paper was to investigate the usefulness of invariant moments for the automatic identification of common galaxy shapes. The obtained results show that using invariant moments for feature extraction in combination with Principal Component Analysis and rough set as feature reduction techniques provide better results than original using invariant moments in combination with principal component analysis. We also conclude that a small set of features (seven features) is sufficient to classify galaxy images and provides a fast classification process and with rough set each image present by (6 features)attributes provide best result with high accuracy. Future work includes testing these methods for classifying more types of galaxy images and dealing with unbalanced data sets. build an automated classification system that can be detect peculiar galaxies. The contribution here is using a simple, fasting, and robust algorithm for galaxies image classification. The experiment results are shown the promising results.

References

1. de la Calleja, J., Fuentes, O.: Automated Classification of Galaxy Images. In: Negoita, M.G., Howlett, R.J., Jain, L.C. (eds.) KES 2004. LNCS (LNAI), vol. 3215, pp. 411–418. Springer, Heidelberg (2004)
2. Lahav, O.: Artificial neural networks as a tool for galaxy classification. In: Data Analysis in Astronomy, Erice, Italy (1996)
3. Pettini, M., Christensen, L., D'Odorico, S., Belokurov, V., Evans, N.W., Hewett, P.C., Koposov, S., Mason, E., Vernet, J.: CASSOWARY20: a wide separation Einstein Cross identified with the X-shooter spectrograph. Monthly Notices of the Royal Astronomical Society 402, 2335–2343 (2010)
4. Frei, Z.: Zsolt Frei Galaxy Catalog, Princeton University, Department of Astrophysical Sciences (1999), `http://www.astro.princeton.edu/frei/catalog.htm` (retrieved 2002)
5. De La Calleja, J., Fuentes, O.: Machine Learning and Image Analysis for Morphological Galaxy Classifcation. Monthly Notices of Royal Astronomical Society 349(1), 87–93 (2004)
6. Mohamed, M.A., Atta, M.M.: Automated Classification of Galaxies Using Transformed Domain Features. IJCSNS International Journal of Computer Science and Network Security 10(2) (2010)
7. Hu, M.: Visual pattern recognition by moment invariants. IEEE Transactions on Information Theory 8, 179–187 (1962)
8. Duda, Stork, D.G.: Pattern Classification. Wiley-Interscience Publication (2001)
9. Hsieh, W.W.: Nonlinear Principal Component Analysis by Neural Network. Univeristy of British Columbia, Appeared in Tellus 53A, 559–615 (2001)
10. Jahanbani, A., Karimi, H.: A new Approach for Detecting Intrusions Based on the PCA Neural Networks. J. Basic. Appl. Sci. Res. 2(1), 672–679 (2012)
11. Pawlak, Z.: Rough sets. International J. Comp. Inform. Science 11, 341–356 (1982)
12. Pawlak, Z.: Rough sets – Theoretical aspects of reasoning about data. Kluwer (1991)
13. Pawlak, Z., Grzymala-Busse, J., Slowinski, R., Ziarko, W.: Rough Sets. Communications of the ACM 38(11), 88–95 (1995)
14. Polkowski, L.: Rough Sets: Mathematical Foundations. Physica-Verlag (2003)
15. Frei, Z.: Zsolt Frei Galaxy Catalog, Princeton University, Department of Astrophysical Sciences (1999), `http://www.astro.princeton.edu/frei/catalog.htm` (retrieved 2002)
16. Banerji, M., Lahav, O., Lintott, C.J., Abdala, F., Schawinski, K., Bamford, S., Andreescu, D., Raddick, M., Murray, P., Jordan, M., Slosar, A., Szalay, A., Thomas, D., Vandenberg, J.: Galaxy Zoo: reproducing galaxy morphologies via machine learning. Monthly Notices of the Royal Astronomical Society 406, 342–353 (2010)
17. Bishop, D.V.M., North, T., Donlan, C.: Nonword repetition as a behavioural marker for inherited language impairment: Evidence from a twin study. Journal of Child Psychology and Psychiatry 37, 391–403 (1996)

Comparison of Crisp, Fuzzy and Possibilistic Threshold in Spatial Queries

Jan Caha, Alena Vondráková, and Jiří Dvorský

Department of Geoinformatics, Faculty of Science, Palacký University Olomouc
17. listopadu 50, 771 46, Olomouc, Czech Republic
{jan.caha,alena.vondrakova,jiri.dvorsky}@upol.cz

Abstract. Decision making is one of the most important application areas of geoinformatics. Such support is mainly oriented on the identification of locations that fulfil certain criterion. The contribution presents the suitability of various approaches of spatial query using different types of Fuzzy thresolds. Presented methods are based on the classical logic (Crisp queries), Fuzzy logic (Fuzzy queries) and Possibility theory (Possibilistic Queries). All presented approaches are applied in the case study. Use these findings may contribute to the better understanding of the nature of the methods used and can help to obtain more accurate results, which have a determining influence on subsequent decision-making process.

Keywords: decision making, Possibility theory, spatial query.

1 Introduction

It is stated that up to 90% of the information has a spatial character [2]. For efficient work with them is therefore necessary not only a system that allows a spatial data store and manage, but it is also necessary to ensure a sufficiently effective tool for spatial querying. Spatial analysis and spatial querying have a great importance in a wide range of research areas and human activities. Frequently discussed area is decision making (in crisis management, land use planning, modelling, etc.) [14]. Precision of data obtained from answers to such spatial querying, as well as a form of this information, have a critical impact on the accuracy of subsequent decisions [3]. What more accurately it will be possible to define the parameters of spatial query and what more precisely can be visualized the result, subsequently can be more confident the final decision. The nature of a variety of phenomena, however, cannot be described as precise number. Therefore, the criteria to define the boundaries cannot be crisp values, since these exact boundaries in a variety of phenomena do not occur naturally [8]. Assistance with decision making based on spatial data is one the main objectives of geoinformatics [2,14]. Such support is mainly oriented on the identification of locations that fulfil certain criterion. Such criterion can be minimal or maximal distance from specific objects, areas with specific slopes and/or orientation, areas with a certain area etc. Regardless on the purpose or type of the spatial query

A. Abraham et al. (eds.), *Innovations in Bio-inspired Computing and Applications*, Advances in Intelligent Systems and Computing 237,
DOI: 10.1007/978-3-319-01781-5_22, © Springer International Publishing Switzerland 2014

there is one common property that those queries should have. That property is flexibility (softness). Generally it is not reasonable to specify thresholds for spatial queries as crisp values since such query can easily fail to identify suitable solution. Thus so called Fuzzy thresholds were introduced to the problematic of spatial decision support. Such queries are more flexible due to their natural softness which allows better ranking of alternatives and comparison of possible solutions. While these Fuzzy thresholds adds considerable amount of flexibility to the spatial queries the issue can be taken even further by introducing Possibilistic thresholds that are based on Possibility theory and that adds yet another level of flexibility and options for better spatial decision making. The main aim of the article is to present possibilities of Possibilistic thresholds and their use in spatial queries.

The structure of article is following: Sections 2 and 3 offers brief summary of Fuzzy Sets and Possibility Theory . Section 4 summarizes information about approaches and possibilities of spatial queries and their use for decision making. Case study on the matter is shown in section 5 and the results are discussed and evaluated in sections 6 and 7.

2 Fuzzy Sets

Fuzzy set is a special case of set that does not have strictly defined criterion of membership. For example set of "large numbers" or "steep slopes" do not have strict threshold but rather a transitional interval where objects have increasing or decreasing value of membership. Fuzzy set is determined by the membership function, which is defined as mapping

$$\mu_{\tilde{A}} : U \longrightarrow [0,1] \tag{1}$$

that indicates that element x of universe U has membership value $\mu_{\tilde{A}}(x)$ from the interval $[0,1]$ [16]. Elements with $\mu_{\tilde{A}}(x) = 0$ do not belong to the set $\tilde{A}$, while elements $\mu_{\tilde{A}}(x) = 1$ completely belong to the set. Other membership values indicates partial membership in the set. Fuzzy set can be expressed as pairs of elements and their membership values

$$\tilde{A} = \{(x, \mu_{\tilde{A}}(x)) \mid x \in U, \mu_{\tilde{A}}(x) \in [0,1]\}. \tag{2}$$

or by exact the definition of membership function [10]. Fuzzy sets are often use to model sets without precise borders or in situation where natural gradualness of results is required [6,12].

3 Possibility and Necessity Measures

Let $\tilde{F}$ be a normalized fuzzy subset of universe U which is characterized by the membership function $\mu_{\tilde{F}}$. This fuzzy sets act as a fuzzy restriction [4]. Let X be a variable from U. Then the *possibility measure* derived from the membership function $\mu_{\tilde{F}}$ by

$$\Pi_{\tilde{F}}(X) = \sup_{x \in X} \mu_{\tilde{F}}(x) \qquad \forall X \subseteq U \tag{3}$$

and $\mu_{\tilde{F}}$ is a possibility distribution underlying $\Pi_{\tilde{F}}$ that can be denoted as $\pi_{\tilde{F}}$ [5]. If X is a fuzzy set $\tilde{X}$ then this equation can be extended to

$$\Pi_{\tilde{F}}(\tilde{X}) = \sup_{x} \min(\mu_{\tilde{F}}(x), \mu_{\tilde{X}}(x)). \tag{4}$$

Let $\overline{X}$ be complement of X and Π a possibility measure then set function $\mathcal{N}$ defined by

$$\mathcal{N}_{\tilde{F}}(X) = 1 - \Pi(\overline{X}) \qquad \forall X \subseteq U \tag{5}$$

is a *necessity measure* [5]. Such necessity measure can be also called certainty measure. If both X and F are fuzzy set $\tilde{X}, \tilde{F}$ then this equation is extended to

$$\mathcal{N}_{\tilde{F}}(\tilde{X}) = 1 - \sup_{x} \min(\mu_{\tilde{F}}(x), 1 - \mu_{\tilde{X}}(x)). \tag{6}$$

There is constraint that specifies that always $\mathcal{N}_{\tilde{F}}(X) \leq \Pi_{\tilde{F}}(X)$ [5].

3.1 Comparing Real Numbers to Possibility Distribution

In order to asses possibility and necessity of value x being bigger or equal then the possibility distribution (threshold) $\tilde{Y}$, there is need a to evaluate following equations [5]:

$$\mu_{[\tilde{Y},\infty)}(x) = \Pi_x((-\infty, \tilde{Y}]) = \sup_{y \leq x} \mu_{\tilde{Y}}(y) \tag{7}$$

$$\mu_{]\tilde{Y},\infty)}(x) = \mathcal{N}_x((-\infty, \tilde{Y}[) = \inf_{y \geq x} (1 - \mu_{\tilde{Y}}(y)) \tag{8}$$

Where y is any value from $\tilde{Y}$. According to those equations possibility and necessity of $x \leq \tilde{Y}$ can be calculated and used for further decision making. Details on the implementation, proofs and process of answering inverse problem are provided in [4,5] and [17].

4 Spatial Queries

Usual aim of spatial queries is to select areas that meet one or more conditions. The condition is usually defined as a value being higher or smaller than given threshold. Such queries cannot quite well introduce any measure of preference in the result, because they are based on classical logical expressions [1,6]. But many decision making situations that involve spatial data are not well suited for such crisp queries, because the crisp query might be far to restrictive for such utilization [15].

The reasons for creating soft thresholds can be summarized as following:

- the concept of the threshold is naturally vague i.e. definition of "steep slopes",
- there are more than one acceptable definitions of threshold Y and there is no indications that any of them is more correct or precise than the others,
- threshold Y is based on the expert opinion that is provided as an interval of values rather then precise value, or there is a need to merge definitions of Y from several such expert opinions,
- results need to be arranged and the concept of intervals with associated suitability value is insufficient or inappropriate for the problem.

These findings are supported by numerous studies [1,6,9].

4.1 Crisp Queries

Crisp query is based on classical logic usually looks like "is variable X higher (or lower) then threshold Y?". In such case it does not matter what is the difference between X a Y, if X is smaller then it is rejected from the resulting set. However there is a clear difference between $X_1 = 0.1$ and $X_2 = 1.799$ when they are compared to the threshold $Y = 1.8$. While X_1 is clearly smaller and should not be included in the set of numbers equal or higher than Y, X_2 is quite another matter. Indeed it is lower then Y but the difference is so small that X_2 is almost indistinguishable from the threshold. This example illustrates the biggest drawback of Crisp Queries which is the rigidity. It can only classify data into two groups (Fig. 1). Sometimes more such thresholds are defined that classify the data into several intervals that indicates growing or decreasing suitability of the alternatives in these categories. Each category has assigned specific value of suitability. Figure 1 shows such classes with intervals $[0, 0.4]$ and suitability value 0, $[0.4, 1.8]$ with suitability 0.33, $[1.8, 2.7]$ with 0.66 and finally interval $[2.7, 3.0]$ with suitability 1. However the problem of rigidity of definition is preserved, because the crisp thresholds are still used in creation of the stepwise classification and all the values on one class have the same amount of suitability (Fig. 1). While this is helpful for the decision maker, the suitability ordering of the values still is not as fluid as it should be.

4.2 Fuzzy Queries

Fuzzy query is based on Fuzzy logic [18] and its main purpose is to extend possibilities of Crisp Queries [12]. For complex decision making processes it is much better to specify the threshold as a fuzzy set [6]. In a case mentioned in the previous chapter the crisp threshold can be approximated with Fuzzy set that has a membership function defined:

$$\mu_{\tilde{Y}}(x) = \begin{cases} 0 & \text{if } x < 0.4 \\ \dfrac{x - 0.4}{2.7 - 0.4} & \text{if } 0.4 \leq x \leq 2.7 \\ 1 & \text{if } 2.7 < x. \end{cases} \tag{9}$$

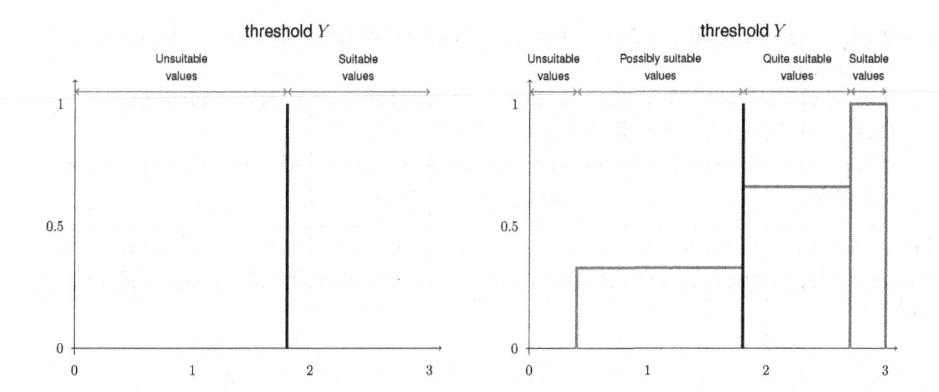

Fig. 1. Left figure - Crisp threshold. Right figure - Several crisp thresholds forming intervals of suitability.

Evaluating if the specific value of X is lower (or higher) then such threshold is then matter of calculation of the membership value to the fuzzy set denoted as $\tilde{Y}$. The main addition is the natural gradualness of the results. Because of that there is no longer any need to create classes with various suitability. Fuzzy thresholds are typically used in a situation when the threshold is described in terms of values that are possible to be used (in the example these are values higher then 0.4) and values absolutely fulfil the criterion (these are values higher then 2.7 in the example). Such description is quite common in geography, because many elements have naturally vague definitions [8]. It is quite apparent that this approach is a natural extension of the Crisp query with several intervals. However this approach is much more complex as it can distinguish two close values, that would otherwise fall into one category, but their suitability value will be also quite similar, yet not the same.

4.3 Possibilistic Queries

Possibilistic Queries are based on Possibility theory [17,5]. In such case the threshold used in the query is modelled as a fuzzy number [10] and it serves as a possibility distribution [17]. Triangular fuzzy numbers are specified by three values, support values - in the previously mentioned example the values would be 0.4 and 2.7, and a kernel value, which is 1.8 (Fig. 2). This allows creation of true soft threshold that is represented by such possibility distribution. This type of query is useful when the criterion is described as values that are absolutely suitable, values that are quite suitable, values that can be used but it would be better to avoid them and values that are completely unsuitable (Fig. 2). This approach extends the fuzzy approach by introducing not only one but two measures of suitability. These Possibilistic thresholds are compared to the crisp values of the data by means Eqs. (7) and (8). The resulting measures identify both possibly and necessary suitable results. Interpretation of the results is then:

- if $\Pi_{\tilde{Y}}(x)) = 0$ and $\mathcal{N}_{\tilde{Y}}(x) = 0$ then x is lower then the threshold and thus unsuitable,
- if $\Pi_{\tilde{Y}}(x)) > 0$ and $\mathcal{N}_{\tilde{Y}}(x) = 0$ then x is possibly (or partly) suitable,
- if $\mathcal{N}_{\tilde{Y}}(x) > 0$ is quite suitable,
- if $\mathcal{N}_{\tilde{Y}}(x) = 1$ then the x is for sure higher number then $\tilde{Y}$ and thus absolutely suitable.

These are four possible variants of outcomes. Details on the implementation, proofs and process of answering inverse problem ($x < \tilde{Y}$) are provided in [4] and [5].

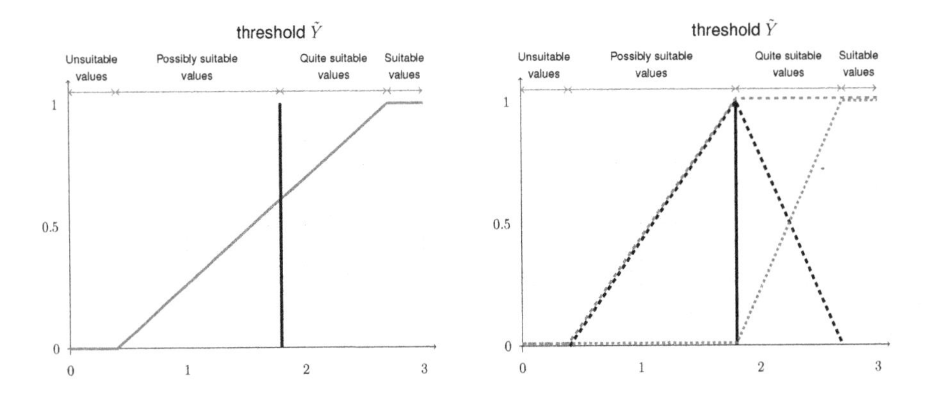

Fig. 2. Left figure - Fuzzy threshold. Right figure - Possibilistic threshold (dashed) with measures of possibility (grey dashed) and necessity (grey dotted).

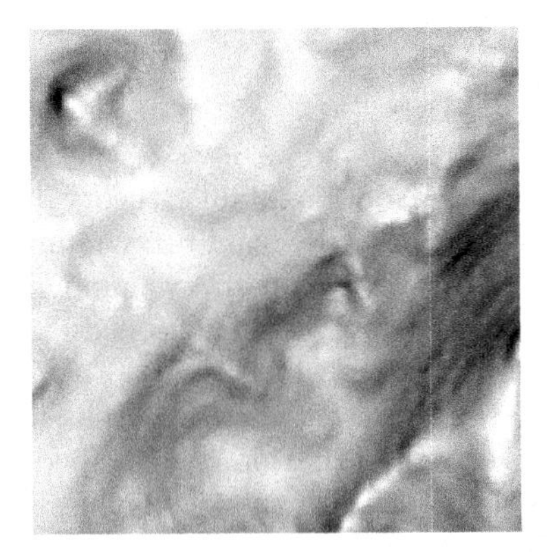

Fig. 3. Slope of the are of the interest. Small values of slope are represented by white colour and the higher values are darker.

5 Case Study

The case study presents rather simple problem. Suppose that we have a surface represented by a field model [13] that has N rows and M columns, with each cell denoted by $\tilde{C}_{i,j}$ where $i = 1, 2, \ldots, N$ and $j = 1, 2, \ldots, M$. For every cell of this surface the slope is calculated using Horn's method [11]. The slope of the testing data is in range $0\% - 110\%$ (Fig. 3). The task is to identify all areas with slope higher then medium slope. All three possible approaches to this task will be shown and compared.

The simplest way is to specify crisp threshold after which the slope is considered medium. Since there is no general definition the value 7% is chosen the threshold by the expert. So the slopes are classified into two classes lower then 7% and higher or equal to 7%. The natural extension of this approach is identify values around the threshold and create special classes for them, to emphasize the areas that might were or were not chosen but differs only by a small margin from the threshold. In this case the classes were specified $0\% - 5\%$, $5\% - 7\%$, $7\% - 12\%$ and $12\% - 110\%$. Such classification obviously offers more information to the user and the decision maker (Fig. 4).

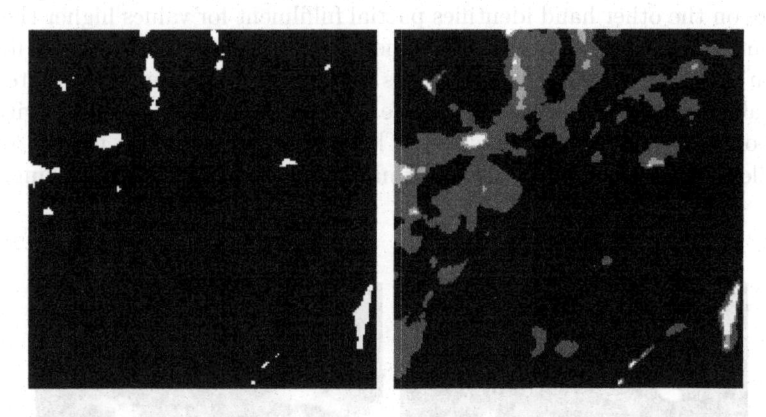

Fig. 4. Crisp classification of the slope with one category (left) and with four categories (right). The white colour indicates unsuitable solutions and the black suitable solutions.

The second mentioned approach is to approximate possible values of the threshold by a fuzzy set. Such fuzzy set can have linear membership function (similar to example on Fig. 2) but with important points at values 5% and 12% (Fig. 5). Such fuzzy set offers linear classification to the crisp classification into four classes from the Fig. 4. The result however is much more smoother and provides even more information then the crisp classification into classes.

The last approach that was mentioned in Section 4 is Possibilistic query. In this case the threshold is modelled by a triangular fuzzy number that has support range $[5\%, 12\%]$ and kernel value 7%. To correctly query such surface there is a need to obtain both measures of possibility and necessity (Fig. 6). As can be seen

Fig. 5. Fuzzy classification of the slope. The white colour indicates unsuitable solutions and the black suitable solutions.

from the figure 6 the possibility identifies much bigger area that has slope value at least 5% and shows complete fulfilment for values higher then 7%. Necessity measure on the other hand identifies partial fulfilment for values higher then 7% and complete fulfilment for values higher then 12%. This is the most flexible solution from all mentioned approaches as it allows the user to operate with limit values (smallest and highest value of the threshold) as well as with the most possible value of the threshold. The use of two measures instead of one then allows obtaining much more information for further decision making.

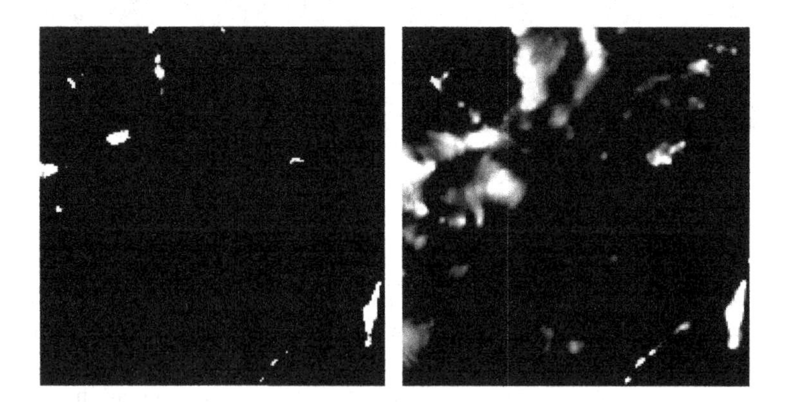

Fig. 6. Classification by the possibilistic threshold. Possibility (left) and necessity (right) values. The white colour indicates unsuitable solutions and the black suitable solutions.

6 Future Work

The topic of Possibilistic queries can be further extended especially by presenting more complex case studies and also by presenting methods for handling results of possibilistic queries in further data analysis. While the theoretical background for the use of Theory of Possibility in decision making and decision support is good [6] there are surprisingly few practical examples and case studies regarding the subject, not only in the geosciences but also in other scientific fields. Considering the potential and benefits that this approach can offer, it definitely should be studied further. For particular case studies there would be monitored relevance of results with regard to the application of different approaches in spatial querying. For effective implementation of the previous findings there is a need to implement appropriate tools into primarily used software and become potential users familiar with this procedure.

7 Conclusion

The contribution described two commonly used ways to query spatial data, Crisp and Fuzzy queries, and present new possible approach to this problematic by introducing Possibilistic queries. These queries are based on the Possibility theory and allows the decision maker to model thresholds by vague numbers. Results are obtained as the measure of possibility, that shows optimistic variant of the resulting selection, and the measure of necessity, that show pessimistic variant of the selection. These measures together allow the decision maker to embrace all types of decisions (Fig. 2), those that were barely selected (low possibility values). Decisions that can be used but are not completely suitable (high possibility values but low necessity values). And even decision that absolutely satisfies the constraint (necessity value of 1), such enlargement of possibilities of spatial queries is much desirable for decision making, as it allows better data handling as well as better spatial decision support.

Acknowledgement. The authors gratefully acknowledge the support by the Operational Program Education for Competitiveness - European Social Fund (projects CZ.1.07/2.3.00/20.0170 and CZ.1.07/2.2.00/28.0078 of the Ministry of Education, Youth and Sports of the Czech Republic).

References

1. Bosc, P., Kraft, D., Petry, F.: Fuzzy sets in database and information systems: Status and opportunities. Fuzzy Sets and Systems 156(3), 418–426 (2005)
2. Burrough, P., McDonnell, R.: Principles of geographical information systems. Oxford University Press, Oxford (1998)
3. Devillers, R., Stein, A., Bédard, Y., Chrisman, N., Fisher, P., Shi, W.: Thirty Years of Research on Spatial Data Quality: Achievements, Failures, and Opportunities. Transactions in GIS 14(4), 387–400 (2010)

4. Dubois, D., Prade, H.: Ranking Fuzzy Numbers in the Setting of Possibility Theory. Information Sciences 30(3), 183–224 (1983)
5. Dubois, D., Prade, H.: Possibility Theory: An approach to Computerized Processing of Uncertainty. Plenum Press, New York (1986)
6. Dubois, D.: The role of fuzzy sets in decision sciences: Old techniques and new directions. Fuzzy Sets and Systems 184(1), 3–28 (2011)
7. Dunn, M., Hickey, R.: The effect of slope algorithms on slope estimates within a GIS. Cartography 27(1), 9–15 (1998)
8. Fisher, P.: Sorites paradox and vague geographies. Fuzzy Sets and Systems 113(1), 7–18 (2000)
9. Fonte, C.C., Lodwick, W.A.: Modelling the Fuzzy Spatial Extent of Geographical Entities. In: Petry, F., Robinson, V.B., Cobb, M.A. (eds.) Fuzzy Modeling with Spatial Information for Geographic Problems, pp. 120–142. Springer, Berlin (2005)
10. Hanss, M.: Applied fuzzy arithmetic: an introduction with engineering applications. Springer, Berlin (2005)
11. Horn, B.: Hill shading and the reflectance map. Proceedings of the IEEE 69(1), 14–47 (1981)
12. Hwang, S., Thill, J.-C.: Modeling Localities with Fuzzy Sets and GIS. In: Petry, F., Robinson, V.B., Cobb, M.A. (eds.) Fuzzy Modeling with Spatial Information for Geographic Problems, pp. 120–142. Springer, Berlin (2005)
13. Janoška, Z., Dvorský, J.: P systems: State of the art with respect to representation of geographical space. In: CEUR Workshop Proceedings - 12th Annual Workshop on Databases, Texts, Specifications and Objects, DATESO 2012, pp. 13–24 (2012)
14. Longley, P., Goodchild, M., Maguire, D., Rhind, D.: Geographical information systems and science, 2nd edn. Wiley, Chichester (2005)
15. Witlox, F., Derudder, B.: Spatial Decision-Making Using Fuzzy Decision Tables: Theory, Application and Limitations. In: Petry, F., Robinson, V.B., Cobb, M.A. (eds.) Fuzzy Modeling with Spatial Information for Geographic Problems, pp. 120–142. Springer, Berlin (2005)
16. Zadeh, L.A.: Fuzzy Sets. Information and Control 8(3), 338–353 (1965)
17. Zadeh, L.A.: Fuzzy sets as a basis for a theory of possibility. Fuzzy Sets and Systems 1, 3–28 (1978)
18. Zadeh, L.A.: Is there a need for fuzzy logic? Information Sciences 178(13), 2751–2779 (2008)

Visualizing Clusters in the Photovoltaic Power Station Data by Sammon's Projection

Martin Radvanský, Miloš Kudělka, and Václav Snášel

VSB Technical University Ostrava, Ostrava, Czech Republic
{martin.radvansky.st,milos.kudelka,vaclav.snasel}@vsb.cz

Abstract. This paper presents results of the finding and the visualization cluster in the hourly recorded data of power from the small photovoltaic power station. Our main aim was to evaluate the use of Sammon's projection for visualizing clusters in the data of power. The photovoltaic power station is sensitive for changes according to the sun's light power. Although one can think that sunny days are the same the power of the sun light is very volatile during a day. When we wanted to analyse the efficiency of the power station, it was necessary to use some kind of clustering method. We propose the clustering method based on social network algorithms and the result is visualized by the Sammon's projection for explorational analysis.

Keywords: clustering, Photovoltaic power station, Sammon's projection.

1 Introduction

Photovoltaics are widely considered to be an expensive method of producing electricity. However, in off-grid situations photovoltaics are very often the most economical solution to provide the required electricity service. The growing market all over the world indicates that solar electricity has entered many areas in which its application is economically viable. Additionally, the rapidly growing application of photovoltaics in grid-connected situations shows that photovoltaics are very attractive for a large number of private people, companies and governments who want to contribute to the establishment of a new and more environmentally benign electricity supply system. A small photovoltaic power stations (PVPS) can be a reliable and pollution-free producer of electricity for home or office where it can help reduce high fees for electricity.

One of the major problems with PVPS is that the amount of power generated by a solar system at a particular site depends on how much of the sun's energy reaches it. In our country, the building of the small home PVPS is supported by government and many of the houses have their own PVPS on the roof. Although it was a very good idea to support families, a lot of really huge PVPS, owned by the local companies were built. Goverment reduced the redemption price of supplied electricity. For a small home, PVPS is necessary to minimize supplied power to the grid and in an efficient way reduce house power consumption. For this purpose it is necessary to know how PVPS generates electricity the during day time, month and a whole year. These parameters are specific to the each PVPS installation.

A. Abraham et al. (eds.), *Innovations in Bio-inspired Computing and Applications*,
Advances in Intelligent Systems and Computing 237,
DOI: 10.1007/978-3-319-01781-5_23,

In this paper we have processed a method for finding interesting clusters (group of days) in the PVPS power data. This helps to understand how particular installation of the PVPS generates power over time. The results forming this clustering can be used for optimizing electricity consumption in the house and it can also be used in the smart house systems.

This paper is organized as follows: Section 2 contains an overview of related work. Section 3 explains the methods used for data visualization. In the section 4, data preprocessing and clustering is described. The following section 5 contains experiments and the last section 6 concludes the paper.

2 Related Work

Many methods exist for finding and visualizing clusters in the PVPS power data. We will briefly describe some of them. The papers described in the next paragraph works with the large PVPS installations and also work with huge amounts of data. Although there exist many times smaller or home PVPS installations the aim of researchers is focused mainly on big PVPS.

Haghdadi in the paper [1] proposes the method for finding the optimal number of clusters and the best clustering method for 4 MW PVPS. He worked with four year hourly cumulated power data. In his article he utilizes several methods an compares their result. He used K–means, K–medoid, Fuzzy C–means and Guastafson–Kessel method applied to the dataset. In his paper results were evaluated by applying more comprehensive clustering validity by the crisp or fuzzy indices. In the paper [2] Haghdadi focuses on sitting and sizing of photovoltaic units with the aim of minimizing the total power loss in a distribution network. Due to fluctuations in output power of photovoltaic and variable nature of the loads in electricity network, for evaluating power loss and the bus voltages, the grid should be simulated for all days of the year and all hours of the day. He worked in the clustering method not only with hourly cumulated power but also with the level of irradiation, the ambient temperature and the wind speed. These three main parameters greatly influence the output power of PVPS. Modeling and estimation of daily global solar radiation data by an artificial neural network is described in the Benghammen paper [3]. He compared six different ANN models and conventional regression model. In his paper he worked with global radiation, diffuse radiation, air temperature and relative humidity. He tried to find a model for description of the different clusters of the similar days based on sunshine duration.

Chen [4] published paper about prediction of the solar radiation based on the fuzzy logic and neural network. His proposed techniques can find a difference of the solar radiation between the different sky conditions. Performance analysis of a small 3 kW grid-connected PVPS is described in the paper [5]. The PVPS are monitored for the experimental validation of the models and of the simulation codes that allow the evaluation of the performance of the single components and of the overall plant. There are experimental results of the efficiencies of the photovoltaic field and of the inverter are presented, as well as other plant data.

Our approach is different from the previously described methods. We are mainly focused on the visualization for exploratory analysis, therefore we need to obtain easily understandable visualization. As an input for Sammon's projection we used different settings. We have built a matrix of dependency between all data points from the data set and this matrix was used as an input for Sammon's projection algorithm by setting different thresholds of the dependency value we obtained clusters in the data. By using this method, we can visualize the networks into euclidean 2D/3D space.

The next section describes the Sammon's projection.

3 Sammon's Projection

One of the methods for projecting a dataset from high–dimensional space to a space with lower dimensionality is the method introduced by Sammon [6] in the 1969. This method is based on preserving inner–point distances. This goal is achieved by minimizing the error criteria that penalizes the difference in distance between points in the high–dimensional original space and the projected low–dimensional space. We are primarily interested in projections into two and three dimensional space because obtained projections can be easily explored and evaluated by human.

Assuming that we have a collection X with m data points $X = (X_1, X_2, ..., X_m)$ where each data point X_i is n dimensional vector $X_i = \{x_{i1}, x_{i2}, ..., x_{in}\}$. At the same time we define a collection Y of m data points $Y = (Y_1, Y_2, ..., Y_m)$ where each data point Y_i is d dimensional vector and $(d < m)$. As d^*_{ij} we denote the distance between vectors X_i and X_j. The distance between corresponding vectors Y_i and Y_j in lower dimensional space is denoted as d_{ij}. Although any distance measure can be used, the distance measure suggested by Sammon is Euclidean metric.

The projection error E (so–called Sammon's stress) measure how well the current configuration of m data points in the d–space fits the m points in the original space.

$$E = \frac{1}{\sum_{i<j}^{m} [d^*_{ij}]} \sum_{i<j}^{m} \frac{[d^*_{ij} - d_{ij}]^2}{d^*_{ij}} \quad (1)$$

In order to minimize the projection error, any minimization technique can be used. Sammon's original paper [6] proposes some widely known methods such as pseudo–Newton (Steppest descent) minimization:

$$y'_{ik}(t+1) = y'_{ik} - \alpha \frac{\frac{\partial E(t)}{\partial y'_{ik}(t)}}{\left| \frac{\partial^2 E(t)}{\partial y'_{ik}(t)^2} \right|} \quad (2)$$

where y'_{ik} is the kth coordinate of the data point's position y'_i in the projected low–dimensional space. Constant α setting 0.3 – 0.4 is given by Sammon as optimal. However, this range is not optimal for all solved problems. Equation 2 may cause a problem at the inflection points where second derivative is very small. Therefore the gradient descent may be used as an alternative minimization method.

4 Preprocessing Data

In this paper we worked with six hundred records of the PVPS measurement of hourly cumulated values of power. Dataset[1] contains daily entries, each of which contains 24 records of power (one record per hour). Each day we can represent vectors of dimension 24. This vector space model can be easily converted to a weighted undirected network. In our case, the vertices of the network are individual days. Weight of edges between any two vertices (days) corresponds to the similarity of vectors that they represent (similar days). For the calculation we used the cosine similarity.

$$CosSim(x,y) = \frac{\sum_i x_i y_i}{\sqrt{\sum_i x_i^2}\sqrt{\sum_i y_i^2}} \tag{3}$$

The result of dataset conversion is weighted undirected network. Sammon's projection, however, requires the calculated distance of each vertex. The input must be a distance matrix. There are different approaches to calculate the distance of the network vertices (e.g. diffusion distance). We used a method based on the calculation of local dependency (see [7]). The dependency takes into account the surroundings of each vertex into depth 2.

4.1 Dependency Measuring

We understand the dependency as a local unsymmetrical feature of a pair of vertices, which have distance of 2 at the most. More distant vertices are considered as being independent, therefore the dependency value is zero.

Let $E(x)$ be the set of all edges adjacent to the vertex x. Let $\mathrm{Adj}(x,y)$ be the set of all edges between the vertex x and any of the neighbours of the vertex y. Clearly, $\mathrm{Adj}(x,y) \subseteq E(x)$. Let $W(e)$ be the weight of an edge e and $W(v_1,v_2)$ is the weight of an edge between vertices v_1 and v_2 ($W(v_1,v_2) = 0$, if there is no such edge).

Let x not be an isolated vertex of the network. The dependency $D(x,y)$ (as we have introduced in [8]) of the vertex x on a vertex y is defined as:

$$D(x,y) = \frac{W(x,y) + \sum_{e_i \in \mathrm{Adj}(x,y)} W(e_i) \cdot R(e_i)}{\sum_{e_i \in E(x)} W(e_i)}, \tag{4}$$

$$R(e_i) = \frac{W(y,v_i)}{W(e_i) + W(y,v_i)}, \tag{5}$$

where $R(e_i)$ is the coefficient of dependency of the vertex x on the vertex y via the common neighbour v_i, therefore $v_i \in e_i$.

For particular situations from the Fig. 1 holds:

1a: $D(x,y) = D(y,x) = 1$,
1b: $D(x,y) = 1, D(y,x) = \frac{1}{2}$,
1c: $D(x,y) = \frac{13}{21}, D(y,x) = \frac{13}{18}$.

[1] Dataset can be downloaded from
`http://www.forcoa.net/resources/ibica2013/`

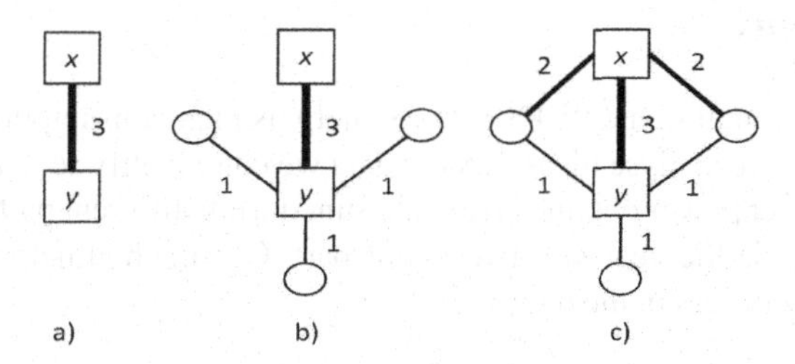

Fig. 1. Examples of dependency between two vertices

The dependency describes a relation of one vertex to another vertex from the point of view of their surroundings. Particular values of the dependency are used as threshold for creating distance matrix. Presented equations infer $D(x,y) \in \langle 0;1 \rangle$. The dependency being equal to zero means that vertices x and y have no common edge or neighbour. The full dependency (dependency equal to one) describes a situation where vertex x has only one common edge with vertex y.

Remark. The dependency is non-zero, if at least one of the following conditions hold:

1. There exists an edge between vertices x and y.
2. Vertices x and y share at least one common neighbour.

4.2 Conversion into Weighted Undirected Network

4.3 Dependency-Based Vertex Distance (DDST)

Distance matrix A for n vertices has order n. Matrix element A_{ij} represents the distance between vertices V_i and V_j. This matrix is symmetric. In order to calculate the distance between any two vertices in the network, we set the largest possible distance between points to 1. This corresponds to totally independent vertices (the vertices have no common neighbor).

The dependency between vertices V_i and V_j is unsymmetrical, but the distance matrix A for the Sammon's projection should be symmetric. Let $D(V_i,V_j)$ be the dependency of a vertex V_i on the vertex V_j. The elements of the distance matrix may be calculated as follows:

1. if $i = j$, then $A_{ij} = 0$
2. else $A_{ij} = A_{ji} = 1 - MAX(D(V_i,V_j),D(V_j,V_i))$

All vertices are therefore represented as points inside the hyper–ball of dimension n and diameter 1. The main diagonal of the matrix contains zeros, which may be ignored during the calculation of the projection. Matrix A serves as the direct input for the Sammon's projection.

5 Experiments

Evaluation of the small home PVPS data can help us plan consumption in a house in a better way. When we analyse measurement data we can identify several clusters during the year and predict power obtained from the sun, during different parts of year. By this sense we can divide the year into two major part. Our method helps us to find some other interesting clusters in the data.

5.1 Input Data Collection

We worked with the data of the measurement from small home PVPS. Power station has maximal power 2200 Watt per hour. The measurement of the hour power was recorded directly from the inverter device. We worked with 608 records in period since 9-26-2011 to 6-5-2013. Data set contained some records when the inverter was in the error state. These records were completely full of zeros or some hourly measurements were cleared.

For our paper we used different colors for the particular period of the year. The months of December, January and Febuary were colored brown. Green color was used for March, April and May. For June, July and August yellow was used. September, October and November were colored by orange color. These groups of months divided the year into four parts where obtained power was different from each other. It is necessary to say, that there is not a crisp border between these parts of the year.

In the figure 2(a) the power obtained from PVPS, during year is depicted. Each curve shows power in one randomly selected day in each month. The variability of the obtained power from power station is depicted there as well. Second figure 2(b) shows output power for each day in June. Power obtained from the power station is very dependent on the weather. Clouds in the sky can change power very dramatically. Although photovoltaic power station needs sun for getting maximal output power, high temperature of the photovoltaic cell reduces output power. The charts of the power are dependent on the particular location of the PVPS and can not be generalized.

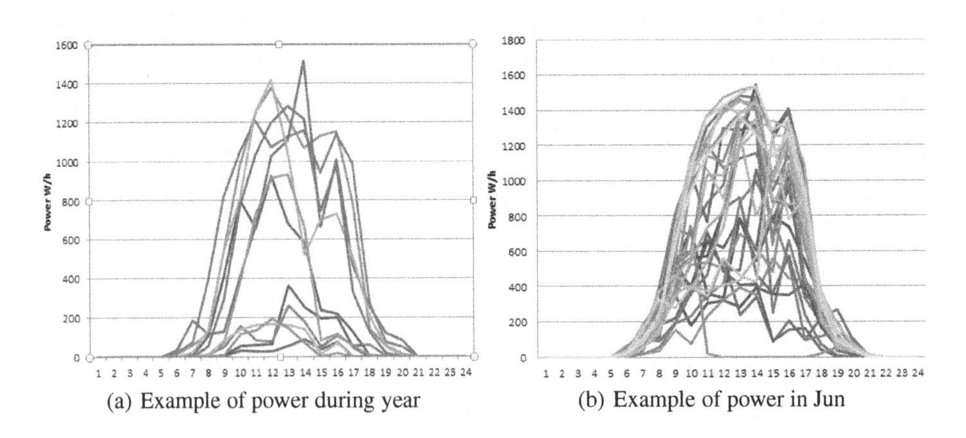

(a) Example of power during year

(b) Example of power in Jun

Fig. 2. Example Power of the PVPS during one year and particular month

Table 1. Number of vertices used in visualization by the level of threshold

Level of threshold	Number of vertices
0.00	608
0.90	568
0.95	551
0.99	300

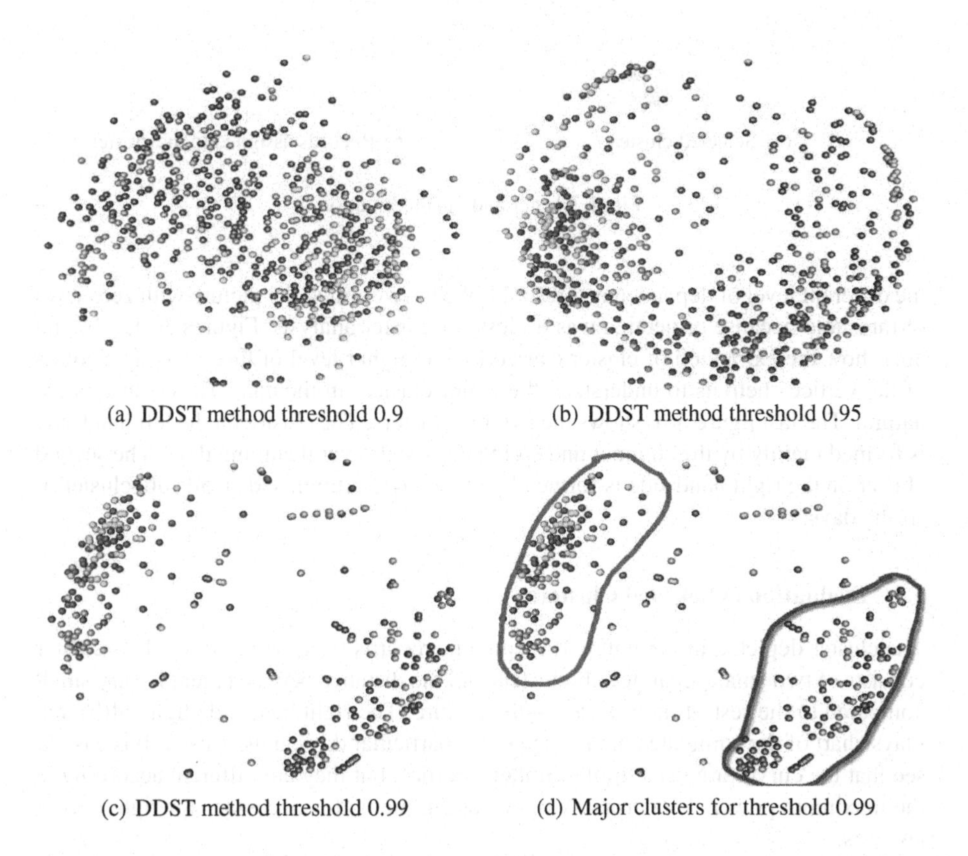

(a) DDST method threshold 0.9 (b) DDST method threshold 0.95

(c) DDST method threshold 0.99 (d) Major clusters for threshold 0.99

Fig. 3. Different level of threshold

5.2 Evaluation of Main Clusters

Recorded data from PVPS was used for computing matrices by the DDST method. These matrices were used as input for Sammon's projection. We prepared several matrices for different levels of threshold. The level of threshold (dependency value) was experimentally evaluated. In the matrices, dependencies were computed between each vertice. For vertices with level of dependency lower than the selected threshold the largest possible distance, was set between them (1). The number of vertices used for visualization is shown in the table 1. In the figure 3 Sammon's projection is presented for

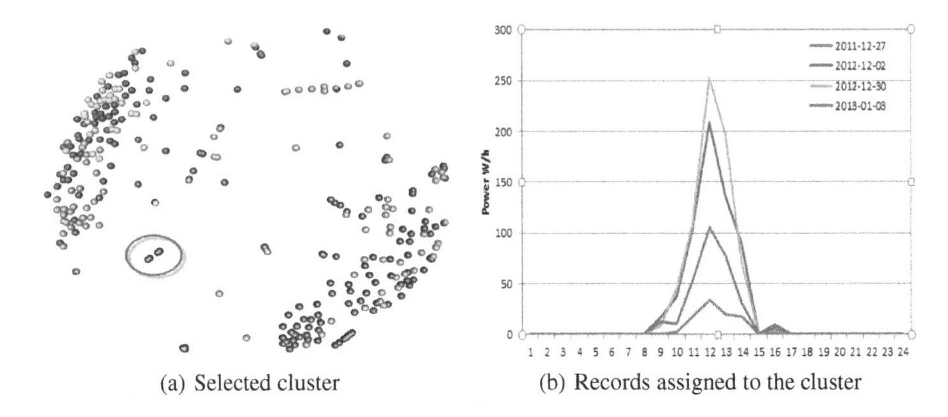

(a) Selected cluster

(b) Records assigned to the cluster

Fig. 4. Cluster and his presentation

the different level of dependency threshold. We have omitted the picture with zero level of threshold because projection was useless for cluster analysis. Figures 3(a), 3(b) and 3(c) show the extraction of cluster caused by the higher level of threshold. The colors of the vertices help us to understand the major clusters in the data. These clusters are natural. The last figure 3(d) shows the two big clusters. The cluster on the left hand side is formed mainly by the summer and spring days and several autumn days. The second cluster on the right hand side is formed by the winter, autumn and small sub–cluster of spring days.

5.3 Evaluation of Selected Clusters

The cluster depicted in the figure 4(a) contains records from winter days. This cluster consists of two small separate sub-clusters and the distance between them is very small compared to the rest of the vertices in the picture. The right hand side figure 4(b) displays chart of the cumulated hourly power for particular days in the cluster. It is easy to see that the curves are very similar in their profiles, but they are different according to the maximum power. This property of clustering method is necessary for right cluster analysis.

Small cluster of the winter days is displayed in the figure 5(a). This is one of the sub-clusters of the winter and autumn days. This cluster contains 7 days in the period under consideration. The curves of the power are also very similar although there are days from the years 2011 and 2012. The right hand side figure 5(b) is identifiable by the difference in the middle part of the chart. These differences are depicted in the Sammon's visualization by the close distance between vertices and line shape.

The final selected cluster for evaluation is our clustering algorithm which is depicted in the figure 6(a). This cluster is situated on the top of the projection space. The vertices are shaped into the line and there is distance between vertices. This shape and space suggest a profile of the curves. There are eleven curves for eleven days in this cluster. These days are mainly from Autumn 2011 and 2012 and there is one day from Winter 2013. The curves are shaped into similar shapes and we can identify a very close shape

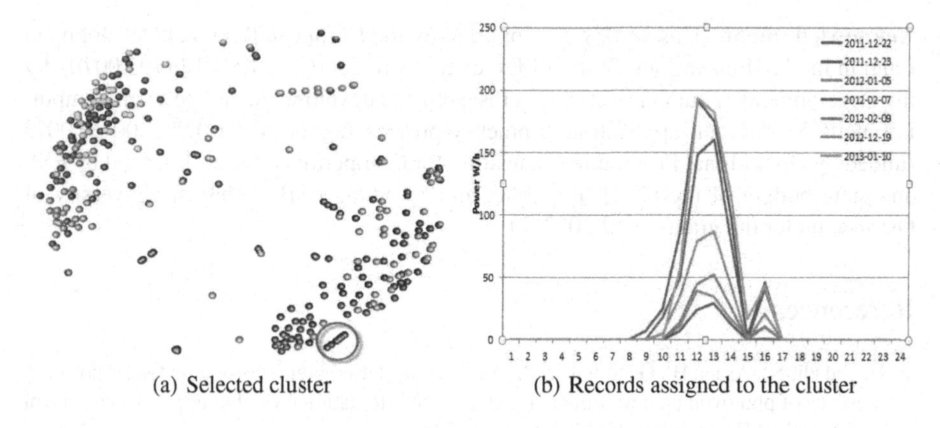

(a) Selected cluster (b) Records assigned to the cluster

Fig. 5. Cluster and his presentation

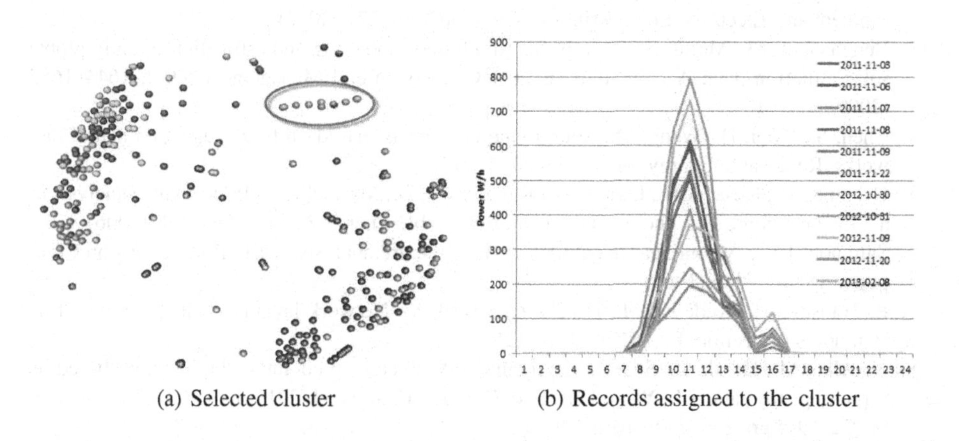

(a) Selected cluster (b) Records assigned to the cluster

Fig. 6. Cluster and his presentation

for vertices which are very close in the visualization. The method of the clustering based on the DDST method together with Sammon's visualization seems to be appropriate for explorational analysis of the data form the PVPS.

6 Conclusion and Future Work

In this paper, we have introduced an approach for finding interesting cluster in the photovoltaic power station power measurement. We have used a method based on the graph theory and vertex dependency. The used method gave us a very interesting hierarchical view of the cluster with the same daily power profile. In our future work we plan to compare our results with the other clustering methods and finding a method for selecting the best value of threshold.

Acknowledgment. This work was supported by the European Regional Development Fund in the IT4Innovations Centre of Excellence project (CZ.1.05/1.1.00/02.0070), by the Development of human resources in research and development of latest soft computing methods and their application in practice project, reg. no. CZ.1.07/2.3.00/20.0072 funded by Operational Programme Education for Competitiveness, co-financed by ESF and state budget of the Czech Republic, and by SGS, VSB-Technical University of Ostrava, under the grant no. SP2013/70.

References

1. Haghdadi, N., Asaei, B., Gandomkar, Z.: A clustering-based preprocessing on feeder power in presence of photovoltaic power plant. In: 2011 10th International Conference on Environment and Electrical Engineering (EEEIC), pp. 1–4 (2011)
2. Haghdadi, N., Asaei, B., Gandomkar, Z.: Clustering-based optimal sizing and siting of photovoltaic power plant in distribution network. In: 2012 11th International Conference on Environment and Electrical Engineering (EEEIC), pp. 266–271 (2012)
3. Benghanem, M., Mellit, A., Alamri, S.: Ann-based modelling and estimation of daily global solar radiation data: A case study. Energy Conversion and Management 50(7), 1644–1655 (2009)
4. Chen, S., Gooi, H., Wang, M.: Solar radiation forecast based on fuzzy logic and neural networks. Renewable Energy 60(0), 195–201 (2013)
5. Cucumo, M., Rosa, A.D., Ferraro, V., Kaliakatsos, D., Marinelli, V.: Performance analysis of a 3 kw grid-connected photovoltaic plant. Renewable Energy 31(8), 1129–1138 (2006)
6. Sammon, J.W.: A nonlinear mapping for data structure analysis. IEEE Transactions on Computers 18, 401–409 (1969)
7. Radvanský, M., Kudělka, M., Horák, Z., Snášel, V.: Network layout visualization based on sammon's projection. In: INCoS. IEEE (2013)
8. Kudelka, M., Horak, Z., Snasel, V., Abraham, A.: Weighted co-authorship network based on forgetting. In: Park, J.J., Yang, L.T., Lee, C. (eds.) FutureTech 2011, Part II. CCIS, vol. 185, pp. 72–79. Springer, Heidelberg (2011)

Rough Power Set Tree for Feature Selection and Classification: Case Study on MRI Brain Tumor*

Waleed Yamany[1,6], Nashwa El-Bendary[2,6], Hossam M. Zawbaa[3,6], Aboul Ella Hassanien[4,6], and Václav Snášel[5]

[1] Fayoum University, Faculty of Computers and Information, Fayoum, Egypt
waleed-2math@yahoo.com
[2] Arab Academy for Science, Technology, and Maritime Transport, Cairo, Egypt
nashwa.elbendary@ieee.org
[3] BeniSuef University, Faculty of Computers and Information, BeniSuef, Egypt
hossam.zawbaa@gmail.com
[4] Faculty of Computers and Information, Cairo University, Egypt
aboitcairo@fci-cu.edu.eg
[5] VSB-Technical University of Ostrava, Czech Republic
[6] Scientific Research Group in Egypt (SRGE)
http://www.egyptscience.net

Abstract. This article presents a feature selection and classification system for 2D brain tumors from Magnetic resonance imaging (MRI) images. The proposed feature selection and classification approach consists of four main phases. Firstly, clustering phase that applies the K-means clustering algorithm on 2D brain tumors slices. Secondly, feature extraction phase that extracts the optimum feature subset via using the brightness and circularity ratio. Thirdly, reduct generation phase that uses rough set based on power set tree algorithm to choose the reduct. Finally, classification phase that applies Multilayer Perceptron Neural Network algorithm on the reduct. Experimental results showed that the proposed classification approach achieved a high recognition rate compared to other classifiers including Naive Bayes, AD-tree and BF-tree.

Keywords: Rough sets, power trees, K-mean clustering, classification.

1 Introduction

Feature selection is one of the most essential problems in the fields of data mining, machine learning, and pattern recognition [16]. The main purpose of feature

* This work was partially supported by Grant of SGS No. SP2013/70, VSB - Technical University of Ostrava, Czech Republic., and was supported by the European Regional Development Fund in the IT4Innovations Centre of Excellence project (CZ.1.05/1.1.00/02.0070) and by the Bio-Inspired Methods: research, development and knowledge transfer project, reg. no. CZ.1.07/2.3.00/20.0073 funded by Operational Programme Education for Competitiveness, co-financed by ESF and state budget of the Czech Republic.

A. Abraham et al. (eds.), *Innovations in Bio-inspired Computing and Applications*, Advances in Intelligent Systems and Computing 237,
DOI: 10.1007/978-3-319-01781-5_24,

selection is to determine a minimal feature subset from a problem domain, while retaining a suitably high accuracy in representing the original features [2]. In real world problems, feature selection is a must due to the abundance of noisy, misleading, or irrelevant features [10]. By removing these factors, learning from data techniques can be of a great use. The motivation of feature selection in data mining, machine learning, and pattern recognition is to reduce the dimensionality of feature space, improve the predictive accuracy of a classification algorithm, and develop the visualization and the comprehensibility of the induced concepts [13].

Medical data, such as MRI brain data, oftentimes contain irrelevant features in addition to missing values and uncertainties [1]. The analysis of medical data was required for dealing with vagueness and inconsistent information as well as manipulation of different levels of representation of data. Rough set theory [5] can transact with uncertainty and vagueness in data analysis. It has been widely applied in many fields, such as data mining [4], machine learning [6], etc. [14]. It considers knowledge as a kind of discriminability. Furthermore, feature reduction algorithms remove redundant information or features and selects a feature subset that has the same discernibility as the original set of features.

In this paper rough set algorithm based on power set tree was applied for feature selection from twenty features based on shape, color, and texture that have been extracted in order to obtain a feature vector for each object to identify the tumor and classify it in 2D brain Magnetic resonance imaging (MRI) images.

The rest of the paper is organized as follows. Section 2 gives an overview of rough set theory as well as power set tree and artificial neural network (ANN) algorithms. Section 3 describes the proposed system model and its phases. Section 4 presents the experimental result. Section 5 addresses conclusions and discusses future work.

2 Background

2.1 Rough Sets

Rough set theory [7], [9], [11], [5] is a fairly new intelligent technique for managing uncertainty that has can be used for the discovery of data dependencies, evaluation of the importance of attributes, discovery of patterns in data, reduction of attributes, and the extraction of rules from databases. Such rules have the potential to reveal new patterns in the data and can also collectively function as a classifier for unseen datasets. Unlike other computational intelligence techniques, rough set analysis requires no external parameters and uses only the information present in the given data.

One of the interesting features of rough sets theory is that it can tell whether the data is complete or not based on the data itself. If the data is incomplete, it suggests more information about the objects to be collected in order to build a good classification model. On the other hand, if the data is complete, rough sets can determine the minimum data needed for classification. This property of rough sets is important for applications where domain knowledge is limited or data collection is very expensive/laborious because it makes sure the data collected is just good enough to build a good classification model without sacrificing the accuracy of the classification model or wasting time and effort to gather extra information about the objects [7], [9], [11], [5].

In rough sets theory, the data is collected in a table, called a decision table (DT). Rows of the decision table correspond to objects, and columns correspond to attributes. In the dataset, we assume that the a set of examples with a class label to indicate the class to which each example belongs are given. We call the class label the decision attributes, and the rest of the attributes the condition attributes. Rough sets theory defines three regions based on the equivalent classes induced by the attribute values: lower approximation, upper approximation and boundary. Lower approximation contains all the objects, which are classified surely based on the data collected, and upper approximation contains all the objects which can be classified probably, while the boundary is the difference between the upper approximation and the lower approximation. So, we can define a rough set as any set defined through its lower and upper approximations. On the other hand, indiscernibility notion is fundamental to rough set theory.

Informally, two objects in a decision table are indiscernible if one cannot distinguish between them on the basis of a given set of attributes. Hence, indiscernibility is a function of the set of attributes under consideration. For each set of attributes we can thus define a binary indiscernibility relation, which is a collection of pairs of objects that are indiscernible to each other. An indiscernibility relation partitions the set of cases or objects into a number of equivalence classes. An equivalence class of a particular object is simply the collection of objects that are indiscernible to the object in question.

2.2 Power Set Tree (PS-tree)

This section recalls some fundamental of definition of power set tree (PS-tree), more details can be found in [3]. Trees provides us with an efficient method to solve many problems. The PS-tree is a tree building to represent the power set in an order style. Let $I = (O, C \cup D, V, f)$ be a decision table,where D is a decision feature and C is a condition features, $C = \{c_1, ..., c_n\}$ P is the power set of $C.PT$ is a PS-tree for I, which is a tree that satisfies the following conditions:

1. The root of PT is a $\langle c_1, ..., c_n, n \rangle$, which denote that the root has n element $c_1, ..., c_n$ and n children.
2. Suppose that $PT_1, PT_2, ..., PT_n$ be n children of root PT. PT_1 has *n-1* member and *n-1* children; PT_n has *n-1* member and zero children. Members of PT_n are inherited from PT by deleting the n member of PT. Each child of the root is a PS-tree.
3. The arrangement of member in PT_1 is unmodified. The order of $PT_2, ..., PT_n$ must be changed.The *n-1* member of PT_n equal the deleted members of $PT_{n-1}, ..., PT_2, PT_1$.

The PS-tree will be used for caching solutions and for testing the subsumption relation between two sets, which could efficiently reduce the time and space complexities of the algorithms.

2.3 Artificial Neural Network (ANN)

Artificial neural networks (ANN) or simply neural networks (NN) have been developed as generalizations of mathematical models of biological nervous systems. In a simplified mathematical model of the neuron, the effects of the synapses are represented by connection weights that modulate the effect of the associated input signals, and the nonlinear characteristic exhibited by neurons is represented by a transfer function. There are a range of transfer functions developed to process the weighted and biased inputs, among which four basic transfer functions widely adopted for multimedia processing [20].

The neuron impulse is then computed as the weighted sum of the input signals, transformed by the transfer function. The learning capability of an artificial neuron is achieved by adjusting the weights in accordance to the chosen learning algorithm. The behavior of the neural network depends largely on the interaction between the different neurons. The basic architecture consists of three types of neuron layers: input, hidden and output layers.

In feed-forward neural networks, the signal flow is from input to output units strictly in a feed-forward direction. The data processing can extend over multiple units, but no feedback connections are present, that is, connections extending from outputs of units to inputs in the same layer or previous layers. There are several other neural network architectures (Elman network, adaptive resonance theory maps, competitive networks, etc.) depending on the properties and requirement of the application [21].

3 Proposed Feature Selection and Classification Approach

The feature selection and classification approach, proposed in this article, composed of four fundamental phases; namely *clustering*, *feature extraction*, *reduct generation*, and *classification*, as follows:

- **Clustering phase:** K-means clustering algorithm is applied on the 2D MRI slices.
- **Feature extraction phase:** Extract the optimum feature subset via using brightness and circularity ratio.
- **Reduct generation phase:** Rough set based on power set tree algorithm is applied to choose the reduct.
- **Classification phase:** Multilayer Perceptron Neural Network algorithm is applied to classify the reduct.

Fig. 1 depicts the building phases of the proposed system. These phases are described in more details in the next subsections along with the steps involved and the characteristics feature for each phase.

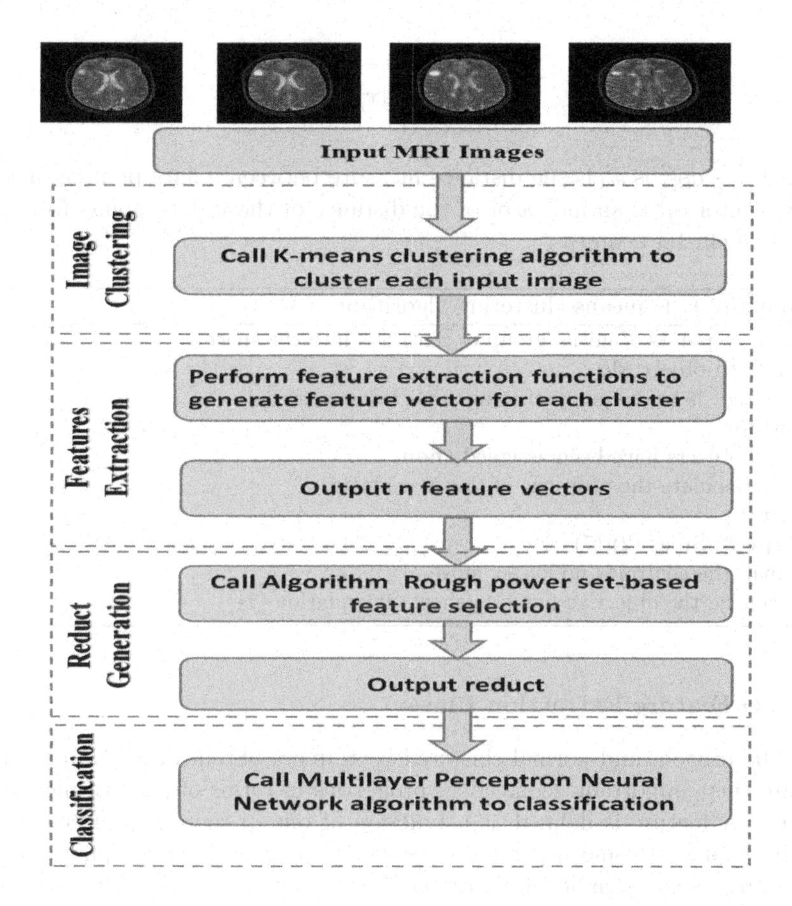

Fig. 1. 2D Brain Tumor Feature Selection and Classification Proposed Scheme

3.1 K-Means 2D Slice Clustering Phase

K-means [18], [12] is a simplest unsupervised learning techniques that solve the well known clustering problem. The algorithm follows a simple way to classify a given dataset through a certain number of clusters (say k-clusters). The basic idea is to assign k-centroids for each cluster. The better way to select the k is to place them as much as possible far away from each other. The next step is to take each point belonging to a given dataset and associate it to the nearest centroid [18]. When no point is pending, the first step is completed and an early groupage is done. At this point, k new centroid needs to be re-calculated as barycenters of the clusters resulting from the previous step. Having these k new centroid, a new binding has to be done between the same dataset points and the nearest new centroid. By repeating this process, the k centroids change their location step by step until no more changes are done. Finally, as shown in algorithm 3.1, the K-means clustering algorithm aims at minimizing an objective function, which is defined as in the equation (1).

$$J = \sum_{j=1}^{x} \sum_{i=1}^{k} \|x_i^j - c_j\|^2 \tag{1}$$

where $\|x_i^j - c_i\|^2$ is a chosen distance measure between a data point x_i^j and the cluster center c_j, is an indicator of the distance of the n data points from their respective cluster centers.

Algorithm 1. K-means clustering algorithm

1: Place K points of initial group of centroids into the space
2: **for** Each object **do**
3: Assign it to the group that has the closest centroid
4: **end for**
5: **if** All objects have been assigned **then**
6: Recalculate the positions of the K centroids
7: **end if**
8: **Repeat**: Steps (2)-(7)
9: **Until**: the centroids no longer move
10: Minimize the objective function shown in equation (1)

3.2 2D Feature Extraction Phase

Once the tangent and normal clustered vectors are obtained, it is important to capture their important geometrical properties in terms of a set of meaningful features. A feature is defined as a function of one or more measurements and are the values of some quantifiable property of an object, computed so that it quantifies some significant characteristics of the object [15], [19]. One of the biggest advantages of feature extraction lies that it significantly reduces the information to represent an image for understanding the content of the image. In this paper, we used only the *circularity ratio* and the *brightness* of the object.

The circularity ratio is the ratio of the area of the shape to the area of a circle having the same perimeter, which is expressed mathematically as in the equation (2)

$$f_{circ} \frac{4 * \pi * A}{P^2} \tag{2}$$

where P and$A$ are the perimeter and area of the object, respectively. f_{circ} is equal to one for a circle and it is less than one for any other shape. The area is obtained using the total number of pixels in each separated cluster.

3.3 PS-tree Reduct Generation Phase

In this phase, rough set algorithm based on power set tree (PS-tree) [3] was applied for feature selection from 22 features based on shape, color, and texture which were extracted to obtain feature vector for each object that characterizes the tumor and identifies it [1]. The PS-tree is an arrangement tree representing the power set, and each possible reduct in rough set is mapped to a node of thee tree. The PS-tree will be used for caching solutions and for testing the subsumption relation between two sets, which could efficiently reduce the time and space complexities of the algorithms. For its exponential size, it is impossible to totally explore the PS-tree. The rules of removing a branch of the search tree from consideration without examining the nodes in the branch are called pruning rules. The specific PS-tree-based pruning is called *rotation* and *backtracking.*

In the PS-tree, its left branches are larger than the right branches. Then, it was more efficient to prune the left branches. If a node is not super reduct, then its children obviously are not reducts. By effect of this property of PS-tree, the unpromising parts of the search space can be pruned. When an extended node in the right of PS-tree is not super reduct, it should be rotated to the left and pruned as a large branch. Simultaneously, the arrangement of elements in the node must be revised according to the definition of PS-tree. Backtracking is a methodical way to search for the solution to a given problem.

Once we had determined an organization for the solution space, this space was searched in a depth-first manner beginning at a start node. If the search operation constructs a solution by depth-first manner, then backtracks to search for a more optimal solution. Rotation technique can prune left branches of the PS-tree and backtracking technique can prune the right branches of the PS-tree. Out of rotation and backtracking, we can eliminate those parts of the solution space which do not have the potential to lead to a solution. Then, a rough set algorithm based on PS-tree-based for feature selection (also called *Rough power set-based feature selection*), presented in algorithm 3.4, extends nodes according to some priority function in an order style. Nodes along the tree's extending fringe are kept in a priority queue and next node to be extended is obtained by virtue the guiding heuristic.

3.4 Multilayer Perceptron Neural Network Classification Phase

Artificial neural networks (ANN) have been developed as generalizations of mathematical models of biological nervous systems. In a simplified mathematical model of a neuron, the effects of the synapses are represented by connection weights that modulate the effect of the associated input signals, while the non-linear characteristic exhibited by neurons is represented by a transfer function. Each neuron is characterized by an activity level (representing the state of polarization of a neuron), an output value (representing the firing rate of the neuron), a set of input connections (representing synapses on the cell and its dendrite), a bias value (representing an internal resting level of the neuron), and a set of output connections (representing a neuron's axonal projections). Each of these aspects of the unit is represented mathematically by real numbers. Thus, each connection has an associated weight (synaptic strength), which determines the effect of the incoming input on the activation level of the unit.

Algorithm 2. Rough power set-based feature selection algorithm

Input: A decision table $I = (O, C \cup D, V, f)$

1: Let $R = C$; Add C to $OPEN - NODES$; $feat - num = |C|$; $child = |C|$
2: **Repeat**
3: Choose the first node $NODE$ in $OPEN - NODES$
4: Let $T = NODE$, $T.feat - num = feat - num$ and $T.child = child$
5: **if** $(\mu_T(D) = \mu_C(D))$ and $(\vee P \subset T, \mu_P(D) \neq \mu_C(D))$ **then**
6: **if** $|T| < |R|$ **then**
7: $R = T$ and *Return*
8: **end if**
9: **else** *Return*
10: **end if**
11: **if** $(\mu_T(D) = \mu_C(D))$ **then**
12: **for** $i = 1$ to $T.child$ **do**
13: Select a feature $a_i \in T$ in sequence T , let $T_i = T - a_i$
14: Let $T_i.feat - num = T.feat - num - 1$
15: Let $T_i.child = T.child - i$
16: **if** $(\mu_{T_i}(D) \neq \mu_C(D))$ **then**
17: Rotate T_i to the left of PS-tree to be pruned
18: **end if**
19: **else** Rotate T_i to the right of PS-tree
20: **end for**
21: **end if**
22: Let $K_1, K_2, ..., K_m, K_{m+1}, ..., K_{child}$ be the nodes rotation where $K_1, K_2, ..., K_m$ are the pruned parts, and $K_{m+1}, ..., K_{child}$ are non- pruned parts.
23: Modify the arrangement of elements in nodes of non-pruned parts according to definition of PS-tree.
24: **for** $j = m + 1$ to $T.child$ **do**
25: **if** $K_j.feat - num - K_j.child < |R|$ **then**
26: Add K_j to to $OPEN - NODES$
27: Delete the node $NODE$ from $OPEN - NODES$
28: Sort all nodes in $OPEN - NODES$ according to the heuristic
29: **end if**
30: **end for**
31: **Until** $OPEN - NODES$ is empty
32: Output R

The neuron impulse is hence computed as the weighted sum of the input signals, transformed by the transfer function. The learning capability of an artificial neuron is achieved by adjusting the weights in accordance with a chosen learning algorithm. The reader may refer to [17] for an extensive overview of the artificial neural networks. The most common methodology for accomplishing this type of learning in an ANN is the multilayered perceptron NN employing back-propagation. A three-layered perceptron is the simplest multi-layered perceptron is with feed-forward connections from the input layer to the hidden layer and from the hidden layer to the output layer [8]. This function estimating NN records pattern pairs using a multi-layered gradient error correction method. It applies its internal representation features of the training data by minimizing a cost function. The cost function is the sum squared error or the summation of the difference between the computed and desired output values for the output neuron across the patterns in the training data.

4 Experimental Results

The dataset used in this article contains different MRI images from 10 patients with cerebral tumors. Figure (2) shows 39 slices and the tumor appeared from slice 29 to slice 33. These MRI images of the brain shows many small irregular areas of increased signal that turned out to be cancer. Figure (3) shows the four clusters results obtained using the K-mean clustering. The brain tumor cluster in this experiments has the maximum brightness and is circular. After applying circularity measure to the clusters that have maximum brightness, we obtain the resulted cluster.

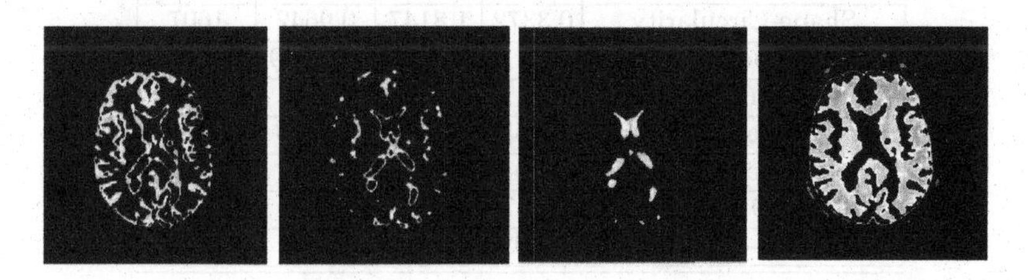

Fig. 2. K-mean clustering results

For the analysis of patients, the information system was defined, where a set of 32 instances was described by a set of features. The set of features includes 22 condition features and one decision feature. The condition features characterize the patients, while the decision feature defines the patients' classification according to the tumor. A list of the chosen twenty two condition features with characteristics of their domains is presented in Table 1.

Table 2 shows results of a comparative analysis for reduct generation of MRI brain tumor dataset using different reduct generation algorithms such as Genetic Search [22, 23], Greedy Stepwise [25], Linear Forward Selection [26], Best First [27], and PS-FS-C [3]. From the results, we can conclude that the GLCM02-Contrast and Shape-Circularity are the minimal reduct, which are the same features that are chosen according to the recommendations of a physician.

Table 3 depicts classification accuracy using different classifiers such as Multilayer Perceptron NN, BFtree, ADtree and Naive Bayes classifiers. As presented in Table 3, the classification accuracy using the Multilayer Perceptron NN was 85%, using best first decision tree (BF-tree) [29] was 83%, using alternating decision tree (AD-tree) [28] was 84% , and using Naive Bayes [24] was 75%, which means that using the Multilayer Perceptron NN outperformed the other other classification algorithms for the experienced dataset.

Table 1. Characteristics of condition features

Feature	Min	Max	Avg	Std
GLCM02-Contrast	0	6870.1	1020.121	1462.853
GLCM02-Correlation	-0.973	1.1	1.1	0.428
GLCM02-Energy	0	0.6	0.126	0.124
GLCM02-Homogeneity	0	0.485	0.247	0.11
GLCM20-Contrast	0	2984.1	669.866	684.255
GLCM20-Correlation	-0.408	0.989	0.666	0.31
GLCM20-Energy	0	0.433	0.104	0.094
GLCM20-Homogeneity	0	0.423	0.255	0.098
GLCM22-Contrast	0	9390	1730.957	2430.185
GLCM22-Correlation	-1.1	0.951	0.056	0.696
GLCM22-Energy	0	1.1	0.166	0.211
GLCM22-Homogeneity	0	0.399	0.184	0.105
Gray-Entropy	1.5	5.515	4.084	1.096
Gray-Mean	6.736	53.6	28.632	14.283
Gray-Variance	97	191.259	129.138	21.582
Shape-Circularity	0.3372	1.8147	0.9642	1607
Shape-Area	1	1557.1	154.099	348.542
Shape-EulerNumber	1	2.1	1.144	0.306
Shape-MajorAxisLength	1.155	72.915	14.935	16.714
Shape-MinorAxisLength	1.155	50.585	8.932	11.441
Shape-Orientation	-89.575	90.1	5.727	58.268
Shape-Solidity	0.402	1.1	0.913	0.171

Table 2. Reduct set results

Algorithms	Feature reduct
Genetic Search	GLCM02-Energy GLCM02-Homogeneity GLCM20-Correlation GLCM20-Energy GLCM22-Contrast
Greedy Stepwise	GLCM02-Energy GLCM02-Homogeneity GLCM20-Correlation GLCM20-Energy GLCM22-Contrast
Linear Forward Selection	GLCM02-Energy GLCM02-Homogeneity GLCM20-Correlation GLCM20-Energy GLCM22-Contrast
Best Frist	GLCM02-Energy GLCM02-Homogeneity GLCM20-Correlation GLCM20-Energy GLCM22-Contrast
Rough power set-based	GLCM02-Contrast Shape-Circularity

Table 3. Classification results: Accuracy, Mean absolute error(MAE), Root mean squared error (RMSE), and Relative absolute error(RAE)

Classifier	Accuracy	MAE	RMSE	RAE
Multilayer Perceptron NN	85 %	0.2192	0.4153	47.0524 %
ADtree	84 %	0.2959	0.3493	63.6448%
BFtree	83 %	0.1846	0.3917	39.4453 %
Naive Bayes	75%	0.3144	0.4309	66.9829%

5 Conclusions and Future Work

This article presents feature selection and classification approach, which consists of four main phases (Clustering phase, Feature extraction, Reduct generation, and Classification). A rough set algorithm based on power set tree was applied for feature selection from twenty features based on shape, color, and texture that have been extracted in order to obtain a feature vector for each object to identify the tumor and classify it in 2D brain Magnetic resonance imaging (MRI) images. The experimental results showing a high recognition rate with the proposed scheme. For future work, we will work on applying different machine learning techniques in order to get more accurate classification results.

References

1. Moftah, H.M., Hassanien, A.E., Shoman, M.: 3D brain tumor segmentation scheme using K-means clustering and connected component labeling algorithms. In: IEEE International Conference in Intelligent Systems Design and Applications (ISDA), Cairo, Egypt, pp. 320–324 (2010)
2. Dash, M., Liu, H.: Feature selection for Classification. Intelligent Data Analysis 1(3), 131–156 (1997)
3. Chen, Y., Miao, D., Wang, R., Wu, K.: A Rough Set Approach to Feature Selection Based on Power Set Tree. Knowledge-Based System 24, 275–281 (2011)
4. Guo, Q.L., Zhang, M.: Implement web learning environment based on data mining. Knowledge-Based Systems 22(6), 439–442 (2009)
5. Pawlak, Z.: Rough sets. International Journal of Computer and Information Science 11(5), 341–356 (1982)
6. Swiniarski, R.W., Skowron, A.: Rough set methods in feature selection and recognition. Pattern Recognition Letters 24, 833–849 (2003)
7. Pawlak, Z.: Rough sets: Theoretical aspects of reasoning about data. Kluwer (1991)
8. Hassanien, A.E., Kim, T.-H.: Breast cancer MRI diagnosis approach using support vector machine and pulse coupled neural networks. Journal of Applied Logic 10, 277–284 (2012)
9. Pawlak, Z., Grzymala-Busse, J., Slowinski, R., Ziarko, W.: Rough Sets. Communications of the ACM 38(11), 88–95 (1995)
10. Jensen, R.: Combining rough and fuzzy sets for feature selection. Ph.D. Thesis, University of Edinburgh (2005)
11. Polkowski, L.: Rough Sets: Mathematical Foundations. Physica-Verlag (2003)

12. Lee, G.N., Fujita, H.: K-means Clustering for Classifying Unlabelled MRI Data. In: Proceedings of the 9th Biennial Conference of the Australian Pattern Recognition Society on Digital Image Computing Techniques and Applications, pp. 92–98 (2007)
13. Liu, H., Motoda, H.: Feature Selection for Knowledge Discovery and Data Mining. Kluwer, Boston (1998)
14. Mi, J.S., Wu, W.Z., Zhang, W.X.: Approaches to knowledge reduction based on variable precision rough set model. Information Sciences 159(3-4), 255–272 (2004)
15. Prastawa, M., Bullitt, E., Gerig, G.: Simulation of brain tumors in MR images for evaluation of segmentation efficacy. Medical Image Analysis 13(2), 297–311 (2009)
16. Cannas, L.M.: A framework for feature selection in high-dimensional domains. Ph.D. Thesis, University of Cagliari (2012)
17. Bishop, C.M.: Neural Networks for Pattern Recognition. Oxford University Press (1995)
18. Latha, P.P., Sharmila, J.S.: A dynamic 3D clustering algorithm for the ligand binding disease causing targets. IJCSNS International Journal of Computer Science and Network Security 9 (February 2009)
19. Dubey, R.B., Hanmandlu, M., Gupta, S.K., Gupta, S.K.: An advanced technique for volumetric analysis. International Journal of Computer Applications 1(1) (2010)
20. Zawbaa, H.M., El-Bendary, N., Hassanien, A.E., Kim, T.-H.: Event Detection Based Approach for Soccer Video Summarization Using Machine learning. International Journal of Multimedia and Ubiquitous Engineering (IJMUE) 7(2) (2012)
21. Yu, B., Zhu, D.H.: Automatic Thesaurus Construction for Spam Filtering Using Revised: Back Propagation Neural Network. Journal Expert Systems with Applications 37(1), 24–30 (2010)
22. Wrblewski, J.: Finding minimal reducts using genetic algorithms. In: Proceedings of Second Annual Join Conference on Information Sciences, Wrightsville Beach, NC, September 28-October 1, pp. 186–189 (1995)
23. Zhai, L.Y., Khoo, L.-P., Fok, S.-C.: Feature extraction using rough set theory and genetic algorithms: an application for the simplification of product quality evaluation. Computers and Industrial Engineering 43, 661–676 (2002)
24. Caruana, R., Niculescu-Mizil, A.: An empirical comparison of supervised learning algorithms. In: Proceedings of the 23rd International Conference on Machine Learning (2006)
25. Samala, R.K., Potunuru, V.S., Zhang, J., Cabrera, S.D., Qian, W.: Comparative Study of Feature Measures for Histopathological Images of the Lung. In: Digital Image Processing and Analysis, Tucson, Arizona United States, June 7-8. Medical Image Processing II (2010)
26. Wilkinson, L., Dallal, G.E.: Tests of significance in forward selection regression with an F-to enter stopping rule. Technometrics 23, 377–380 (1981)
27. Pearl, J.: Heuristics: Intelligent Search Strategies for Computer Problem Solving, p. 48. Addison-Wesley (1984)
28. Pfahringer, B., Holmes, G., Kirkby, R.: Optimizing the Induction of Alternating Decision Trees. In: Cheung, D., Williams, G.J., Li, Q. (eds.) PAKDD 2001. LNCS (LNAI), vol. 2035, pp. 477–487. Springer, Heidelberg (2001)
29. Shi, H.: Best-first decision tree learning. University of Waikato (2007)

The Nelder-Mead Simplex Method with Variables Partitioning for Solving Large Scale Optimization Problems*

Ahmed Fouad Ali[1], Aboul Ella Hassanien[2], and Václav Snášel[3]

[1] Suez Canal University, Dept. of Computer Science,
Faculty of Computers and Information, Ismailia, Egypt
Member of Scientific Research Group in Egypt
[2] Cairo University, Faculty of Computers and Information, Cairo, Egypt
Chair of Scientific Research Group in Egypt
[3] VSB-Technical University of Ostrava, Czech Republic
vaclav.snasel@vsb.cz

Abstract. This paper presents a novel method to solve unconstrained continuous optimization problems. The proposed method is called SVP (simplex variables partitioning). The SVP method uses three main processes to solve large scale optimization problems. The first process is a variable partitioning process which helps our method to achieve high performance with large scale and high dimensional optimization problems. The second process is an exploration process which generates a trail solution around a current iterate solution by applying the Nelder-Mead method in a random selected partitions. The last process is an intensification process which applies a local search method in order to refine the the best solution so far. The SVP method starts with a random initial solution, then it is divided into partitions. In order to generate a trail solution, the simplex Nelder-Mead method is applied in each partition by exploring neighborhood regions around a current iterate solution. Finally the intensification process is used to accelerate the convergence in the final stage. The performance of the SVP method is tested by using 38 benchmark functions and is compared with 2 scatter search methods from the literature. The results show that the SVP method is promising and producing good solutions with low computational costs comparing to other competing methods.

1 Introduction

Nelder and Mead devised a local search method for finding the local minimum of a function of several variables, the method is called the Nelder-Mead method

* This work was partially supported by Grant of SGS No. SP2013/70, VSB - Technical University of Ostrava, Czech Republic., and was supported by the European Regional Development Fund in the IT4Innovations Centre of Excellence project (CZ.1.05/1.1.00/02.0070) and by the Bio-Inspired Methods: research, development and knowledge transfer project, reg. no. CZ.1.07/2.3.00/20.0073 funded by Operational Programme Education for Competitiveness, co-financed by ESF and state budget of the Czech Republic.

A. Abraham et al. (eds.), *Innovations in Bio-inspired Computing and Applications*, 271
Advances in Intelligent Systems and Computing 237,
DOI: 10.1007/978-3-319-01781-5_25, © Springer International Publishing Switzerland 2014

[13]. The method is one of the most popular derivative-free nonlinear optimization methods. A simplex is a triangle, for function of two variables and the method is a pattern search that compares function values at the three vertices of a triangle. The worst vertex is rejected and replaced with a new vertex. A new triangle is formed and the search is continued. The process generates a sequence of triangles for which the function values at the vertices get smaller and smaller. The size of the triangles is reduced and the coordinates of the minimum point are found. The algorithm is stated using the term simplex (a generalized triangle in N dimensions) and will find the minimum of a function of N variables. Four scaler parameter must be specified to define a complete Nelder-Mead method; coefficients of reflection ρ, $\rho > 0$, expansion χ, $\chi > 1$, contraction γ, $0 < \gamma < 1$ and shrinkage σ, $0 < \sigma < 1$ as shown in Figure 1. The Nelder-Mead algorithm steps are described in Algorithm 1 and all its parameters are defined in Table 1. In this paper, we proposed a new method based on the simplex Nelder-Mead method. The proposed method is called SVP (simplex variable partitioning). The main goal of the SVP method is construct an efficient method to solve unconstrained large scale optimization problems. SVP starts with a random initial solution, the iterate solution is divided into pre-specified number of partition. In order to generate a trail solution around the iterate solution the Nelder-Mead method has been applied in a random selected partitions. The trail solution with the best objective function is always accepted. In the final stage a local search method is applied in order to accelerate the search instead of letting the algorithm running for several iterations without much significant improvement of the objective function values. The SVP method is compared with 2 main scatter search methods by using 38 benchmark functions with different properties (uni-model, multi-model, shifted, rotated). The numerical results show that SVP method is a promising method and faster than other methods. The rest of the paper is organized as follows. The next section survey the related work on high dimension and large scale optimization problems. Section 3 describes the proposed SVP method. The performance of the SVP method and its numerical results are reported in Sections 4, 5. The conclusion of this paper is summarized in Section 6.

2 Related Work

Many researches have been attracted to apply their works to solve the global optimization problems, this problems can expressed as follows.

$$\begin{aligned} &Minimize \quad f(x) \\ &Subject\ to \quad l \leq x \leq u, \end{aligned} \tag{1}$$

where $f(x)$ is a nonlinear function, $x = (x_1, \ldots, x_n)$ is a vector of continuous and bounded variables, $x, l, u \in \Re^n$.

Although the efficiency of there works when applied with lower and middle dimensional problems e.g $D < 100$, they suffer from the curse of dimensionality when applied to large scale and high dimensional problems. Some efforts have been done to overcome this problem. The quality of any proposed method to solve the large scale and optimization problem is the capability of performing the wide exploration and the deep exploration processes. These two processes have been invoked in many works through different strategies, see for instances [3], [6], [7], [12]. These two processes have been considered in the SVP method through three strategies as follows. The dimension reduction process which the search space can be divided into smaller partitions. The exploration process where the trail solutions are generated around the iterate solution. In the variable neighborhood search [14], the search space is treated as nested zones and each one is searched through iterative solutions [1], [10]. Finally the intensification process by applying a local search method with the elite solution obtained from the pervious stage [4], [5], [7]. Invoking these strategies together in the SVP method is the main difference between it and other related methods existing in the literature.

Table 1. Parameters used in Algorithm 1

Parameters	Definitions
x_r	Reflection point
x_e	Expansion point
x_{oc}	Outside contraction point
x_{ic}	Inside contraction point
ρ	Coefficients of reflection
χ	Coefficients of expansion
γ	Coefficients of contraction
σ	Coefficients of shrinkage

3 The Proposed SVP Method

In this section a proposed method is presented for solving large scale optimization problems. The proposed method is called SVP (simplex variable partitioning). SVP starts with a random initial solution and consists of n variables. The solution is divided into η partitions, each partition contains ξ variables (if the number of variables n is not a multiple of ξ, a limited number of dummy variables my be add to the last partition). At a fixed number of iteration (SVP inner loop), a random partition is selected, and a trail solution is generated by applying the simplex Nelder-Mead method in the selected partition. The overall trail solution is accepted if it's objective function value is better than the previous solution. Otherwise the trial solution is rejected. The scenario is repeated until the termination criteria satisfied (SVP outer loop). In order to refine the best solution, SVP method applies a local search method as final intensification process. The definitions of the used parameters in SVP method is reported in Table 2. In the next subsections we give more descriptions of SVP method.

Algorithm 1. The Nelder-Mead method

1. Set x_i denote the list of vertex in the current simplex, $i = 1, \dots, n+1$.
2. Order Order and re-label the $n+1$ vertices from lowest function value $f(x_1)$ to highest function value $f(x_{n+1})$ so that $f(x_1) \leq f(x_2) \leq \dots \leq f(x_{n+1})$.
3. Reflect. Compute the reflection point x_r by
$x_r = \bar{x} + \rho(\bar{x} - x_{(n+1)})$,
where $\bar{x}$ is the centroid of the n best points,
$\bar{x} = \sum(x_i/n), i = 1, \dots, n$.
if $f(x_1) \leq f(x_r) < f(x_n)$ **then**
    replace x_{n+1} with the reflected point x_r.
    go to Step 7.
end if
4. Expand.
if $f(x_r) < f(x_1)$ **then**
    Compute the expansion point x_e by $x_e = \bar{x} + \chi(x_r - \bar{x})$.
end if
if $f(x_e) < f(x_r)$ **then**
    replace x_{n+1} with x_e and go to Step 7.
else
    replace x_{n+1} with x_r and go to Step 7.
end if
5. Contract.
if $f(x_r) \geq f(x_n)$ **then**
    perform a contraction between $\bar{x}$ and the better of x_{n+1} and x_r.
end if
Outside contract.
if $f(x_n) \leq f(x_r) < f(x_{n+1})$ **then**
    Calculate $x_{oc} = \bar{x} + \gamma(x_r - \bar{x})$.
    if $f(x_{oc}) \leq f(x_r)$ **then**
        replace x_{n+1} with x_{oc}
        go to Step 7.
    else
        go to Step 6.
    end if
end if
Inside contract.
if $f(x_r) \geq f(x_{(n+1)}$ **then**
    Calculate $x_{ic} = \bar{x} + \gamma(x_{n+1} - \bar{x})$.
end if
if $f(x_{ic}) \geq f(x_{(n+1)}$ **then**
    replace x_{n+1} with x_{ic}
    go to Step 7.
else
    go to Step 6.
end if
6. Shrink. Evaluate the n new vertices
$x' = x_1 + \sigma(x_i - x_1), i = 2, \dots, n+1$.
Replace the vertices $x_2, \dots, x_{n+1}$ with the new vertices $x'_2, \dots, x'_{n+1}$.
7. Stopping Condition. Order and re-label the vertices of the new simplex as $x_1, x_2, \dots, x_{n+1}$ such that $f(x_1) \leq f(x_2) \leq \dots \leq f(x_{n+1})$.
if $f(x_{n+1}) - f(x_1) < \epsilon$ **then**
    stop, where $\epsilon > 0$ is a small predetermined tolerance.
else
    go to Step 3.
end if

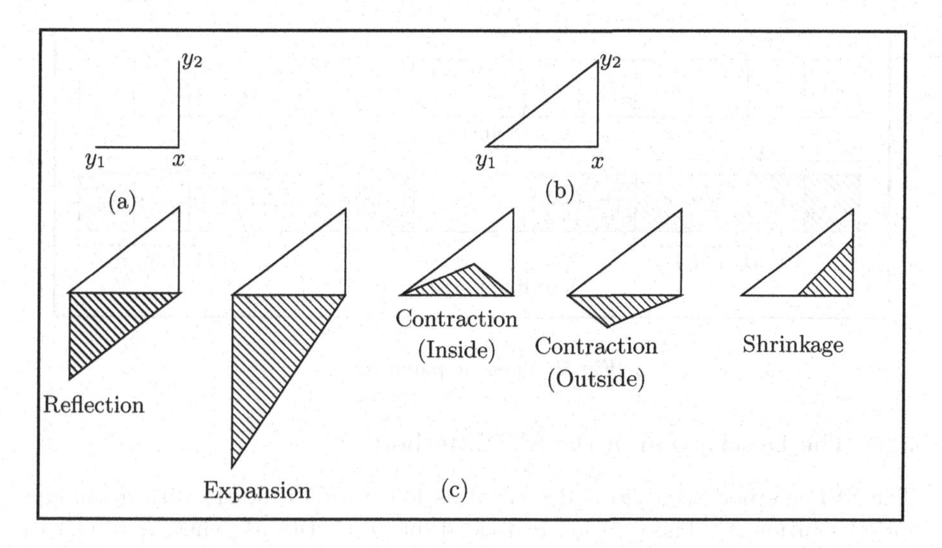

Fig. 1. Nelder-Mead search strategy in two dimensions

Table 2. Parameters used in Algorithm 2

Parameters	Definitions
n	No. of variables
ξ	Partition size
η	No. of partitions
ϕ	Random selected partition
x^0	Initial solution
$x\prime$	Best trail solution
$Maxitr$	Maximum number of iterations
N_{elite}	No. of best solution for intensification

3.1 Variable Partitioning and Trail Solutions Generating

SVP method starts with an iterate solution which divided into η small partitions, where $\eta = n/\xi$. Searching the partitioned subspaces is controlled by applying the search in a limited number of subspaces in the current iteration. This allows the SVP method to intensify the search process in each iteration. However, choosing different subspaces in consequent iterations maintains the search diversity. Moreover, searching a limited number of subspaces prevents SVP from wandering in the search space especially in high dimensional spaces. The variable partitioning process with $\xi = 4$ is shown in Figure 2. At a fixed number of iterations (SVP inner loop), a random partitions are selected in order to generate a trail solutions by applying the Nelder-Mead method in each selected partition ϕ as shown in Algorithm 1. If the overall trail solution objective function is better the previous solution, then the trail solution is selected to become the current solution. The operation of generating new trail solutions is repeated until stoping criteria satisfied (SVP outer loop).

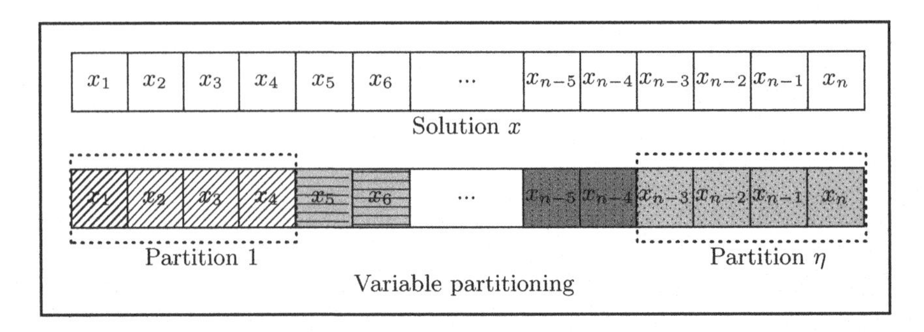

Fig. 2. Variable partitioning

3.2 The Description of the SVP Method

The SVP method scenario is described as follows. SVP starts with a random initial solution x^0. The solution is divided into η partitions, where $\eta = n/\xi$. In order to generate a trail solutions, SVP method uses two type of iterations as follows.

- **SVP inner loop**
  SVP uses a given number of iterations in order to generate a trail solutions by applying the simplex Nelder-Mead method as shown in Algorithm 1 in a random selected partitions. The number of the selected partitions is depend on the number of the applied iterations in the inner loop. The higher the number of the applied iterations, the higher the value of the cost function.
- **SVP outer loop**
  The main termination criterion in SVP is the number of the external loop which is the maximum number of iterations. The number of the best overall solutions is depends on the value of the external iterations.

Finally a local search method is applied as a final intensification process in order to refine the best solution N_{elite} which is obtained in the previous search stage. The structure of the SVP method with the formal detailed description is given in Algorithm 2, all variables of Algorithm 2 and it's definitions are reported in Table 2.

4 Numerical Experiments

In order to test the efficiency of the proposed SVP method and present the comparison results between it and other competing methods. SVP uses three sets of instances, LM-HG1, LM-HG2, CEC05 instances. The LM-HG1 instances consist of 10 uni-model and shifted functions, these functions have been used by Hvttum and Glover [9]. The LM-HG2 instances consist of 16 multi-model and shifted functions, these functions are based on functions in Laguna and Marti [11]. The CEC05 instances consist of 12 multi-model function based on classical

Algorithm 2. The proposed SVP method

Generate an initial solution x^0 randomly.
Set initial values for ξ, η, $Maxitr$, N_{elite}.
Set $x = x^0$.
repeat
  $I := 0$. ▷ Counter for inner loop.
  $K := 0$. ▷ Counter for outer loop.
  repeat
    Divide the solution x into η partitions,
    where $\eta = n/\xi$.
    Pick a random partition ϕ from η partitions of x.
    Apply Nelder-Mead Algorithm into the selected
    partition ϕ, as shown in **Algorithm 1** .
    Generate neighborhood trail solutions around x.
    $I := I + 1$.
  until $I \leq \eta$.
  Set $x\prime$ equal to the best trial solution.
  if $f(x\prime) \leq f(x)$ **then**
    $x = x\prime$. ▷ Accept the trial solution.
  end if
  $K := K + 1$.
until $K \leq Maxitr$. ▷ Stoping criteria satisfied.
Apply local search method starting from N_{elite} on the best solution in the previous stage. ▷ Intensification search.

function after applying some modifications (shifted, rotated, biased and added), these instances are described in detail in Suganthan [15]. The objective function values of all sets of instances are known. All details about the three mentioned sets of instances are described in [8]. The names and main feature of LM-HG1, LM-HG2, CEC05 functions are summarized in Table 4, Table 5 and Table 10, respectively. The SVP method was programmed in MATLAP. The results of the SVP method and the other competing benchmark methods are averaged over 10 runs. The parameter setting of the SVP method are reported in Table 3. The numerical results of all LM-HG1 instances are reported in Tables 6 - 8 where the numerical results of LM-HG2 and CEC05 instances are reported in Table 9, Table 11 respectively.

Table 3. Parameter setting

Parameters	Values
ξ	4
η	n/ξ
$Maxitr$	$n/4$
N_{elite}	1

Table 4. LM-HG1 test functions

f	Function name	f	Function name	f	Function name
f_1	Branin	f_5	Zakharov	f_9	Stair-Ros
f_2	Booth	f_6	Trid	f_{10}	Stair-LogAbs
f_3	Matyas	f_7	Sum Squares		
f_4	Rosenbrock	f_8	Sphere		

Table 5. LM-HG2 test functions

g	Function name	g	Function name	g	Function name	g	Function name
g_1	B2	g_5	Beale	g_9	Perm(o.5)	g_{13}	Powell
g_2	Easom	g_6	SixH.C.Back	g_{10}	Perm(10)	g_{14}	Dixon&Price
g_3	Gold.&Price	g_7	Schhwefel	g_{11}	Rastrigin	g_{15}	Levy
g_4	Shubert	g_8	Colville	g_{12}	Griewank	g_{16}	Ackley

4.1 Performance Analysis

In this subsection we analyze the performance of the SVP method as follows.

The Efficiency of Variables Partitioning. We compared the SVP method with variable partitioning process, with the basic simplex Nelder-Mead method (BSNM) in order to check the efficiency of variable partitioning process. The same parameters and termination criteria are used in both methods. The results are shown in Figure 3. In Figure 3, the dotted line represents the results of the basic simplex Nelder-Mead method, the solid line represents the results of the SVP method. Two functions f_3, f_7 are selected with dimensions 512, 1000 by plotting the number of iterations versus the function values. Figure 3 shows that the function values are rapidly decreases as the number of generations increases for the SVP method results than those of the basic simplex Nelder-Mead method.

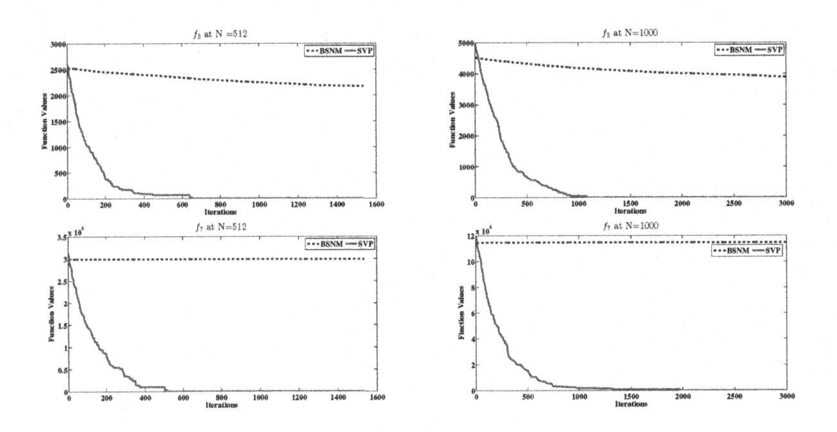

Fig. 3. Basic simplex Nelder-Mead algorithm Vs. SVP algorithm

The Performance of Final Intensification. The SVP method uses the MATLAB function "fminunc" as a local search method in the final intensification process. The final intensification can accelerate the convergence in the final stage instead of letting the algorithm running for several iterations without much significant improvement of the objective function values as shown in Figure 4. Figure 4 represents the general performance of the SVP method and the effect of the final intensification by selecting two functions f_2, g_{12} with different properties and plotting the values of objective functions versus the number of iterations. Figure 4 shows that the objective values are decreases as the number of function iterations increases. The behavior in the final intensification phase is represented in Figure 4 by dotted lines, in which a rapid decrease of the function values during the last stage is clearly observed.

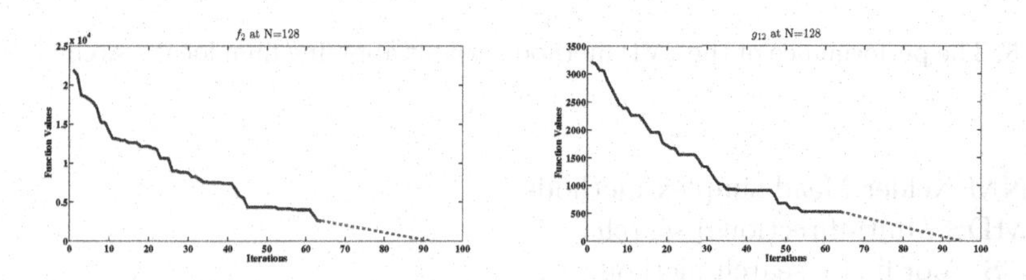

Fig. 4. The effects of final intensification process

The Performance of the Final Local Search Method. The performance of the SVP method without applying the last final intensification process on two functions f_1, f_8 with dimensions 512, 1000 is shown in Figure 5. Figure 5 represents the number of iterations versus the function values. The function values in Figure 5 are rapidly decreased and reached to its objective values without applying the final intensification process. We can conclude from Figure 5 that the MATLAB function "fminunc" is not doing the majority work of the SVP method.

4.2 The Proposed SVP Method and the Other Competing Methods

The SVP method is compared with two scatter search methods, whereas the two scatter methods were recently developed by [9] for solving high dimensional problems. The first scatter search method is called scatter search with randomized subset combination (SSR), where the second scatter search method is called scatter search with clustering subset combination (SSC). Also we compare our SVP method with a combination between the main two scatter search methods and six direct search methods. The SVP method uses the same termination criteria such as the gap between a heuristic solution x and the optimal solution x^* is $\mid f(x) - f(x^*) \mid < \epsilon$, where $\epsilon = 10^{-8}$, the other termination criterion is the maximum number of function evaluation is 50,000. The main local search methods are listed as follows.

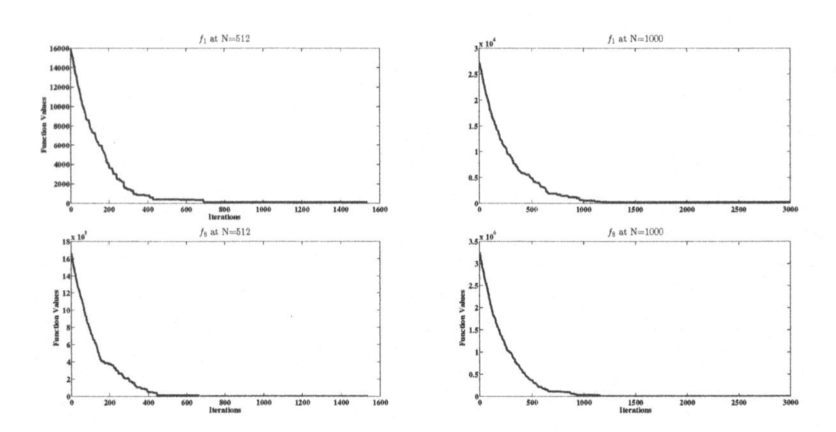

Fig. 5. The performance of the SVP method without applying final local search method

- **NM** Nelder-Mead simplex methods
- **MDS** Multi-directional search
- **CS** Coordinate search method
- **HJ** Hook and Jeeves method
- **SW** Solis and Wet's algorithm
- **ROS** Rosenbrock's algorithm

The description of these methods are reported in [8].

Comparison Results on LM-HG1 Instances. The SVP method was tested on the LM-HG1 instances in Havattum and Glover [9]. The dimensions of functions are 16, 32, 64, 128, 256 and 512. Each execution is repeated 10 times with different initial solutions. All tested methods have the same maximum number of function evaluations (50,000). We report the largest dimension n for which the method successfully found the optimal solution in all 10 runs. The comparison results between the SVP method and the other methods for all LM-HG1 instances are reported in Table 6. The best results are highlighted in bold. Table 6 shows that the SVP method outperforms in most of all functions except f_9. Also we can reach to the global minimum of these instances for $n > 512$ by increasing the number of function evaluation values as shown in Table 8. In case of at least one run fails to reach to the global minimum of the function, the ratio of successful run is recorded in parentheses.

Table 6 shows the comparison results between the SVP method and two scatter search methods. Now we evaluate the SVP method with the combined scatter search (SS) method and the direct search method as a improvement methods (IM). In addition to the combined SS method with the improvement direct search methods, there are extra two methods. SS method which is the scatter search function without improvement methods and STS method which is a hybrid scatter tabu search method. All details of STS method are reported in [2]. Table 7

Table 6. Largest successful dimension for SVP and other improvement methods on the LM-HG1 instances

f	SSR	SSC	SVP
f_1	**512**	**512**	**512**
f_2	**512**	**512**	**512**
f_3	256	256	**512**
f_4	32	**64**	**64**
f_5	32	32	**64**
f_6	16	16	**128**
f_7	**512**	**512**	**512**
f_8	**512**	**512**	**512**
f_9	32	**64**	32
f_{10}	**512**	**512**	**512**

Table 7. Largest successful dimension for the SVP method and the other combined SS with the improvement methods on the LM-HG1 instances

f	SS	STS	SS+NM	SS+MDS	SS+SW	SS+HJ	SS+ROS	SS+CS	SS+SSC	SS+SSR	SVP
f_1	2	16	4	4	8	8	16	16	16	16	**512**
f_2	2	8	4	4	16	128	8	128	128	128	**512**
f_3	2	16	8	4	32	128	32	256	256	256	**512**
f_4	0	4	2	2	2	4	4	4	4	2	**64**
f_5	2	8	4	4	8	16	8	16	8	8	**32**
f_6	2	8	4	4	8	32	8	16	16	16	**32**
f_7	2	16	4	4	16	32	32	32	32	32	**512**
f_8	2	32	4	4	64	32	64	32	32	32	**512**
f_9	0	4	2	2	2	4	2	2	4	2	**32**
f_{10}	0	4	0	0	0	128	2	128	256	256	**512**

Table 8. Function Evaluations (feval) of the SVP method with 1000 Dimension

f	f_1	f_2	f_3	f_4	f_5	f_6	f_7	f_8	f_9	f_{10}
feval	112,949	70,802	61,357	(0.0)	(0.0)	545,698	196,552	56,253	(0.0)	203,485

shows that the SVP method is outperforms all combined scatter search with the improvement methods and the hybrid scatter tabu search method in all LM-HG1 instances.

Comparison Results on LM-HG2 Instances. Now we test the performance of the SVP method with the multi-model functions in LM-HG2 sets. The committing methods is the same methods in Table 7. The terminations criteria is the same as in the LM-HG1 functions. The results in Table 9 show that the performance of the SVP method in the LM-HG2 is better then the other competing methods, although the functions in LM-HG2 seem to be more difficult and only smaller dimensions can be solved as seen in functions g_2, g_8, g_9, g_{10}, g_{11}, g_{16}. The best results are highlighted in bold.

Table 9. Largest successful dimension for the SVP method and the other combined SS and improvement methods on the LM-HG2 functions

g	SS	STS	SS+NM	SS+MDS	SS+SW	SS+HJ	SS+ROS	SS+CS	SS+SSC	SS+SSR	SVP
g_1	0	2	2	2	4	8	4	16	16	**32**	**32**
g_2	0	2	2	2	2	2	4	2	**4**	**4**	**4**
g_3	0	8	2	2	4	8	8	8	8	8	**16**
g_4	0	8	0	2	4	8	2	2	2	2	**16**
g_5	2	8	4	4	4	8	**16**	**16**	8	**16**	**16**
g_6	2	16	2	4	16	32	16	32	**64**	**64**	**64**
g_7	0	8	2	2	2	4	0	2	2	2	**128**
g_8	0	4	0	0	4	**8**	4	**8**	**8**	4	4
g_9	2	**4**	2	2	2	2	2	2	2	2	**4**
g_{10}	0	4	2	2	4	4	4	4	4	4	**8**
g_{11}	0	**16**	0	0	2	4	2	2	2	2	4
g_{12}	0	0	0	0	0	2	2	2	0	32	**256**
g_{13}	4	16	4	4	16	64	16	64	128	128	**256**
g_{14}	2	8	2	2	8	8	8	8	16	16	**32**
g_{15}	2	256	4	4	16	128	128	**512**	128	64	64
g_{16}	0	8	0	0	0	**32**	8	**32**	**32**	**32**	4

5 CEC05 Benchmark Test Instances

In order to evaluate the proposed SVP method, a new set of a modified benchmark functions with various properties provided by CEC2005 special session [15] are used in this section. Most of these functions are the shifted, rotated, expanded, and combined variants of the classical functions. These modifications make them to be more hard, and resistant to simple search. We used 12 functions h_1-h_{12} as shown in Table 10. Specifically, these functions span a diverse set of problem features, including multi-modality, ruggedness, noise in fitness, ill-conditioning, non-separabilty, interdependence(rotation), and high-dimensionality. Most of these functions are based on classical benchmark functions such as Rosenbrocks, Rastrigins, Swefels, Griwank and Ackleys function. The applied maximum evaluation function value as a termination criterion is increased to 100,000. Table 11 shows the CEC2005 functions h_1 - h_{12} with its objective function values for 10-100 dimensions. The exact global minimum is highlighted in bold. Results in Table 11 show that the SVP method obtains the exact global minimum of functions h_1 and h_2 at 100 dimension, where obtained the exact global minimum for function h_7, h_{12} at 80 dimension and for function h_6, h_{10} the exact global minimum at 40, 20 dimensions respectively. For functions h_3, h_4, h_5, h_8, h_{11} the SVP method fails to obtain the exact global minimum for any dimension and the obtained function values of these functions are reported in Table 11.

Table 10. CEC05 benchmark test functions

h	Function name	Bounds	Global minimum
h_1	Shifted Sphere	[-100,100]	-450
h_2	Shifted Schwefel's 1.2	[-100,100]	-450
h_3	Shifted rotated high conditioned elliptic	[-100,100]	-450
h_4	Shifted Schwefel's 1.2 with noise in fitness	[-100,100]	-450
h_5	Schwefel's 2.6 with global optimum on bounds	[-100,100]	-310
h_6	Shifted Rosenbrock's	[-100,100]	390
h_7	Shifted rotated Griewank's without bounds	[0,600]	-180
h_8	Shifted rotated Ackley's with global optimum on bounds	[-32,32]	-140
h_9	Shifted Rastrigin's	[-5,5]	-330
h_{10}	Shifted rotated Rastrigin's	[-5,5]	-330
h_{11}	Shifted rotated Weierstrass	[-0.5,0.5]	90
h_{12}	Schwefel's 2.13	[-100,100]	-460

Table 11. Mean number of function values with 10~ 100 dimensions

h	10	20	30	40	50	60	70	80	90	100
h_1	**-450**	**-450**	**-450**	**-450**	**-450**	**-450**	**-450**	**-450**	**-450**	**-450**
h_2	**-450**	**-450**	**-450**	**-450**	**-450**	**-450**	**-450**	**-450**	**-450**	**-450**
h_3	9235.12	6432.56	5429.27	4312.7	7134.2	5426.3	3458.45	4869.57	6778.24	4879.125
h_4	7143.61	52,848.2	72,582.4	86,931	315,056	370,425	416,241	504,957	565,784	590,668
h_5	194,431	201,456	293,84	312,58	614,58	640,15	714,26	785.12	798,147	812,58
h_6	**390**	**390**	**390**	**390**	465.23	485.45	486.78	512.56	540.25	545.13
h_7	**-180**	**-180**	**-180**	**-180**	**-180**	**-180**	**-180**	**-180**	-179.993	-179.78
h_8	-120	-120	-120	-120	-120	-120	-119.99	-119.98	-119.99	-119.98
h_9	**-330**	**-330**	**-330**	**-330**	512.716	643.05	714.95	740.12	815.56	845.89
h_{10}	**-330**	**-330**	-237.47	-215.45	112.48	145.26	242.23	361.68	412.25	415.23
h_{11}	101.56	120.731	123.908	126.56	131.25	152.36	158.69	162.54	165.98	167.57
h_{12}	**-460**	**-460**	**-460**	**-460**	**-460**	**-460**	**-460**	**-460**	-457.15	-441.25

6 Conclusion

In this paper, the simplex Nelder-Mead method which is called SVP (simplex variable partitioning) has been proposed to solve large scale optimization problems. The use of variable partitioning process assists effectively the SVP method to achieve promising performance specially with high dimensional test functions. Moreover, applying the Nelder-Mead method in different partitions helps the SVP method to achieve a wide exploration and a deep exploitation before stopping the search by generating a trail solution around the iterate solution. Finally the intensification process has been inlaid in the SVP method to accelerate the search process. The numerical experiments on 38 test benchmark functions with different properties have been presented to show the efficiency of the proposed SVP method. The comparison with other competing methods indicate that the SVP method is promising and it is cheaper than other methods.

References

1. Conn, A.R., Gould, N.I.M., Toint, P.L.: Trust-region methods. MPS-SIAM series on optimization. SIAM, Philadelphia (1987)
2. Duarte, A., Mart, R., Glover, F., Gortazar, F.: Hybrid scatter tabu search for unconstrained global optimization. Ann. Oper. Res. 183, 95–123 (2011a)
3. Garcia, S., Lozano, M., Herrera, F., Molina, D., Snchez, A.: Global and local real-coded genetic algorithms based on parentcentric crossover operators. Eur. J. Oper. Res. 185, 1088–1113 (2008)
4. Hedar, A., Fukushima, M.: Hybrid simulated annealing and direct search method for nonlinear unconstrained global optimization. Optim. Methods Softw. 17, 891–912 (2002)
5. Hedar, A., Fukushima, M.: Minimizing multimodal functions by simplex coding genetic algorithm. Optim. Methods Softw. 18, 265–282 (2003)
6. Hedar, A., Fukushima, M.: Directed evolutionary programming: towards an improved performance of evolutionary programming. In: Proceedings of Congress on Evolutionary Computation, CEC 2006, IEEE World Congress on Computational Intelligence, Vancouver, Canada, pp. 1521–1528 (July 2006)
7. Hedar, A., Ali, A.F.: Tabu search with multi-level neighborhood structures for high dimensional problems. Appl. Intell. 37, 189–206 (2012)
8. Hvattum, L.M., Duarte, A., Glover, F., Mart, R.: Designing effective improvement methods for scatter search: an experimental study on global optimization. Soft Comput 17, 49–62 (2013)
9. Hvattum, L.M., Glover, F.: Finding local optima of high dimensional functions using direct search methods. Eur. J. Oper. Res. 195, 31–45 (2009)
10. Kolda, T.G., Lewies, R.M., Torczon, V.J.: Optimization by direct search: new perspectives on some classical and modern methods. SIAM Rev. 45, 385–482 (2003)
11. Laguna, M., Mart, R.: Experimental testing of advanced scatter search designs for global optimization of multimodal functions. J. Glob. Optim. 33, 235–255 (2005)
12. Linhares, A., Yanasse, H.H.: Search intensity versus search diversity: a false trade off? Appl. Intell. 32, 279–291 (2010)
13. Nelder, J.A., Mead, R.: A simplex method for function minimization. Comput. J. 7, 308–313 (1965)
14. Mladenovic, N., Drazic, M., Kovac, V., Angalovic, M.: General variable neighborhood search for the continuous optimization. Eur. J. Oper. Res. 191, 753–770 (2008)
15. Suganthan, P.N., Hansen, N., Liang, J.J., Deb, K., Chen, Y.-P., Auger, A., Tiwari, S.: Problem definitions and evaluation criteria for the CEC 2005 special session on real-parameter optimization. Technical report, Nanyang Technological University, Singapore (2005)

SVM-Based Classification for Identification of Ice Types in SAR Images Using Color Perception Phenomena

Parthasarty Subashini[1], Marimuthu Krishnaveni[1], Bernadetta Kwintiana Ane[2], and Dieter Roller[2]

[1] Department of Computer Science, Avinashilingam University, Coimbatore, India
{mail.p.subashini, krishnaveni.rd}@gmail.com
[2] Inst. of Computer-aided Product Development Systems, Univ. of Stuttgart, Stuttgart, Germany
{ane,roller}@informatik.uni-stuttgart.de

Abstract. In rise of global temperatures, the formation of ice in freshwater like rivers and lakes are apparent to high condition which has to be significantly monitored for the importance of forecasting and hydropower generation. For this research, Synthetic Aperture Radar (SAR) based images gives good support in mapping the variation between the remote sensing data analysis. This paper presents an approach to map the different target signatures available in the radar image using support vector machine by providing limited amount of reference data. The proposed methodology takes a preprocess expansion of transforming the grayscale image into a synthetic color image which is often used with radar data to improve the display of subtle large-scale features. Hue Saturation Value based sharpened Synthetic Aperture Radar images are used as the input to supervised classifier in which evaluation metrics are considered to assess both the phase of the approach. Based on the evaluation, Support Vector Machine classifier with linear kernel has been known to strike the right balance between accuracy obtained on a given finite amount of training patterns and the facility to generalize to undetected data.

Keywords: SAR images, sharpening techniques, supervised classification, SVM kernel, κ-coefficient.

1 Introduction

Remotely-sensed data are used majorly in ice classification application in which it is initiated to convert data into meaningful information. Ice formation in fresh water takes the fundamental role of change in hydrology and climate. Satellite-derived estimates of ice coverage are found to be the eminent research technology to have ice detection run-off models. Automated processing and retrieval systems are required to be able to develop the quantities of SAR data available for freshwater ice monitoring applications. Hence, the classification of ice types plays a key task which is not simple in executing. In particular, in machine learning the supervised learning algorithms

A. Abraham et al. (eds.), *Innovations in Bio-inspired Computing and Applications*, Advances in Intelligent Systems and Computing 237,
DOI: 10.1007/978-3-319-01781-5_26,

require large sets of training samples which is the vital problem of satellite data. Focus on synthetic aperture radar (SAR) technology, SAR is capable of producing high resolution images using microwave by which it can capture cloud, smoke, fog and moisture area. Depending on the signal frequency, SAR has the capability to penetrate below the surface and shallow water. But still interpretation of images from ERS-2 and RADARSAT-2 are highly complex, since it operates at one frequency and one polarization for sending and receiving [1-2]. Azimi-Sadjadi and Zekavat [3] used a hierarchical arrangement of support vector machines to classify six different cloud types in infrared GOES-8 imageries. Several studies have been done that artificial intelligence can be an unconventional method suitable to class problems than the statistical approaches. As there is no consistent approach applied for mapping process, this paper tries to present a systemic mapping approach with the methodology of color enhancement, preprocessing and classification. As the characteristics of each image and the circumstances for each study differs in its own way [4], a suitable classifier is identified according to the experimentation and on the evaluation metrics. The organization of the paper is as follows. Section 1 explains the preface of the theme carried out in this work. Section 2 describes the methodology in which the framework is derived for remote sensing data analysis. The image enhancement and its subjective evaluation are given in Section 3. Section 4 gives the context of classification and the visual assessment of the images based on the supervised learning classifiers. Section 5 illustrates the experimentation and the findings for the conclusion. Section 6 concludes the research work carried out based on the experimental analysis.

2 Methodological Approach for SAR Data Analysis

Mapping and identification of ice types by categorizing a promising classifier for the data analysis is the objective of the research work. Hence the boundary between the targets is achieved especially in satellite images. This could be a complex task as the number of surface cover types will always be evident in the captured SAR images. A comprehensive process is followed to achieve visual interpretation in subsequent classification phase as indicated in Figure1. Details of the SAR images used for experiments are given in Figure 2.

3 Image Enhancement

To bring out adequate information and to improve the fidelity of the image, color enhancement is done using synthetic color image method. This improves the interpretation effort in the classification phase. Edge preserving is done using sharpening techniques to highlight edges and fine details in an image. Sharpened images provide clear image to view visually and assist for the classification. Hue Saturation Value (HSV) and Brovey spectral sharpening techniques are two methods used for image sharpening. Hence a simple pre-processing component is initiated for mapping by which the data exhibit higher sharpness with spectral quality.

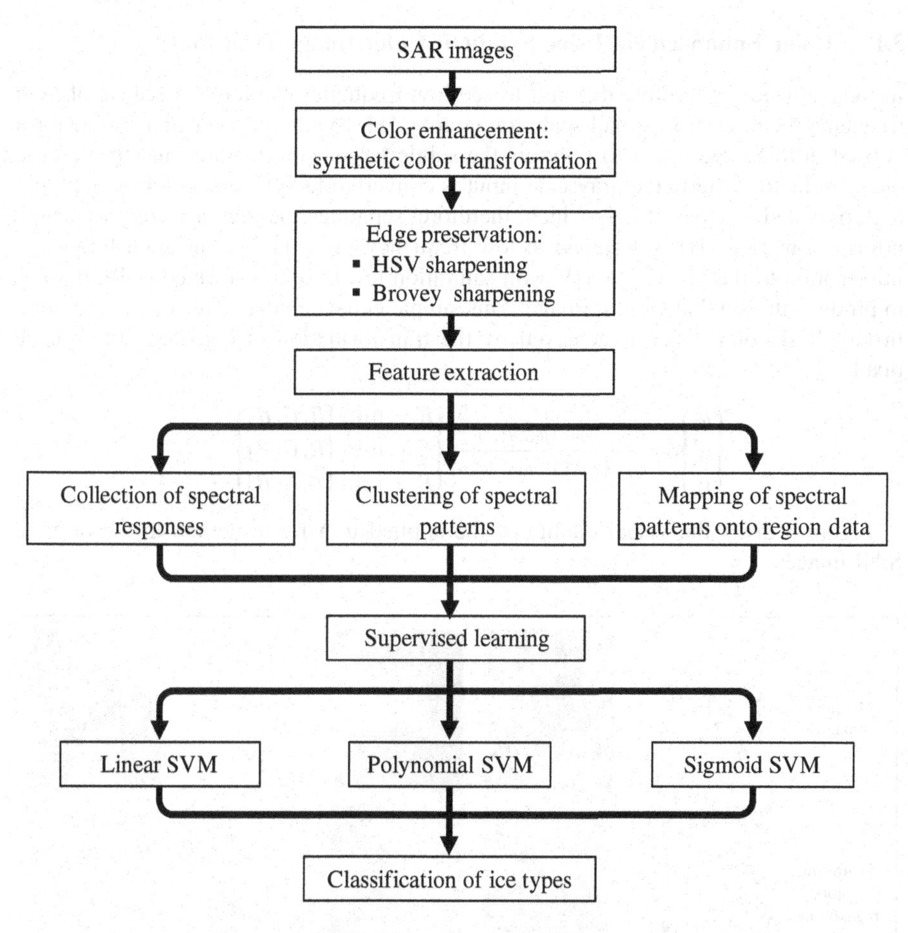

Fig. 1. Mapping methodology

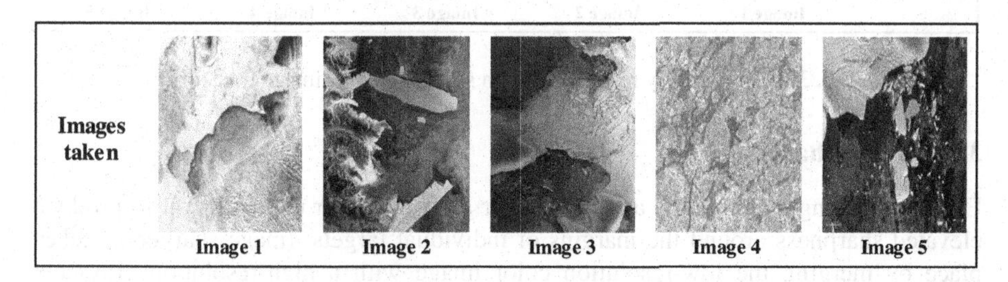

Fig. 2. Image 1 represents Visible Satellite pictures of the progression of the ice cover on Lake Erie. Image 2 represents Wilkins ice shelf with Charcot Island, a satellite image from terra SAR-X Scan SAR on April 2009. Image 3 represents the entire northern coast of Alaska and parts of the Russian Canadian coast, a satellite image from RADARSAT-1 SAR. Image 4 represents bottle-shaped B-15A iceberg adjacent to the land fast Aviator Glacier ice tongue, a satellite image from Envisat Advanced Synthetic Aperture Radar on 16 May 2005. Image 5 represents the Sulzberger Ice shelf along the Antarctic coast, Ross Sea, a satellite image from Envisat on March 11, 2011.

3.1 Color Enhancement Using Synthetic Color Image Transform

In radar images, it is often difficult to see low frequency variations because of high frequency features from small scale topography [5]. Synthetic color image transform is used in this experiment to improve the subtle large scale features and to preserve the edge factors. Here, the grayscale input is converted to synthetic color by applying high pass and low pass filters which, therefore, separates the high and low frequency information. Hue (H) is assigned to low frequency information, and high frequency information to the value (V). HV with saturation level are transformed to RGB so as to produce the synthetic color image. This simple enhancement of the color saturation in the RGB color space is achieved by the transformation of equation (1) on each pixel

$$\begin{Bmatrix} R' \\ G' \\ B' \end{Bmatrix} = \frac{\max\{R,G,B\}}{max\{R,G,B\}-min\{R,G,B\}} \begin{Bmatrix} R - \min\{R,G,B\} \\ G - \min\{R,G,B\} \\ B - \min\{R,G,B\} \end{Bmatrix} \tag{1}$$

Figure 3 gives the visual quality of the applied transform method to the original SAR images.

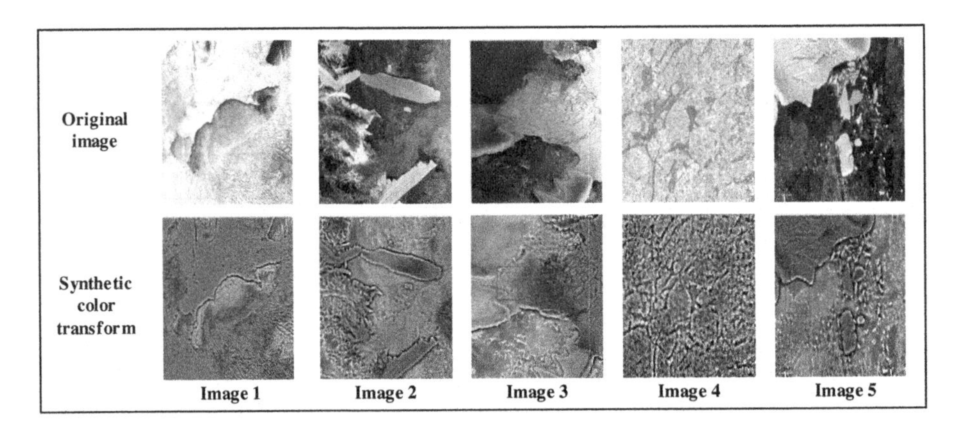

Fig. 3. Visual quality results based on synthetic color image transform

3.2 Edge Sharpening

Edge sharpening technique is applied further after the color enhancement to produce elevated sharpness around the margins of individual targets. Image sharpening takes place by merging the low-resolution color image with a high-resolution grayscale image and re-sampling the output to the high-resolution pixel size. Two image sharpening techniques are experimented to find the suitable one based on the visual assessment and evaluation metrics.

3.2.1 Hue Saturation Value (HSV) Sharpening

Based on the color perception phenomenon three main quantities known as hue, saturation and value are considered. Hue is defined as an angle in the range [0, 2π] relative to the Red with read at angle 0, green at 2π/3, blue at 4π/3 and red again at 2π [6]. These mentioned parameters will reflect the same as color intensity, equal intensity and perceived intensity respectively. HSV sharpening is used to transform HSV color space and replace the value band with the high-resolution. This technique re-samples the H and S bands by using the nearest neighbor method and the output result is shown with high pixel size. The numerical functions used for computing H in degrees are

$$U = \cos\,(H * \pi/180) \tag{2}$$

$$W = \sin\,(H * \pi/180) \tag{3}$$

$$Th = \begin{bmatrix} 1 & 0 & 0 \\ 0 & U & -W \\ 0 & W & U \end{bmatrix} = \begin{bmatrix} 1 & 0 & 0 \\ 0 & \cos\,(H * \pi/180) & -\sin\,(H * \pi/180) \\ 0 & \sin\,(H * \pi/180) & \cos\,(H * \pi/180) \end{bmatrix} \tag{4}$$

and saturation matrix is given in matrix (5)

$$T_s = \begin{bmatrix} 1\,0\,0 \\ 0\,S\,0 \\ 0\,0\,S \end{bmatrix} \tag{5}$$

Finally, the value transformation is a simple scaling of the color which can be represented by matrix

$$T_v = \begin{bmatrix} V\,0\,0 \\ 0\,V\,0 \\ 0\,0\,V \end{bmatrix} \tag{6}$$

3.2.2 Brovey Sharpening

This method is applied to sharpen the image using high spatial resolution data. It is done by using the mathematical combination of the color image and high resolution data. Brovey transform can be expressed as

$$\mathrm{DN}fusedMS_i = \frac{DN_{b_i}}{DN_{b_1} + DN_{b_2} + \cdots + DN_{b_n}} DN_{PAN} \tag{7}$$

Each band in the image taken is multiplied by a ratio of the high resolution data and divided by the sum of the color bands. The nearest neighbour method is used as the convolution technique for re-sampling the three-color bands to the high-resolution pixel size similar to HSV sharpening [7]. This method provides good spatial quality, but poor spectral quality. The comparative subjective evaluation of both HSV and Brovey spectral techniques are shown in Figure 4.

4 Support Vector Machine Based Classification

In remote sensing data, instance classifiers classify the coarse classes appropriately, but more detailed classification is required for ice mapping. Hence, it is much importance to possess more detail spectral information. In order to subdivide the classes, knowledge about spatial distribution is taken to find the difference in the mass of the object class correspondingly. The training labels derived are from four classes such as cloud, thin-ice, middle-ice and water. In neural network approach, a given unknown pixel or a region is classified into one of the predefined classes. This is one of [8]. To avoid unknown class and to sharpen boundaries between the classes, support vector machine (SVM) is used. Three kernels of SVM are studied in order to find high possibility of accurate results.

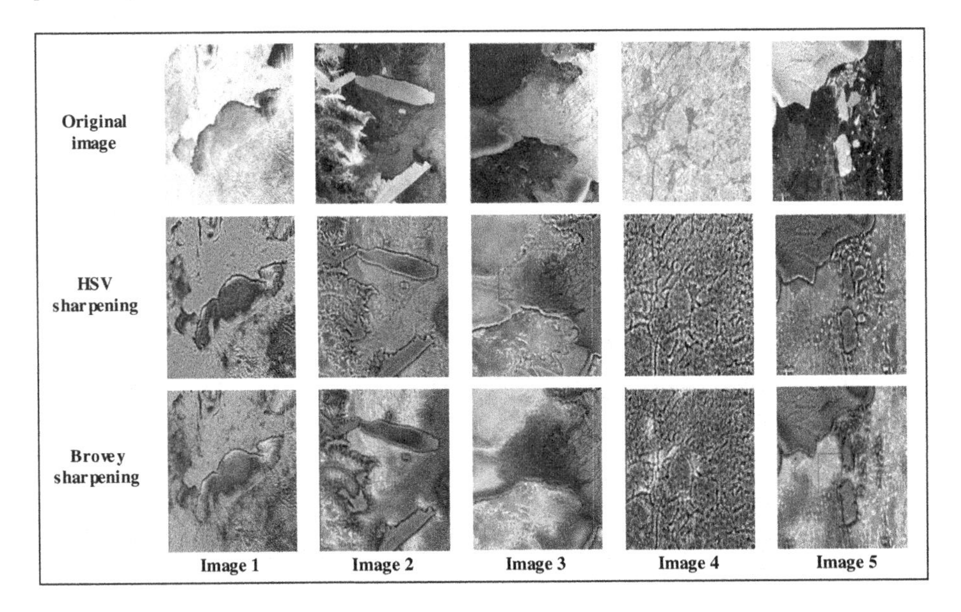

Fig. 4. Subjective comparison results of edge sharpening techniques

Each supervised learning algorithm develops a model of data which is used to classify original labels in the correct category. SVM is specifically constructed to minimize a statistical bound on the generalization error resulting in models that can extrapolate to new examples quite well [9]. In addition to its potential, a randomized procedure is followed for quick learning phase. SVM is applied by breaking the problem down into a number of binary problems. In this study, four classes are taken in which {4(4-1)}/2 classifiers are trained to differentiate between each pair of classes. Prior to classification process, pixels of SAR image targets are acquired and used as training vectors that are of high-quality statistics. This study is analyzing three types of SVM kernels, i.e., linear, polynomial and sigmoid. These kernels are represented as $(\vec{u} \cdot \vec{v}) = \vec{u} \cdot \vec{v}$ for linear, $K(\vec{u} \cdot \vec{v}) = 2(\vec{u} \cdot \vec{v} + a)$ for polynomial and $K(\vec{u} \cdot \vec{v}) = tanh(\gamma \cdot \vec{u} \cdot \vec{v} + a)$ for sigmoid functions.

It is found that the linear SVM outperforms of the other types of SVM classifier. Figure 5 gives the visual assessment of the linear SVM classifier. Figure 6 depicts the color distribution of the images in RGB values. Meanwhile, measurement of the SVM classifiers performance is described in Section 5.

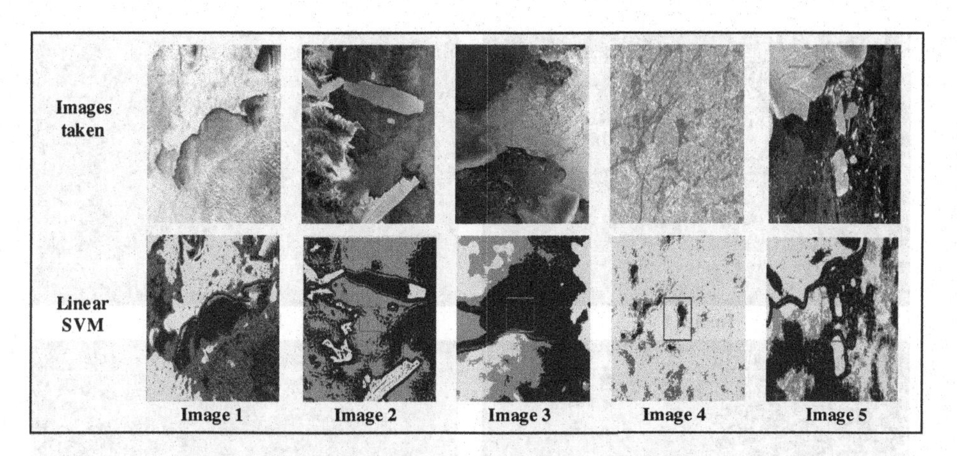

Fig. 5. Classification results using linear SVM algorithm

5 Performance of SVM Classifier

The performance of SVM classifiers is evaluated using its standard metrics, i.e., overall accuracy and κ-coefficient. Measurement is based on synthetic aperture radar images which indicate the four classes of ice types.

5.1 Accuracy Assessment

The accuracy of a classification is assessed by comparing the classification results with the reference data, which has accurately reflected the types of ice cover in the SAR image. The accuracy of the classifier is measured by counting the number of pixels classified as the same in the satellite image and on the reference data, and then dividing these by the total number of pixels.

5.2 The κ-Coefficient

The κ-coefficient is a measure of integrator agreement which is given as

$$k = \frac{P_0 - P_e}{1 - P_e} \tag{8}$$

Kappa (κ) is a positive value with its magnitude reflecting the strength of the integrator agreement and it becomes negative when the observed agreement is less than the chance agreement. To bring the best and suitable kernel for SVM, different kernels have been studied to find the reliable one. Table 1 describes the overall accuracy and κ-coefficient for each type of SVM kernels.

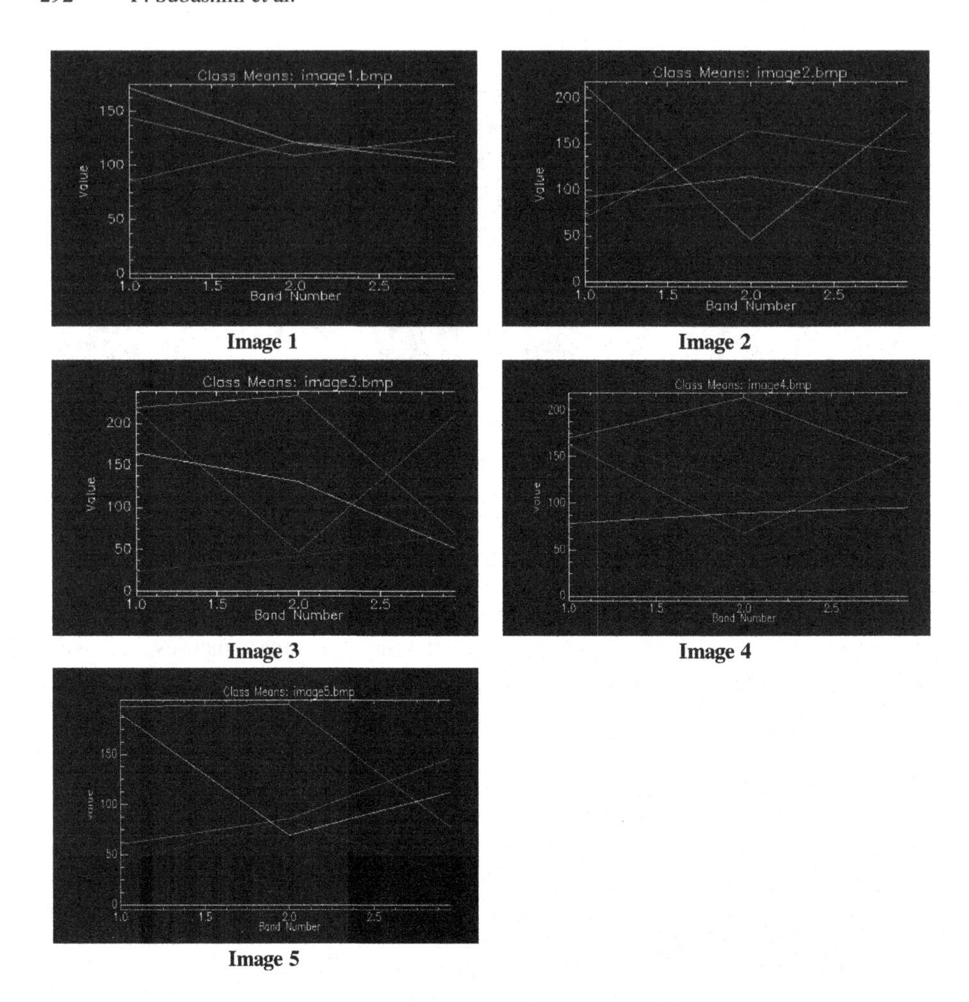

Image 1 **Image 2**

Image 3 **Image 4**

Image 5

Fig. 6. Distribution of RGB color values

Table 1. Comparison of overall accuracy and κ-coefficients between SVM kernels

Images	Linear SVM		Polynomial SVM		Sigmoid SVM	
	Accuracy	κ-value	Accuracy	κ-value	Accuracy	κ-value
Image 1	97.91	0.97	82.45	0.77	72.30	0.64
Image 2	90.00	0.86	80.35	0.74	84.00	0.77
Image 3	84.21	0.80	81.00	0.76	990.81	0.87
Image 4	83.95	0.79	77.27	0.71	80.00	0.74
Image 5	86.00	0.81	83.33	0.78	78.18	0.72

6 Conclusion

This paper describes the freshwater ice class identification based on machine learning method with remote sensing technology. Experiment is done using ENVI 4.2 software to determine the effective classifier which could be reliable for recognition and mapping of ice cover types. It is proven that image enhancement has significant contribution in determining the performance of classifiers. In addition, color enhancement and sharpening technique provide benefits and flexibility to explore important features in the SAR image. The advantage of using a supervised classifier states that it can be trained interactively and the performance is high on satellite data. This is specifically important in situations where the mapping of boundaries is very much crucial. The future improvement of this research work is to optimize the suitable kernel that will allow SVM to extract the spectral features more efficiently.

References

[1] Kaleschkel, L., Kern, S.: ERS-2 SAR Image Analysis for Sea Ice Classification in the Marginal Ice Zone. In: Proceedings of IEEE Geoscience and Remote Sensing Symposium (IGARSS 2002), vol. 5, pp. 3038–3040 (2003)
[2] Begoin, M., Richter, A., Weber, M., Kaleschke, L., Tian-Kunze, X., Stohl, A., Theys, N., Burrows, J.P.: Satellite Observations of Long Range Transport of a Large BrO Cloud in the Arctic. Atmos. Chem. Phys. 10, 6515–6526 (2010)
[3] Azimi-Sadjadi, M.R., Zekavat, S.A.: Cloud Classification Using Support Vector Machines. In: Proceedings of IEEE Geoscience and Remote Sensing Symposium (IGARS 2000), vol. 2, pp. 669–671 (2000)
[4] Kumar, M.: Digital Image Processing.: Satellite Remote Sensing and GIS Applications in Agricultural Meteorology. In: Proceedings of Training Workshop, Dehra Dun, India (2003)
[5] Daily, M.: Hue-Saturation-Intensity Split-Spectrum Processing of Sea-SAT Radar Imagery. Photogrammetric Engineering and Remote Sensing 49(3), 349–355 (1983)
[6] Sural, S., Qian, G., Pramanik, S.: Segmentation and Histogram Generation Using the HSV Color Space for Image Retrieval. In: IEEE International Conference on Image Processing, pp. 589–592 (2002)
[7] Yuhendra, Sumantyo, J.T.S., Kuze, H.: Performance Analyzing of High Resolution Pansharpening Techniques: Increasing Image Quality for Classification using Supervised Kernel Support Vector Machine. Research Journal of Information Technology 3(1), 12–23 (2011)
[8] Gill, R.S.: SAR Ice Classification Using Fuzzy Screening Method. Danish Meteorology Scientific Report (2012) ISBN: 987-7478-466-8
[9] Mazzoni, D., Garay, M.J.: An Operational MISR Pixel Classifier Using Support Vector Machines. Remote Sensing of Environment 107, 149–158 (2007)

Evaluation of Electrocardiographic Leads and Establishing Significance Intra-individuality

Marek Penhaker, Monika Darebnikova, Frantisek Jurek, and Martin Augustynek

VSB - Technical University of Ostrava,
Faculty of Electrical Engineering and Computer Science,
Department of Cybernetics and Biomedical Engineering,
Ostrava, Czech Republic
{marek.penhaker,monika.darebnikova,
frantisek.jurek,martin.augustynek}@vsb.cz

Abstract. The theme of this work was to design and program implementation process electrocardiogram Frank leakage system. Especially when this realization were preprocessed ECG real data, which means the modification of data filtration. The work focuses on the optimal detection of ventricular complex electrocardiographic signals. The main focus of the work is the calculation and graphical presentation vectorcardiogram orthogonal leads. Using a statistical analysis of the results is evaluated for each VCG planes.

Keywords: Vectocardiogram, Frank Leads, Processing.

1 Introduction

Vectorcardiography (VCG) is a diagnostic method that deals with graphic motion capture instantaneous power summation vector in space and time. This is another form of recording electrical manifestations of heart from the body surface. Its aim is to gain immediate projection of the vector changes (changes in direction and magnitude) in the three orthogonal anatomical planes.

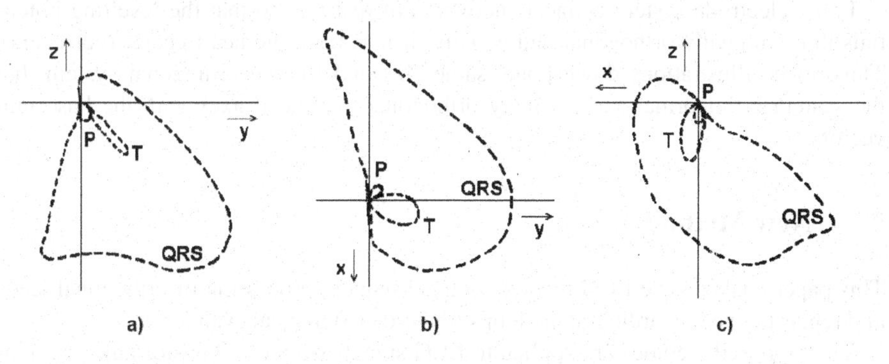

Fig. 1. Vectorcardiographic loops in a plane: a) sagittal, b) the front, c) horizontal [1]

A. Abraham et al. (eds.), *Innovations in Bio-inspired Computing and Applications*, Advances in Intelligent Systems and Computing 237,
DOI: 10.1007/978-3-319-01781-5_27, © Springer International Publishing Switzerland 2014

Unlike standard electrocardiogram signal leakage system which is customary to display in the time domain, it is common to express the information contained in the orthogonal signals spatially in three aforementioned descriptive levels (frontal, sagittal and horizontal). The projections of the instantaneous voltage vectors have the character vectorcrdiography closed loops. If we follow the movement of the instantaneous power summation vector in space and time, we get three loops. These loops correspond vectorcrdiography wave P, QRS and T wave such a display is called vector cardiogram.

Loop P (atrial depolarization) is usually not often obvious and therefore it is necessary to increase it. Most importantly loop QRS complex (ventricular depolarization), which has the physiological state of ovoid shape, its long axis follows the space saving electric axis of the heart. Loop T wave is smaller (ventricular repolarization), a shape similar to a loop QRS complex. Normally rotating loop vector cardiogram at time counterclockwise (CCW). Caution must be put into practice when doctors plotting angles clockwise (CW). Isoelectric line of zero potential is recorded in the electrical record Wednesday. In pathological conditions, the shape, direction and speed of rotation VCG loops changes. When specific pathologies may even happen that the loop will not be closed.

Electrocardiogram and vector cardiogram record the same information, the resulting display for the two methods differ. In today's clinical practice is mainly used classical recording an electrocardiogram, which in certain cases may not be entirely sufficient for accurate localization of certain heart diseases. To illustrate the electrical activity of the heart are therefore also important three-dimensional tracking, which allows leakage orthogonal systems.

1.1 Principle Frank Corrected System

The basic principle of orthogonal lead systems is the use of multiple electrodes unified into a single lead. The voltage of the individual electrodes disposed on the patient's body is transferred to the resistive network. From this network are then derived signals of Lead Ux, Uy and Uz. Examples of such solutions are leaky systems McFee, Schmidt SVEC III system, or the most widespread Frank's system.

Frank electrode system connects networks of resistors so that the resulting potentials three mutually orthogonal and equally distant from the heart center (corrected). Thus, to be illustrations downspouts same size and formed orthogonal system, but the system orthonormal axes, namely orthogonal system of axes with the base unit vectors.

2 New Method

This paper analyzes the ECG records with subsequent processing of orthogonal leads and rendering vectorcardiographic loops for further statistical evaluation.

The proposed methods are applied to ECG signals from the "PhysioBank". National Metrology Institute PTB in Germany provides the user server and PhysioNet ECG database, which are available ECG records. Each ECG recording contains 15 simultaneously measured signals, which are obtained from the 12-Lead system (I, II, III, aVR, aVL, aVF, V1-V6), while Frank leakage from the system (vx, vy, vz). Individual systems are digitized at a sampling rate of 1000Hz. For each elektrocardiographic record is accompanied by a header file (. hea) containing information about the patient and his clinical summary - age, gender, diagnosis, information on history, etc.

3 Realization

In the next step of modifying data was corrected isoelectric line slot signal. The zero reference point, which is determined by isoelectric line, is found in the segment PQ. This is the end wall depolarization (P wave) and the beginning of the Q wave. The PQ segment was chosen cut-out on which was the average value and then was subtracted from the signal.

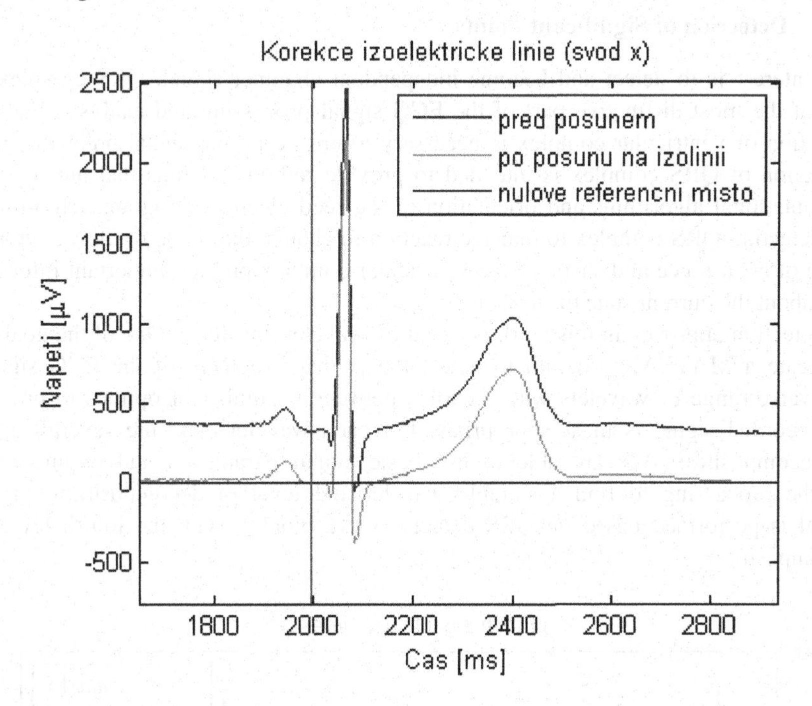

Fig. 2. Correction isoelectric line and the representation of the zero point of reference in the segment PQ

On Fig. 2 can be observed using a purple line at which point in time is shifted original signal to the desired isoelectric line. As already mentioned, this distance is calculated by averaging the values at the desired time Fig. 3.

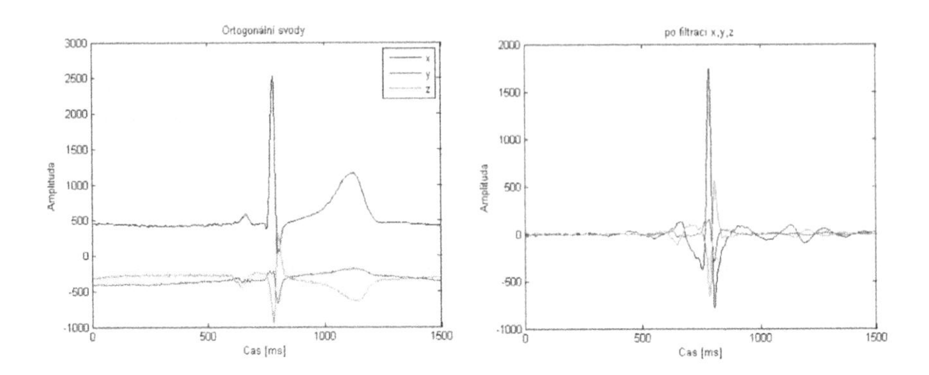

Fig. 3. Modifying data was corrected isoelectric line slot signal

3.1 Detection of Significant Points

Our interest is to detect and become independent chamber signal. QRS complex is one of the most distinctive part of the ECG signal processing and analysis. For the detection of ventricular complex is necessary to ensure proper signal preprocessing. Detection of QRS complex is intended to provide reference points that are used to computational algorithms and orientation of the feed electrocardiogram. Algorithms for detecting QRS complex to manage reaction to change the shape of the waveform. Time of occurrence and shape of the ventricular complex provide important information about the current state of the heart.

Detection approach in this work is based on wavelet transform. One of the goals is to create a MATLAB program to detect the primary segment of the ECG signal. A diverse range of wavelets with a choice between a number of options to process the recorded signal is most appropriate. For each wavelet there are several levels of decomposition. We are able to use basic signal parameters and requirements for the processing to find a suitable wavelet and level of decomposition. Finish signal is performed based on QRS detector wave bior1.5, with the fourth level of decomposition.

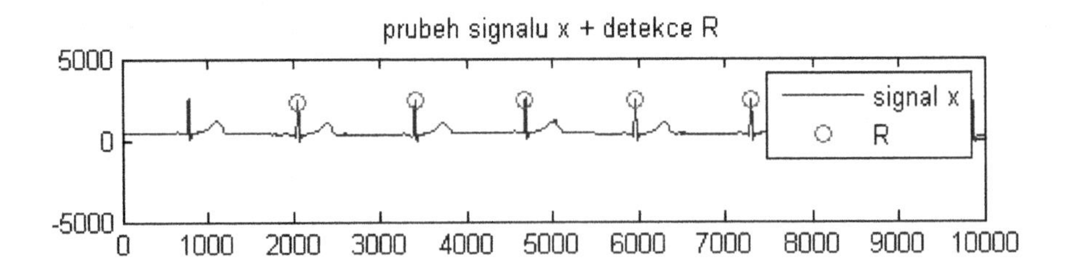

Fig. 4. Detection of positive vibrations R

On Fig. 4 is clear removal detection at the first and last R-wave, thus we prevent the entry of errors in processing due to an improper start recording. Detection of QRS complex serves as the basis for automatic calculation of heart rate, classification of cardiac cycles, or is used in algorithms for compression of ECG data. In a few cases the R wave mistakenly detected as a flicker reminiscent of the QRS complex, which could prove to be a false positive detection.

3.2 Visualization of VCG Curves

Clarifying the ECG is associated with the detection of one cycle of the signal and subsequent transformation of the selected cycle. For the detection of the isolated cycle was proposed algorithm based on the detection of R wave. The cardiac cycle using red lines bordering the QRS complex, manually selected distance from the detected R wave right and left. Information about the time displayed is obtained during each cycle, which is calculated from the index of the first and last time displayed element vector and display the current VCG loop (see Fig. 5-6).

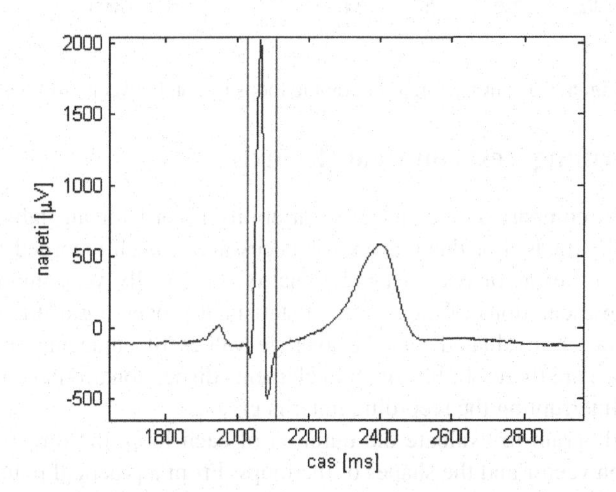

Fig. 5. Detection of one cycle showing the red lines

With all loaded ECG was calculated average curve. To calculate the average wave of the entire record are synchronized sections of each course by the position of R wave. The length of each crop is determined based on the border manually start and end of the QRS complex. Calculation of the average time domain wave was applied to each of the leads x, y and z thus determined stretch is performed at each cardiac cycle, creating trunked ventricular complexes which finally averaged.

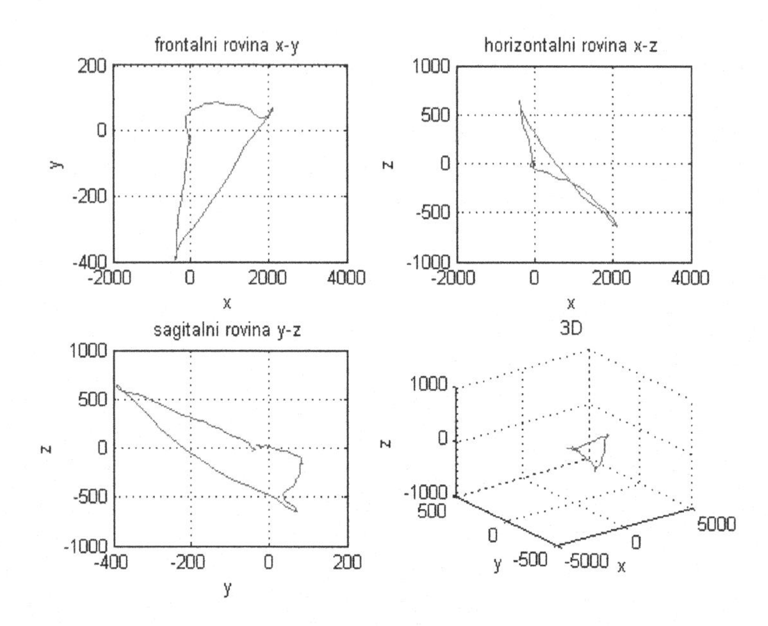

Fig. 6. Current vectorcardiographic loops region between red lines

4 Performing Tests on Real Records

Intra-individual comparison (uniqueness variability record for an individual) differs from other individuals. For this reason it is necessary to each recorded cardio logical recording rate in terms of personal individuality. Practically very important options are graphical presentations of the results of statistical processing of the file. For each processed record is evaluated and the average values of adjacent samples trunked Frank leads. We are therefore information elements dispersion around the mean value, see Fig. 46th Interrupting the recording timer is given.

Vector cardiogram to evaluate the duration of each loop, the direction (sense) of cardiac rotation vector and the shapes of the loops. From a practical point of view, the most important vector cardiogram in the frontal plane.

4.1 Statistical Processing Rate Variance

Major step towards obtaining the necessary values are set aside an appropriate section of the ECG QRS complex nearby. Thus trunked ventricular complexes correspond to the time series, which are then averaged.

Individual results in graphical form as shown in the following figures, which shows the maximum and minimum variance (dark green and green), average (blue). It is therefore determined the average EKG wave and its derivative ratio between the height R oscillation maximum and minimum values at the same time. Graphic display at first glance may seem biased, this is due to a narrowing of the signal amplitude or voltage axis.

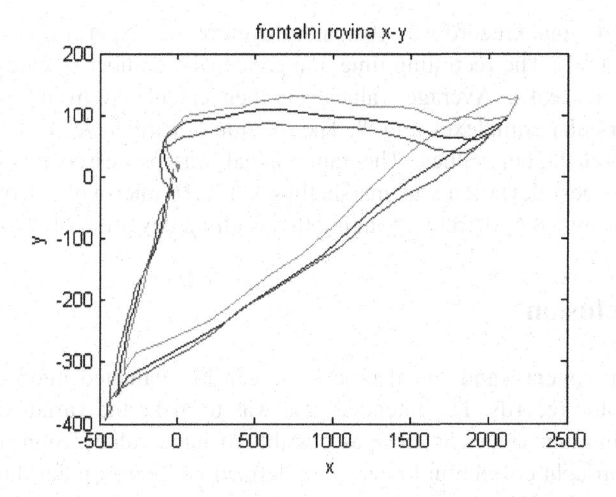

Fig. 7. The variance of the maximum and minimum waveform from the average curves in planes

Selected sections were then with sufficient time resolution plated. In such a predetermined position each time a sample is calculated by the dispersion characteristics of a minimum and maximum course from the average curve. It is a derivation of the relationship between voltage levels fixed average EKG wave (blue) and voltage limit wave (green) Fig. 7.

From these indicators can be calculated by comparing the absolute value of the difference to determine the most. The sample orthogonal lead x largest deviation is 110μV, which corresponds to an approximate difference of 4.4%.

- Lead x: 110μV (minimum curve 54 in the sample), which is 4.4%
- Lead y: 63.7 microvolts (maximum curve 57 of the sample), which makes 13.7%
- Leakage from 135.1 microvolts (minimum curve 67 sample), which is 10.7%

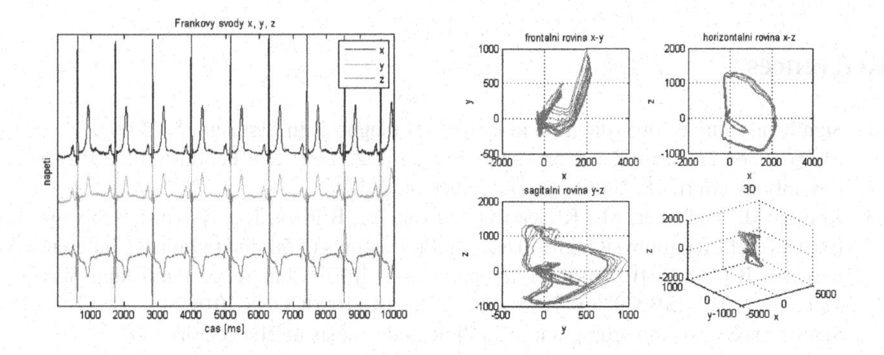

Fig. 8. Frank leads and VCG loops of projections

Length treated female record was 10s, and, therefore, the experiment consider recording more than 10s. The recording time, the greater the number of cardiac cycles, and the detected R peaks. Average values are then calculated from a larger number trunked ventricular complexes lFig. 8. The rotation of cardiac vector and shapes VCG loops can reach higher values. The same signal was processed record length of 1 minute, the largest deviation was marshalling x 170.16 microvolts. Converted to percentage terms by 6.8%, thereby changing the resulting deviation of 2.4%.

5 Conclusion

After the experiments, and signal processing can be evaluated inter-individual electrocardiographic records. The intended goal was to make the variances of the maximum and minimum values as close as possible average values. Significant reduction in variance brought correction to zero in a defined isoelectric point during each lead. As can be seen from the previous section, between the curves is significant statistical difference especially in lead y.

The results of the statistical test based on the correction ECG signal isoelectric line to the relative variance of the signal by about 7.5%. The largest absolute deviation is observed in the analysis of lead x around the peak R wave. During the downspout y reflected the maximum variance in the segment between R and S oscillate. The assembly of these was the results with greater individuality. Evaluation of the individual patient is always stated under the graphical analysis of the record.

Acknowledgment. The work and the contributions were supported by the project SP2013/35 "Biomedical engineering systems IX" and TACR TA01010632 "SCADA system for control and measurement of the process in real time". The paper has been written within the framework of the IT4 Innovations Centre of Excellence project, reg. no. CZ.1.05/1.1.00/02.0070 supported by the Operational Programme 'Research and Development for Innovations' funded by Structural Funds of the European Union and the state budget of the Czech Republic.

References

1. Significance of vectorcardiogram in kardiology diagnosis in 21st century (2011), `http://eminencevcg.com/significance.html` (cit. February 12, 2012)
2. PhysioBank (April 18, 2006) (cit. December 14, 2010)
3. Krawiec, J., Penhaker, M., Krejcar, O., Novak, V., Bridzik, R.: System for Storage and Exchange of Electrophysiological Data. In: Proceedings of 5th International Conference on Systems, ICONS 2010, April 11-16, pp. 88–91. IEEE Conference Publishing Services, Menuires (2010) ISBN 978-0-7695-3980-5, doi:10.1109/ICONS.2010.23
4. Significance of vectorcardiogram in kardiology diagnosis in 21st century (2011), `http://eminencevcg.com/significance.html` (cit. February 12, 2012)
5. Malmivuo, J., Plonsey, R.: Bioelectromagnetism - principles and applications of bioelectric and biomagnetic fields, p. 482. Oxford University Press, New York (1995)

6. Burch, G.E.: The history of vectorcardiography. Medical History (5), 103–131 (1985)
7. Frank, E.: An Accurate, Clinically practical system for spatial vectorcardiography. Circulation 13(5), 737–749 (1956)
8. Pérez Riera, A.R., Uchida, A.H., Filho, C.F., Meneghini, A., Ferreira, C., Schapacknik, E., Dubner, S., Moffa, P.: Significance of vectorcardiogram in the cardiologicaldiagnosis of the 21st Century. Clin. Cardiol. 30(7), 319–323 (2007)
9. Winter, B.B., Webster, J.G.: Reduction of Interference Due to Common Mode Voltage in Biopotential Amplifiers. IEEE Transactions on Biomedical Engineering, BME 30(1), 58–62 (1983)
10. Penhaker, M., Imramovsky, M., Tiefenbach, P., Kobza, F.: Medical diagnostics instruments - learning texts, p. 98. VSB-TU Ostrava, Ostrava (2004)
11. Huhta, J.C., Webster, J.G.: 60-Hz Interference in Electrocardiography. IEEE Transactions on Biomedical Engineering, BME 20(2), 91–101 (1973)
12. Winter, B.B., Webster, J.G.: Driven-right-leg circuit design. IEEE Transactions on Biomedical Engineering, BME 30(1), 62–66 (1983)

Author Index